T0177856

DOMAIN WALLS

Series on Semiconductor Science and Technology

Series Editors

R. J. Nicholas   University of Oxford
H. Kamimura   University of Tokyo

Series on Semiconductor Science and Technology

Domain Walls: From Fundamental Properties to Nanotechnology Concepts

Dennis Meier, Jan Seidel, Marty Gregg, and
Ramamoorthy Ramesh

OXFORD

UNIVERSITY PRESS

OXFORD
UNIVERSITY PRESS

Great Clarendon Street, Oxford, OX2 6DP,
United Kingdom

Oxford University Press is a department of the University of Oxford.
It furthers the University's objective of excellence in research, scholarship,
and education by publishing worldwide. Oxford is a registered trade mark of
Oxford University Press in the UK and in certain other countries

© Oxford University Press 2020

The moral rights of the authors have been asserted

First Edition published in 2020
Impression: 1

All rights reserved. No part of this publication may be reproduced, stored in
a retrieval system, or transmitted, in any form or by any means, without the
prior permission in writing of Oxford University Press, or as expressly permitted
by law, by licence or under terms agreed with the appropriate reprographics
rights organization. Enquiries concerning reproduction outside the scope of the
above should be sent to the Rights Department, Oxford University Press, at the
address above

You must not circulate this work in any other form
and you must impose this same condition on any acquirer

Published in the United States of America by Oxford University Press
198 Madison Avenue, New York, NY 10016, United States of America

British Library Cataloguing in Publication Data

Data available

Library of Congress Control Number: 2019957865

ISBN 978–0–19–886249–9

DOI: 10.1093/oso/9780198862499.001.0001

Printed and bound by
CPI Group (UK) Ltd, Croydon, CR0 4YY

Links to third party websites are provided by Oxford in good faith and
for information only. Oxford disclaims any responsibility for the materials
contained in any third party website referenced in this work.

Preface

Technological evolution and revolution are both driven by the discovery of new functionalities, new materials and the design of yet smaller, faster, and more energy-efficient components. Progress is being made at a breathtaking pace, stimulated by the rapidly growing demand for more powerful and readily available information technology: high-speed internet and data-streaming, home automation, tablets and smartphones are now "necessities" for our everyday lives. Consumer expectations for progressively more data storage and exchange appear to be insatiable.

Oxide electronics is a promising and relatively new field, that has the potential to trigger major advances in information technology. Oxide materials offer a multitude of applications including spintronics, thermoelectrics and power harvesting, which arise from the broad spectrum of tunable phenomena they exhibit, including magnetism, multiferroicity, and superconductivity.

Oxide interfaces are particularly intriguing. Here, low local symmetry combined with an increased susceptibility to external fields leads to unusual physical properties distinct from those of the homogeneous bulk. In this context, but not limited to oxides, ferroelectric domain walls have attracted recent attention as a completely new type of functional interface. In addition to their functional properties, such walls are spatially mobile and can be created, moved, and erased on demand. This unique degree of flexibility enables domain walls to take an active role in future devices and hold great potential as multifunctional 2D systems for nanoelectronics. With domain walls as reconfigurable electronic 2D components, a new generation of adaptive nanotechnology and flexible circuitry becomes possible, that can be altered and upgraded throughout the lifetime of the device. Thus, what started out as fundamental research, at the limit of accessibility, is finally maturing into a promising concept for next-generation technology.

This book provides a state-of-the-art overview about the significant progress that has been made in ferroelectric domain wall research over the last decade and evaluates emerging application possibilities in information technology. Bringing together world-leading scientists from complementary disciplines, the book gives a broad overview of how domain walls can be used as functional nano-objects with distinct physical properties; it also illustrates how domain walls have shifted from being muses for scientific curiosity into becoming key objects of interest to technology developers. Different chapters also highlight the close relationship between the progress and the development of cutting-edge experimental and theoretical analysis tools.

Going beyond the currently available literature, the book identifies major questions and challenges that will influence research on domain walls, refining the reader's picture of the state of the art.

Sharing our excitement about ferroelectric domain walls with you, we and our co-authors are hoping that you will enjoy reading this comprehensive work and become curious to find out how far we can go, in the years to come, to establish a new technology paradigm.

Dennis Meier,
Jan Seidel,
J. Marty Gregg, and
Ramamoorthy Ramesh

Contents

1

Physical Properties inside Domain Walls

Basic Principles and Scanning Probe Measurements

G. Catalan and N. Domingo

Catalan Institute of Nanoscience and Nanotechnology (ICN2), CSIC and BIST, Campus UAB, Bellaterra 08193, Barcelona, Catalonia.
ICREA-Institucio Catalana de Recerca i Estudis Avançats, Barcelona, Catalonia.

1.1 Introduction

Although domain wall properties are material specific, two features are common to all of them.

First: by symmetry, a domain wall cannot just stop in the middle of a crystal. Any domain wall ends at an interface (the surface of the crystal, grain boundary), in another domain wall (forming a needle domain), or on itself (forming a bubble domain) (Figure 1.1). Thus, despite being nanoscopically thin, they can be macroscopically long, providing a continuous path between different interfaces of a crystal irrespective of how big the crystal is. This "topologically protected" percolation path is most useful for transport applications (Lee and Salje 2005; Seidel et al. 2009).

Second: domain walls are mobile; they shift their position as domains grow or shrink in response to external fields. This mobility sets domain walls apart from other types of interfaces, and is a useful feature that can be exploited in devices where the wall is moved into and out of a reading unit, as in the "racetrack memory" concept proposed by Stuart Parkin and co-workers (Parkin et al. 2008), or the domain wall logic devices explored by the group of Russell Cowburn in Cambridge (Allwood et al. 2005). This mobility property means that domain walls need not be regarded as just a transport medium (a connector) for, say, electrical currents, but also as a "container" of information that can itself be moved into and out of the reading head, carrying with it whatever wall-specific physical property is of interest, such as, e.g. internal magnetization or polarization.

G. Catalan and N. Domingo, *Physical Properties inside Domain: Basic Principles and Scanning Probe Measurements* In: *Domain Walls: From Fundamental Properties to Nanotechnology Concepts.* Edited by: Dennis Meier, Jan Seidel, Marty Gregg, and Ramamoorthy Ramesh, Oxford University Press (2020). © G. Catalan and N. Domingo.
DOI: 10.1093/oso/9780198862499.003.0001

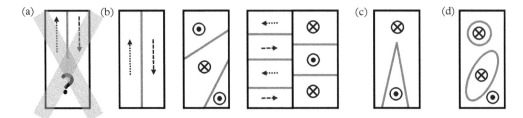

Figure 1.1 *Domain wall configurations. (a) Domain walls ending in the middle of a crystal lead to unresolved configurations. Domains walls ending at (b) surfaces or interfaces, and (c) in another domain wall forming needle domains or (d) on itself, forming bubble domains.*

The field of domain wall nanoelectronics (Catalan et al. 2012a) is predicated on the premise that the distinct physical properties of domain walls offer new conceptual possibilities for devices. The first part of this chapter will deal with the basic physics of domain wall properties, and in particular the cross-coupling that allows domain walls to display properties and order parameters different from those of the parent bulk material. The second part will deal with scanning probe techniques for measuring some of these domain wall properties, and specifically atomic force microscopy (AFM). Together with transmission electron microscopy, discussed in Chapter 10, AFM is one of the most important tools we currently have to probe and manipulate the individual position and physical properties of domain walls; although this book contains many chapters that discuss AFM probes, e.g. to inject domain walls and control their motion (Chapter 13), here we will focus on two recent developments that allow investigating hitherto overlooked properties of domain walls: their magnetotransport and their mechanical response.

1.2   Domain Wall Structure and Thickness

1.2.1   Domain Wall Thickness

Domain walls are not the ground state of any material (with the possible exceptions of incommensurate materials and relaxors), so they cost energy. The minimization of this energy cost dictates their thickness and structure. First, there is a nearest neighbor interaction; in ferromagnets, this is called "exchange energy," but the concept applies to any ferroic. The exchange energy penalizes differences between adjacent unit cells. In other words: it penalizes gradients of the order parameter. The associated energy cost of the gradient of the order parameter, U_{grad}, for a generic ferroic with an order parameter $\Theta = \Theta(x)$ (which can be polarization, or spontaneous strain, or magnetization) is

$$U_{grad} = \frac{k}{2} \left(\frac{\partial \Theta}{\partial x} \right)^2 \tag{1.1}$$

where k plays the role of the "exchange constant." This energy contribution is quadratic as it cannot depend on whether the gradient is positive or negative: the wall energy must be invariant under space inversion, as it obviously does not depend on whether you cross the wall from left to right or from right to left.

Minimization of the gradient of the order parameter to minimize the energy favors broadening of domain walls (broader walls means smaller gradients), but this broadening comes at the expense of having more material misaligned with respect to the ideal ordered state, and this also costs energy. Since the Landau free energy, F, of the homogeneous ferroic state is

$$F = \frac{a}{2}\Theta^2 + \frac{b}{2}\Theta^4 + O\left(\Theta^6\right), \tag{1.2}$$

where a and b are constants and Θ is the order parameter, the equilibrium thickness of the domain wall δ can be found by variational minimization of the total energy ΔG, including gradient and homogeneous terms, integrated across the domain wall thickness:

$$\Delta G = \int_{-\infty}^{\infty} \frac{a}{2}\Theta^2 + \frac{b}{2}\Theta^4 + \frac{k}{2}\left(\frac{\partial\Theta}{\partial x}\right)^2 \tag{1.3}$$

with the center of the wall at $x = 0$. Here, k is a constant and the boundary conditions are $\Theta(\infty) = -\Theta(-\infty) = \Theta_0$, with Θ_0 being the homogeneous monodomain state. The solution for this equation is (Mitsui and Furuichi 1953; Zhirnov 1959)

$$\Theta = \Theta_0 \tanh\left(\frac{x}{\lambda}\right) \tag{1.4}$$

with

$$\Theta_0 = \sqrt{-\frac{a}{b}} \tag{1.5}$$

and the correlation length λ

$$\lambda = 2\Theta_0^{-1}\sqrt{\frac{2k}{b}} = \sqrt{\frac{2k}{-a}}. \tag{1.6}$$

From the polarization profile defined by Equation (1.4), the domain wall thickness δ can in good approximation be defined as twice the correlation length, $\delta = 2\lambda$. Moreover, taking into account that the second derivative of the free energy, in this case with respect to the polarization P, yields the permittivity χ:

$$\chi = \frac{\partial^2 F}{\partial P^2} = -\frac{1}{2a}, \tag{1.7}$$

the domain wall thickness becomes

$$\delta = 2\sqrt{\frac{2k}{-a}} = 4\sqrt{\chi k}.$$

(1.8)

An intuitive simplification that brings in quantitatively similar results consists in replacing the hyperbolic tangent by a linear polarization profile across the wall (Catalan et al. 2012a):

$$\Theta(x) = \Theta_0 \frac{x}{\delta/2} \quad (-\delta/2 < x < \delta/2).$$

(1.9)

In this approximation, the gradient of the order parameter is equal to the total change of the order parameter, $2\Theta_0$, divided by the wall thickness, δ, so the energy density U (per unit volume) associated with the gradient is

$$U_{gradient} = {}^1\!/_2 \, k\left(2\Theta_0\big/\delta\right)^2.$$

(1.10)

The other term of the energy density is the standard electrostatic (or magnetostatic, or elastic) energy, which is proportional to the square of the order parameter:

$$U_{quadratic} = {}^1\!/_2 \, \chi^{-1}\Theta(x)^2.$$

(1.11)

The total energy density per unit area of the wall (σ) is obtained by integrating the volume energy density across the domain wall thickness:

$$\sigma = \int_{-\delta/2}^{\delta/2} \left[{}^1\!/_2 \, k\left(2\Theta_0\big/\delta\right)^2 + {}^1\!/_2 \, \chi^{-1}\Theta(x)^2 \right] dx = 2k\frac{\Theta_0{}^2}{\delta} + \frac{1}{6}\chi^{-1}\Theta_0{}^2\delta.$$

(1.12)

Minimizing this energy with respect to the domain wall thickness, .

$$\frac{\partial\sigma}{\partial\delta} = 0 = -2k\frac{P_0{}^2}{\delta^2} + \frac{1}{6}\chi^{-1}P_0{}^2,$$

(1.13)

we obtain an expression that is surprisingly close to Equation (1.8) despite the simplifications:

$$\delta = 2\sqrt{3}\sqrt{k\chi}.$$

(1.14)

All in all, the aforementioned equations tell us that:

i) The domain wall thickness is *inversely* proportional to the anisotropy of the order parameter (the quadratic term that leads the free energy term in the Landau expansion). This term is generally smaller for magnets than for ferroelectrics and ferroelastics, for which structural coupling is intrinsically strong. In consequence, ferroelectric and ferroelastic domain walls are thinner than magnetic walls. The orders of magnitude for the most common ferroic domain walls from thinnest to thickest would be ferroelectric domain walls, with $\delta \sim 1$ nm, followed by ferroelastic domain walls with $\delta \in (1 - 10)$ nm, and finally magnetic domain walls with typical thicknesses in the range $\delta \in (10 - 100)$ nm.

ii) The domain wall thickness is *directly* proportional to the susceptibility of the order parameter (Equation (1.8)). As the phase transition is approached, the susceptibility will tend to diverge, and hence so will the domain wall thickness. This is intuitive: near the phase transition, the anisotropy energy (conceived as the depth of the double well responsible for the ferroic state) is reduced, and so is the penalty for departing from the polarized state, resulting in broadened domain walls. Another way to look at this is to think of the domain wall as a layer of paraphase sandwiched between ferroic domains; as the actual paraphase is approached (as the temperature increases), the domain walls increase their thickness (Chrosh and Salje 1999) and eventually, at the transition temperature T_C, occupy the entire material.

1.2.2 Internal Symmetry of Domain Walls

The internal symmetry of domain walls can be deduced by group theory analysis (Fousek and Janovec 1969; Fousek 1971), by thermodynamic/Landau theory (Marton et al. 2010), or by first-principles calculations (Padilla et al. 1996; Lubk et al. 2009; Dieguez et al. 2013) (see also Chapters 3 and 4, respectively). Symmetry-based analyses, though quantitatively moot, are powerful for predicting the structurally allowed properties of the domain walls, and in this respect the seminal works of Privratska and Janovec (1997, 1999) should be credited as the first to point out the possibility of emergent magnetization being present inside the ferroelectric domain walls of magnetoelectric multiferroic materials. An update to those studies has been provided by Privratska (2007). Quantitative analyses of domain wall structure use the free energy as the starting point, and impose boundary conditions to obtain the domain wall profile and properties (Mitsui and Furuichi 1953; Zhirnov 1959). Daraktchiev et al. (2010) and Marton et al. (2010) solve the problem of calculating local symmetry inside the domain wall by first calculating the free energy of the bulk ferroic state, and then calculating the least energy path connecting ferroic minima, i.e. the potential wells of the free energy surfaces. The trajectory of this connecting path in phase space defines the state of the domain wall, and the symmetry in the middle of this path defines the state at the center of the domain wall. By way of illustration, the trajectory that connects the <100> minimum with the

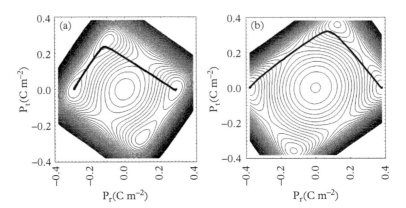

Figure 1.2 *Domain wall trajectories (in phase space) for orthorhombic 120° walls (left) and rhombohedral 180° walls in BaTiO₃, superimposed with the corresponding free energy surfaces. The vertical axis is the polarization along the tetragonal <001> direction, and the horizontal axis is the polarization along the rhombohedral <111> direction. Note that the domain wall trajectory is not straight, and as the polarization switches it takes a detour through a local minimum in which the polar symmetry of the wall is different from that of the domains.*
Source: Figure from Marton et al. 2011, copyright by the American Physics Society.

<010> minimum in a tetragonal ferroelectric such as BaTiO₃ corresponds to a 90° domain wall, and it is straightforward to see that this trajectory passes through a <110> point. In the center of the domain wall, then, the local polar symmetry of the material must be orthorhombic or monoclinic (polarization along <110> instead of tetragonal). Figure 1.2 provides equivalent examples for domain walls in the orthorhombic and rhombohedral phases of BaTiO₃, extracted from the work of Marton et al. (2010, 2011).

Another physically intuitive approach is to treat the domain wall as a slab of paraelectric phase epitaxially sandwiched between the two opposite domains on either side (Catalan 2012b). Let us take, for example, a tetragonal ferroelectric such as BaTiO₃, polarized along z and with a 180° domain wall in the y-z plane with thickness parallel to the x direction. In this case, the spontaneous strain in the domains is

$$s_x = s_y = Q_{12}P_0^2, s_z = Q_{11}P_0^2. \tag{1.15}$$

where P_0 is the spontaneous polarization, and Q_{12} and Q_{11} are, respectively, the transverse and longitudinal electrostrictive coefficients. Notice that, in general, the transverse coefficient is negative $Q_{12} < 0$, and the longitudinal one is positive $Q_{11} > 0$. Since the wall is coherently clamped between rigid domains, the strains s_z and s_y must remain unchanged, while the component perpendicular to the domain wall plane, s_x, is allowed to relax. Given that the electrostrictive contraction is suppressed at the wall (that is, $Q_{12} \to 0$ at the domain wall), s_x will relax to a bigger lattice parameter, and thus the lattice will expand perpendicular to the wall, leading to a cubic-like like structure such as the one of the corresponding paraphrase for this case. This can have consequences for

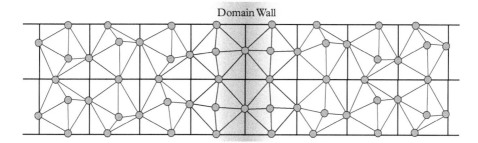

Figure 1.3 *In perovskite oxides, octahedral tilts appear in order to fit the oxygen octahedral inside the perovskite pseudocubic unit cell. If the unit cell is expanded—as must happen inside a ferroelectric domain wall—then the tilts will be reduced.*

Source: *Figure from Catalan 2012b, copyright by Taylor and Francis, 2012.*

the mechanical and thermal properties; longer bonds generally mean softer bonds and slower phonons, so one might expect the domain walls to be mechanically softer and to have different thermal conductivity compared to the bulk. The internal strain of the wall can also affect other important structural parameters, such as perovskite octahedral rotations, as illustrated in Figure 1.3.

1.3 Order Parameter Coupling

The previous analysis shows how the suppression of the main order parameter (e.g. polarization) inside the wall affects its properties and other parameters. Let us dwell further on this concept with respect to multiferroic (or quasi-multiferroic) materials. The phenomenological treatment of this problem starts with a Landau thermodynamic potential with two coupled order parameters (polarization P and magnetization M for magnetoelectrics) (Daraktchiev et al. 2008, 2010). Depending on the symmetry of the material, there can be lower-order couplings, but one that is universally allowed is biquadratic. In this case, the free energy is given by

$$\Delta G = \frac{\alpha}{2}P^2 + \frac{\beta}{4}P^4 + K(\Delta P)^2 + \frac{a}{2}M^2 + \frac{b}{4}M^4 + A(\Delta M)^2 + \gamma P^2 M^2 \qquad (1.16)$$

Here it is important to notice two things. First, the last term is always positive: the order parameters are both squared and thus positive, and the coupling term γ must be positive if the material has a paraphase. Second, because this term is positive, it increases the total energy of the system. Accordingly, in order to minimize the energy, if one of the two order parameters is nonzero, the other one should be zero; if both of them were nonzero at the same time, then the energy would be higher.

Let us imagine that the system is already in the ferroelectric state ($P \neq 0$). We can now group together the terms multiplying M^2, so that the coefficient of the magnetic order parameter is rewritten as

$$\Delta G \sim \left(\frac{a}{2} + \gamma P^2\right) M^2. \tag{1.17}$$

Typically, a depends on temperature as $a_0(T\text{-}T_C)$ (Curie Weiss law), where a_0 is a constant. Here, T_C is the Curie temperature, where the system would become magnetically ordered. Notice that there can be a range of temperatures $T < T_C$ where $a/2$ is negative (which would favor magnetic order), but γP^2 is positive and its absolute value is greater than $a/2$. This condition is fulfilled for $T^* < T < T_C$, where

$$T^* \equiv \left(Tc - \frac{2\gamma P^2}{a_0}\right) \tag{1.18}$$

is the renormalized (ferroelectrically modulated) magnetic ordering temperature. From this treatment, we can make two important observations:

1. When $T^* < T < T_C$, the material will be magnetically disordered in the ferroelectric domains, but magnetically ordered inside the domain walls, because the γP^2 term is suppressed. The resulting polarization and magnetization profiles will look as in Figure 1.4. Needless to say, the energy contribution from order parameter coupling will affect the thickness of the domain walls (Goltsev et al. 2003) and, by extension, the domain size scaling properties (Catalan et al. 2008).

2. It follows from the above treatment that there can be phase transitions inside domain walls at a different critical temperature (T_C) from that of the bulk (T^*). The concept of phase transitions inside domain walls was first put forward for

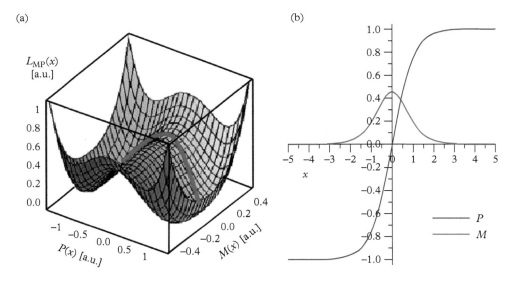

Figure 1.4 *Calculated polarization and magnetization profiles across a ferroelectric domain wall in a magnetoelectric with biquadratic coupling between* P *and* M. *Adapted from Daraktchiev et al. 2010. Copyright by the American Physics Society.*

magnetic systems by Lajzerowicz and Niez (1970) and is supported by the first-principles calculations of Wojdel and Iñiguez (2014) and the thermodynamic calculations of Stepkova et al. (2012).

The key physical insight is that when a component of the primary order parameter goes through zero (as happens inside a domain wall), it permits the emergence of the secondary order parameter that was otherwise frustrated by the primary one (Houchmandzadeh et al. 1991; Tagantsev et al. 2001; Daraktchiev et al. 2010).

1.4　Physical Properties of Domain Walls

1.4.1　Measuring Physical Properties at Domain Walls by AFM

If domain walls are predicted to have intrinsic physical properties distinguishing them from the contiguous domains, the experimental challenge is to measure them independently: they are too thin to be accessible by optical microscopy, and their volume fraction is too small for bulk spectroscopic techniques. Transmission electron microscopy is a suitable instrument for measuring the domain wall structure (Bursill et al. 1983; Bursill and Peng 1986; Floquet et al. 1997; Foeth et al. 1999; Jia et al. 2008) and chemical composition with atomic scale precision (Chapter 10), and even monitor the real-time dynamics of domain switching (Nelson et al. 2011); however, sample preparation is delicate and destructive. By contrast, AFM has emerged as a mainstream nanotool playing the role of an authentic nanolaboratory to characterize domain walls morphologically and functionally in a non-destructive manner, as well as to move them by applying localized electric fields or mechanical stresses (Lu et al. 2012).

The proper combination of cantilever- and tip-designs operated in appropriate modes enables the detection of forces below the piconewton and permits the characterization of physical and chemical properties. The tip of an atomic force microscope can be used as a top mobile electrode for polarization and transport measurements; a magnetic sensor for magnetostatic measurements; and a mechanical deformation sensor, able to operate from the low interacting force regime to a high indentation regime, to examine mechanical properties.

Given the versatility of AFM, it is also possible to combine several modes simultaneously or sequentially to investigate the coupling between different order parameters. Going beyond its multifunctional characterization options with nanoscale resolution, another notable distinguishing feature of AFM is its ability to manipulate physical magnitudes to deliver voltage, stress, magnetic fields, or heat to the domain wall to make it react. Because of these possibilities, AFM has become a unique tool to visualize, measure, and move domain walls at will, achieving nanometric lateral resolution (Shilo et al. 2004) and sub-microsecond dynamic sensitivity via pump-probe techniques (Gruverman et al. 2011).

AFM-based techniques that can be applied, alone or in combination (see Figure 1.5) to study domain walls in ferroic materials are, for example, piezoresponse force microscopy (PFM), which allows the ferroelectric polarization to be measured; magnetic

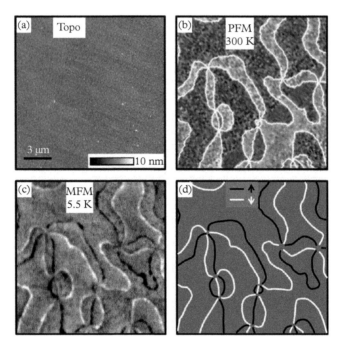

Figure 1.5 *Correlation of DW magnetism over a vortex network in multiferroic domain walls of hexagonal ErMnO₃.. (a) AFM topography image, (b) is the room temperature PFM image showing up and down domains, and (c) shows the MFM image taken at a low temperature and under a magnetic field of 0.1 T, scanning at a constant height of 40 nm above the surface. The sketch in (d) corresponds to the domain walls' net magnetic moments over the entire vortex network.*

Source: Adapted from Geng et al. 2012. Copyright by the American Chemical Society.

force microscopy (MFM), for the study of local magnetization in the domain walls; electric force microscopy (EFM) and Kelvin probe force microscopy (KPFM) for the study of electrostatic fields and surface potentials; conductive–AFM (C-AFM) for charge transport measurements; and contact resonance frequency (CRF) or force spectroscopy techniques for mechanical characterization. In addition, the analysis of any of the exposed physical magnitudes by AFM can be done as a function of in situ changes in temperature, external electric magnetic fields, light, or environmental control (see, e.g. Chapters 11 and 13 for examples).

1.4.2 Polarization

The internal symmetry of domain walls is different from that of the domains they separate and therefore, by von Neumann's principle, the symmetry-allowed functional properties inside domain walls can also be different from those in the domains. The most obvious examples pertain to polarity. Domain walls separating opposite polarities (i.e. 180° walls in ferroelectrics or ferromagnets) must go to zero polarization (along the easy axis) in their center –although the manner in which the polar inversion is achieved can be very different, ranging from rigid rotations (Néel and Bloch walls) to change in magnitude without rotation (see Figure 1.8).[1]

Conversely, in antipolar materials (antiferroelectrics, antiferromagnetics), any antiphase boundary must, by definition, contain a parallel pair of dipoles, and thus the domain walls are polar (Li 1956). These two extreme opposites of no polarization inside a ferroelectric wall and polarization inside an antiferroelectric wall are illustrated in Figure 1.6.

The case of ferroelastics is more subtle but still amenable to symmetry-based analyses. In particular, any ferroelastic domain wall must break the mirror symmetry along the spontaneous strain directions it separates. Consequently, all ferroelastic domain walls must be piezoelectric, even if the ferroelastic material is macroscopically centrosymmetric (Janovec et al. 1999). The strong strain gradients associated with ferroelastic domain walls can also lead to the emergence of ferroelectric polarization at the wall, as observed for the nonpolar $SrTiO_3$ (Zubko et al. 2007) and $CaTiO_3$ (Gonçalves-Ferreira et al. 2008; Van Aert et al. 2012). One consequence of domain wall piezoelectricity is that a high concentration of domain walls in a sample can result in a macroscopically enhanced piezoelectric coefficient (Wada et al. 2006; Hlinka et al. 2009).

Piezoelectricity at the domain walls can be detected by PFM, exploiting the converse piezoelectric effect: the AFM tip used as a top mobile electrode in contact with the samples applies an AC electric field and simultaneously senses the sample mechanical expansion and contraction. The disappearance of polarization inside the 180° domain walls of ferroelectrics has other physical consequences. The first and most obvious one is that, since there is no polarity inside the wall, there cannot be piezoelectricity either. Consequently, the amplitude of the signal in PFM images will be zero at the domain walls where the polarity switches sign, i.e. at the 180° domain walls (see Figure 1.7).

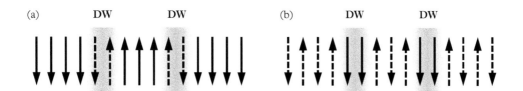

Figure 1.6 *Sketches of domain walls in (a) polar and (b) antipolar materials. In the first case, the polarization disappears inside the walls of polar materials; conversely, polarization appears inside the walls of antipolar materials. In both cases, the polar properties of the walls are exactly the opposite to those of the domains.*

---

[1] Even though the polarization projection along the easy axis must be zero at the center of the wall, it can have components along the other axes; that would be the case for Néel walls and Bloch walls as shown in Figure 1.8.

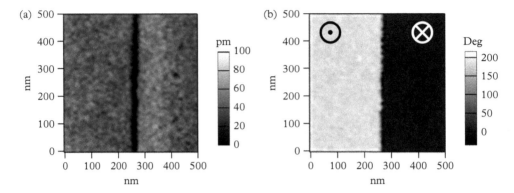

Figure 1.7 *Amplitude and phase PFM images of 180° domain walls. The phase of the induced oscillation is correlated with the sign of the ferroelectric polarization. The lack of amplitude signal at the domain wall separating two opposite domains indicates the lack of piezoelectricity as the center of the wall is nonpolar, as illustrated in Figure 1.6a.*

1.4.3 Charge Transport

The switching of polarization across a ferroelectric domain wall can either happen through progressive change of the magnitude of the polarization (Ising walls), or through rotation—either within the plane of the wall (Bloch walls) or perpendicular to it (Néel walls) (Lee et al. 2009), as depicted in Figure 1.8.

Ising walls and Bloch walls are electrostatically neutral, but Néel walls are not, because they present a discontinuity of the displacement field perpendicular to the wall. According to Poisson's equation, $\nabla D = \rho_f$, where ρ_f is the free charge density, and D is the displacement field, which in a ferroelectric is $D = \varepsilon_0 E + P$. Any Néel-type wall in a ferroelectric must therefore accrue a free charge density that is equal to the polarization discontinuity perpendicular to the wall plane.

Poisson's free charge will also accumulate at the walls when there is a head-to-head or tail-to-tail component of the polarization, as this also implies a discontinuity of the displacement field (Xiao et al. 2005; Gureev et al. 2011). Such head-to-head or tail-to-tail configurations may appear spontaneously in so-called "improper" ferroelectrics (where the ferroelectric order parameter is a "slave" to a dominant primary order parameter—Choi et al. 2010; Meier et al. 2012) (see Chapter 6 for details). Charged domain walls can also be forced to appear even in proper ferroelectrics, such as $BaTiO_3$, by pitting the strain order parameter against the polar order parameter (frustrated poling; Sluka et al. 2013). The trick in either case is to force the appearance of electrostatically unfavorable charged domain walls by acting on a structural order parameter, such as strain, that is coupled to the polarization.

Increased charge density, however, does not imply increased conductivity. Ferroelectrics are semiconductors. For n-type semiconductors, electrons increase conductivity, while holes neutralize the majority carriers, thus decreasing conductivity. Accordingly, for n-type ferroelectrics, head-to-head domain walls will be more

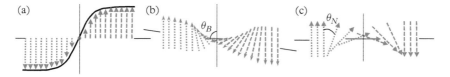

Figure 1.8 *Different types of domain walls: (a) Ising type, (b) Bloch type, and (c) Néel type.*

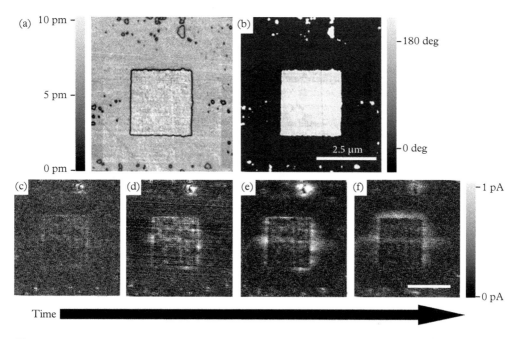

Figure 1.9 *Upward out-of-plane written ferroelectric square domain on an as-grown downward-polarized BiFeO₃ (111) thin film. The domain was written by applying a negative voltage with an AFM tip. The polarization structure is obtained from (a) the amplitude and (b) the phase PFM images. The bottom images from (c) to (f) show the evolution of the domain wall conductivity over different sequential scans. Note that with each consecutive scan, the conductive region becomes broader and more "diffuse," indicative of the sideways diffusion of the conductive defects (likely oxygen vacancies).*

conductive than the bulk, whereas tail-to-tail walls will be less conductive, the converse being true for p-type semiconductors. This phenomenon was theoretically predicted by Eliseev et al. (2011) and experimentally confirmed by Meier et al. (2012).

Domain wall conductivity can also be modified by chemical doping (when the doping goes preferentially into the domain wall, this is referred to as "decoration"). This was famously demonstrated by Aird and Salje's detection of superconductivity along domain walls in sodium-doped WO_3 (Aird and Salje 1998). As well as being dopants, ions can be charge carriers themselves. A signature of ionic conductivity is its slow dynamics, as illustrated in Figure 1.9. It is also worth mentioning that domain wall decoration does not necessarily imply increased domain wall conductivity, as the reverse—increased domain wall resistivity—has also been observed (Bartels et al. 2003). Likewise, it is also worth mentioning that domain wall decoration is sometimes intentional, but it can also happen spontaneously. In oxides, domain walls are known to attract oxygen vacancies (He and Vanderbilt 2003), and these will act as charge donors inside the wall (Farokhipoor and Noheda 2011).

The coupling of ferroelasticity and strain gradients at the domain walls can also impact their conductivity properties. Strain at the domain walls in strongly correlated electron oxides can deform the octahedral tilting in ABO_3 perovskite structures (B-O-B bond angle distortion) and bond lengths (B-O bond length distortion), changing the orbital overlapping. Since orbital overlapping is the structural mechanism governing the bandwidth of these materials, this is observed to enhance electrical conductivity, as observed in Figure 1.10.

1.4.4 Magnetism at Domain Walls

The strain contained in ferroelastic domain walls affects the orbital overlapping in ferroelastic magnetic materials and can enhance the magnetic moment at the domain walls, as observed in $LaSrMnO_3$ and shown in Figure 1.10 (Balcells et al. 2015). In this case, the magnetic tip of an MFM can be used to monitor specifically the magnetic moment of the domain walls, which tends to orient following the magnetic field signal emitted by the MFM tip. This leads to the observation of magnetic domain walls as bright lines in MFM images when the tip is scanned sufficiently close to the surface.

The coupling of magnetism and ferroelectricity in multiferroics is a promising source of magnetoelectric coupling that can be exploited to manipulate spins with electric fields. Most multiferroics are antiferromagnets with a vanishing magnetic moment, but the symmetry breaking at the domain walls can lead to confined magnetic moments in these topological structures. For example, spin rotation across a magnetic Néel wall can induce, via Dzyaloshinskii–Moriya interaction, an electric polarization (Logginov et al. 2008). Also, strain gradients can cause polarization via flexoelectricity (Zubko et al. 2013), and in turn this polarization may cause spin canting via magnetoelectricity.

An example of magnetization emerging inside the domain walls of an antiferromagnetic multiferroic is hexagonal $ErMnO_3$, an improper ferroelectric showing a collective domain wall magnetism over its vortex structure (Geng et al. 2012). Low-temperature MFM images shown in Figure 1.5 are able to map the net magnetization at the domain walls.

1.4.5 Magnetotransport

Any magnetic semiconductor (and, by extension, most magnetoelectric multiferroics, given that they tend to be wide-bandgap semiconductors) is likely to display some form of magnetoresistance. This is the case with the domain walls of multiferroic $BiFeO_3$ (He et al. 2012; Domingo et al. 2017). It is important to note, however, that while all magnetic materials can be magnetoresistive, magnetism is not a prerequisite

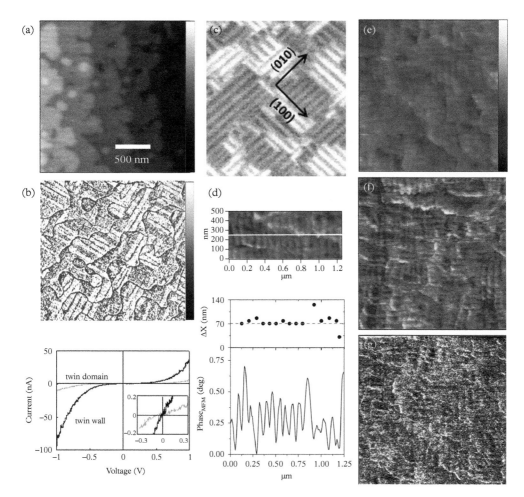

Figure 1.10 *Correlation between strain and enhanced magnetic and conductive properties at domain walls of LaSrMnO₃ domain walls. (a) Topography AFM image of the film and (b) C-AFM image showing an enhanced conductivity at the domain walls, observed as thin black lines, together with the corresponding I(V) curves at the domains and domain walls. The ferroelastic twin walls can be observed as patterns in OC-SEM images in (c). Finally, the magnetic properties of the strained twin walls are shown in images d–g; (d) shows an MFM image together with the associated profile section demonstrating the correlation of the magnetic pattern with that of the twin walls. The right column shows the MFM images at different distances from the surface: (e) d = 35 nm, (f) d = 20 nm, and (g) d = 10 nm. As the magnetic tip approaches the surface, it interacts more strongly with the magnetic moment of the domain walls until it is finally able to switch it, enhancing the magnetic contrast.*

for magnetoresistance, because the Hall Effect can modify the current paths in any thin disordered semiconductor (Parish and Littlewood 2003). Since "thin disordered semiconductor" is a good description of any ferroelectric domain wall, it follows that it may in principle be possible to measure magnetoresistance in the domain walls of any ferroelectric even if it is not multiferroic, a hypothesis that remains to be tested. A distinguishing feature of this "geometric magnetoresistance" is that it is positive (resistance increases with increasing magnetic field). This is the opposite of what is found in magnetic conductors, for which field-induced spin alignment reduces the magnetic scattering of the charge carriers. Another feature is that the effect should be maximum when the magnetic field is perpendicular to the transport plane, i.e. perpendicular to the domain wall. Both of these features have been detected in in-situ magnetotransport measurements of single domain walls of $BiFeO_3$ (Domingo et al. 2017), suggesting a geometric origin for their room-temperature magnetoresistance (see Figure 1.11).

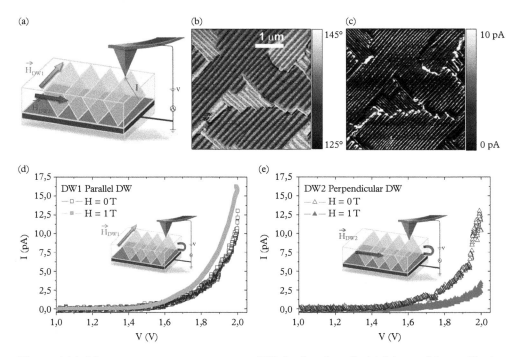

Figure 1.11 *Magnetotransport measurements on $BiFeO_3$ domain walls. (a) Scheme of the combined C-AFM setup with an external magentic field. (b) The amplitude PFM image shows two families of a1/a2 domains with perpendicular relative orientation with respect to the external applied magnetic field. (c) All the domain walls are observed to be conductive, but their level of conductivity changes when a magnetic field is applied either parallel as in (d) or perpendicular to the longitudinal domain walls as shown in (e).*

Source: Adapted from Domingo et al. 2017.

Transport measurements of domain walls can be achieved by different methods. The simplest way is to place microelectrodes on top of a crystal and monitor the conductivity while controlling the number and orientation of the domain walls between the electrodes. This measurement leads to an average of the transport properties over a certain distribution of domains. In contrast, C-AFM can measure the local transport properties at the nanoscale, and enjoys an extra degree of freedom to play with the geometrical symmetry of the transport measurement. Magnetotransport measurements at the nanoscale can be achieved by a combination of C-AFM and the in situ application of an external magnetic field. This allows to capture subtle coupling effects that might otherwise get cancelled when averaged over microscopic dimensions. In the example shown in Figure 1.11, this allows distinguishing the local effect of the orientation of each domain wall with respect the applied magnetic field; this, it turns out, is important, because the sign of the magnetoresistance is orientation-dependent, and it averages out to zero in macroscopic measurements containing many randomly oriented walls.

1.4.6 Mechanical Response

The mechanical response of domain walls can also be locally probed by scanning force microscopy. Among all the available operation modes, a good choice to accomplish this is CRF measurements based on tracking the mechanical resonance frequency of the cantilever in contact with the sample. By applying an oscillatory mechanical force to the cantilever and monitoring the resonance frequency, we can detect relative differences in stiffness between the domain and the domain wall. Our measurements for different ferroelectrics (Stefani et al. 2020) consistently show that the resonance frequency of 180° domain walls is slower (and thus the domain wall is softer) than the domains themselves (Figure 1.12). This contrast in mechanical response may seem surprising, given that there is no mechanical difference between 180° domains themselves. (Note that there is also a weak CRF contrast between domains due to flexoelectric coupling (Cordero-Edwards et al. 2017))

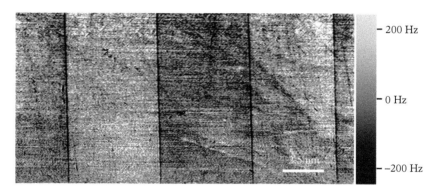

Figure 1.12 *Contact resonance frequency image of a pattern of stripe domains on a periodically poled LiNbO$_3$ single crystal. The domain walls stand out as dark lines as depicted by a decrease in the contact resonance frequency of the cantilever when the tip is scanning over the wall. This feature can be correlated with a decrease in the domain wall stiffness with respect to that of the adjacent domains.*

The reason for the mechanical contrast of the walls is, once again, the coupling between order parameters, in this case, strain and polarization, which are necessarily coupled in any piezoelectric and thus in any ferroelectric. Inducing a deformation in a ferroelectric changes its polarization (ferroelectrics are piezoelectric), and this change in polarization has an electrostatic energy cost. The domain wall, itself, however, is not polar, and therefore it is not piezoelectric, meaning that there is no electrostatic cost for its deformation. Stating that the deformation of the domain wall has a reduced energy cost is just a physicist's way of saying that deforming a ferroelectric wall is easy. In other words, as observed, the wall will respond more softly to the atomic force microscope indentation than the domains.

1.5   Summary and Conclusions

Domain walls behave as embedded layers whose properties can be understood as those of a thin film of paraphase sandwiched coherently between the domains. Thus, the effect of cancelling the order parameter can be modeled by standard thermodynamic arguments. Perhaps the key message from the theoretical section is that the switching off of the order parameter has consequences not only in terms of new symmetry-allowed (or symmetry-forbidden) properties, but also it has a knock-on effect on secondary parameters that might be coupled to or suppressed by the primary ordering. Thus, for example, the suppression of polarization means that the inside of ferroelectric domain walls is not piezoelectric, and this implies meaning that straining them will not generate any polarization—so their mechanical response will be different (softer) than that of the domains. In magnetoelectrics, the emergence or suppression of ferroelectric polarization inside a domain wall also has consequences for the wall's magnetization. In turn, changes in magnetization inside the wall can affect resistivity via magnetoresistance.

The resistivity itself is also affected by the manner in which polarization switches from one side of the wall to the other; hence, head-to-head polarization or tail-to-tail polarization (which, despite being energetically costly, can happen in multiferroic domain walls where the polarization is a "slave" to a primary order parameter) must result in charge accumulation or depletion at the wall, and thus, in dramatically modified electronic properties. Though we have not covered them here, the change in electronic properties will also affect related properties such as the photovoltaic response at the wall (see Chapter 9). The bottom line is that the panoply of properties that domain walls can display is as varied—or more—as that of the multifunctional materials where they reside, but this complexity of behavior can be understood with simple physics: treat the wall as a very thin film, and write the thermodynamic potential including gradient and coupling terms.

The rich variety of domain wall behaviors is matched by the versatility of AFM, arguably the most important tool in any domain wall research laboratory. The possibility of using functionalized cantilevers in different measurement modes and in different in situ environments allows measuring—and manipulating—domain wall responses to different stimuli.

The measurement possibilities of AFM are as varied as the functionalities displayed by domain walls, but perhaps the best-known and most-used one is PFM, which allows identifying the exact location of domain walls, and/or creating them by delivering a high voltage capable of switching a ferroelectric domain. PFM is by now a well-established technique that is described in more detail in other chapters, so in this chapter we have described in more detail a couple of more innovative measurement modes that allow, respectively, the measurement of domain wall magnetoresistance and domain wall mechanical properties. These are just two examples that illustrate the experimental possibilities offered by the creative combination of domain walls and alternative scanning probe microscopy modes. We hope that the readers will be inspired to try their own ideas. It is a fun game.

....

REFERENCES

Aird A., Salje E. K. H., "Sheet superconductivity in twin walls: experimental evidence of WO_{3-x}," *J. of Phys. Cond. Mat.* 10, L377 (1998).

Allwood D. A., Xiong G., Faulkner C. C., Atkinson D., Petit D., Cowburn R. P., "Magnetic domain-wall logic," *Science* 309, 1688 (2005).

Balcells L. L., Paradinas M., Bagués N., Domingo N., Moreno R., Galceran R., Walls M., Santiso J., Konstantinovic Z., Pomar A., Casanove M. -J., Ocal C., Martínez B., Sandiumenge F., "Enhanced conduction and ferromagnetic order at (100)-type twin walls in $La_{0.7}Sr_{0.3}MnO_3$ thin films," *Phys. Rev. B* 92, 075111 (2015).

Bartels M., Hagen V., Burianek M., Getzlaff M., Bismayer U., Wiesendanger R., "Impurity-induced resistivity of ferroelastic domain walls in doped lead phosphate," *J. Phys.: Condens. Mat.* 15, 957–962 (2003).

Bursill L. A., Peng J. L., "Electron microscopic studies of ferroelectric crystals," *Ferroelectrics* 70, 191 (1986).

Bursill L. A., Peng J. L., Feng D., "HREM study of (100) ferroelectric domain-walls in potassium niobate," *Philos. Mag. A* 48, 953 (1983).

Catalan G., "On the link between octahedral rotations and conductivity in the domain walls of $BiFeO_3$," *Ferroelectrics* 433, 65–73 (2012b).

Catalan G., Béa H., Fusil S., Bibes M., Paruch P., Barthélémy A., Scott J. F., "Fractal dimension and size scaling of domains in thin films of multiferroic $BiFeO_3$," *Phys. Rev. Lett.* 100, 027602 (2008).

Catalan G, Seidel J., Ramesh R., Scott J. F., "Domain wall nanoelectronics," *Rev. Mod. Phys.* 84, 119–156 (2012a).

Choi T., Horibe Y., Yi H. T., Choi Y. J., Wu W., Cheong S.-W., "Insulating interlocked ferroelectric and structural antiphase domain walls in multiferroic $YMnO_3$," *Nat. Mat.* 9, 253 (2010).

Chrosh J., Salje E. K. H., "Temperature dependence of the domain wall width in $LaAlO_3$," *J. Appl. Phys.* 85, 722 (1999).

Cordero-Edwards K., Domingo N., Abdollahi A., Sort J., Catalan G., "Ferroelectrics as Smart Mechanical Materials," *Adv. Mater.* 29, 1702210 (2017).

Daraktchiev M., Catalan G., Scott J. F., "Landau theory of ferroelectric domain walls in magneto-electrics," *Ferroelectrics* 375, 122 (2008).

Daraktchiev M., Catalan G., Scott J. F., "Landau theory of domain wall magnetoelectricity," *Phys. Rev. B* 81, 224118 (2010).

Diéguez O., Aguado-Puente P., Junquera J., Íñiguez J., "Domain walls in a perovskite oxide with two primary structural order parameters: first-principles study of $BiFeO_3$," *Phys. Rev. B* 87, 024102 (2013).

Domingo N., Farokhipoor S., Santiso J., Noheda B., Catalan G., "Domain wall magnetoresistance in $BiFeO_3$ thin films measured by scanning probe microscopy," *J. Phys.: Condens. Mat.* 29, 334003 (2017).

Eliseev E. A., Morozovska A. N., Svechnikov G. S., Venkatraman Gopalan, Shur V. Ya., "Static conductivity of charged domain walls in uniaxial ferroelectric semiconductors," *Phys. Rev. B* 83, 235313 (2011).

Farokhipoor S., Noheda B., "Conduction through 71° domain walls in $BiFeO_3$ thin films," *Phys. Rev. Lett.* 107, 127601 (2011).

Floquet N., Valot C. M., Mesnier M. T., Niepce J. C., Normand L., Thorel A., Kilaas R., "Ferroelectric domain walls in $BaTiO_3$: fingerprints in XRPD diagrams and quantitative HRTEM image analysis," *J. Physique III* 7, 1105 (1997).

Foeth M., Sfera A., Stadelmann P., Buffat P. -A., "A comparison of HREM and weak beam transmission electron microscopy for the quantitative measurement of the thickness of ferroelectric domain walls," *J. Electron Microsc.* 48(6), 717–723 (1999).

Fousek J., "Permissible domain walls in ferroelectric species," *Czech. J. Phys.* 9, 955 (1971).

Fousek J., Janovec V., "The orientation of domain walls in twinned ferroelectric crystals," *J. Appl. Phys.* 40, 135 (1969).

Geng Y., Lee N., Choi Y. J., Cheong S.-W., Weida Wu, "Collective magnetism at multiferroic vortex domain walls," *Nano Lett.* 12, 6055 (2012).

Goltsev A. V., Pisarev R. V., Lottermoser Th., Fiebig M., "Structure and interaction of antiferromagnetic domain walls in hexagonal $YMnO_3$," *Phys. Rev. Lett.* 90, 177204 (2003).

Gonçalves-Ferreira L., Redfern S. A. T., Artacho E., Salje E. K. H., "Ferrielectric twin walls in $CaTiO_3$," *Phys. Rev. Lett.* 101, 097602 (2008).

Gruverman A., Wu D., Scott J. F., "Piezoresponse force microscopy studies of switching behavior of ferroelectric capacitors on a 100-ns time scale," *Phys. Rev. Lett.* 100, 097601 (2008).

Gureev M. Y., Tagantsev A. K., Setter N., "Head-to-head and tail-to-tail 180° domain walls in an isolated ferroelectric," *Phys. Rev. B* 83, 184104 (2011).

He L., Vanderbilt D., "First-principles study of oxygen-vacancy pinning of domain walls in $PbTiO_3$," *Phys. Rev. B* 68, 134103 (2003).

He Q., Yeh C.-H., Yang J.-C., Singh-Bhalla G., Liang C.-W., Chiu P.-W., Catalan G., Martin L. W., Chu Y.-H., Scott J. F., Ramesh R., "Magnetotransport at domain walls in $BiFeO_3$," Phys. Rev. Lett. 108, 067203 (2012).

Hlinka J., Ondrejkovic P., Marton P., "The piezoelectric response of nanotwinned $BaTiO_3$," *Nanotechnology* 20, 105709 (2009).

Houchmandzadeh B., Lajzerowicz J., Salje E. K. H., "Order parameter coupling and chirality of domain walls," *J. Phys.: Condens. Mat.* 3, 5163 (1991).

Janovec V., Richterová L., Privratska J., "Polar properties of compatible ferroelastic domain walls," *Ferroelectrics* 222, 331 (1999).

Jia C. L., Mi S. B., Urban K., Vrejoiu I., Alexe M., Hesse D., "Atomic-scale study of electric dipoles near charged and uncharged domain walls in ferroelectric films," *Nat. Mater.* 7, 57 (2008).

Lajzerowicz J., Niez J. J., "Phase transition in a domain wall," *J. Phys. Lett. Paris* 40, L165 (1979).

Lee D., Behera R. K., Wu P., Xu H., Li Y. L., Sinnott S. B., Phillpot S. R., Chen L. Q., Gopalan V., "Mixed Bloch-Néel-Ising character of 180° ferroelectric domain walls," *Phys. Rev. B* 80, 060102(R) (2009).

Lee W. T., Salje E. K. H., "Chemical turnstile," *Appl. Phys. Lett.* 87, 143110 (2005).

Logginov A. S., Meshkov G. A., Nikolaev A. V., Nikolaeva E. P., Pyatakov A. P., Zvezdin A. K., "Room temperature magnetoelectric control of micromagnetic structure in iron garnet films," *Appl. Phys. Lett.* 93, 182510 (2008).

Lu H., Bark C.-W., Esque de los Ojos D., Alcala J., Eom C. B., Catalan G., Gruverman A., "Mechanical writing of ferroelectric polarization," *Science* 336, 59–61 (2012).

Lubk A., Gemming S., Spaldin N. A., "First-principles study of ferroelectric domain walls in multiferroic bismuth ferrite," *Phys. Rev. B* 80, 104110 (2009).

Marton P., Rychetsky I., Hlinka J., "Domain walls of ferroelectric $BaTiO_3$ within the Ginzburg-Landau-Devonshire phenomenological model," *Phys. Rev. B* 81, 144125 (2010).

Marton P., Rychetsky I., Hlinka J., "Erratum: domain walls of ferroelectric $BaTiO_3$ within the Ginzburg-Landau-Devonshire phenomenological model," *Phys. Rev. B* 84, 139906(E) (2011).

Meier D., Seidel J., Cano A., Delaney K., Kumagai Y., Mostovoy M., Spaldin N. A., Ramesh R., Fiebig M., "Anisotropic conductance at improper ferroelectric domain walls." *Nat. Mater.* 11, 284 (2012).

Mitsui T., Furuichi J., "Domain structure of rochelle salt and KH_2PO_4," *Phys. Rev.* 90, 193–202 (1953).

Nelson C. T., Gao P., Jokisaari J. R., Heikes C., Adamo C., Melville A., Baek S.-H., Folkman C. M., Winchester B., Gu Y., Liu Y., et al., "Domain dynamics during ferroelectric switching," *Science* 334, 968–971 (2011).

Padilla J., Zhong W., Vanderbilt D., "First-principles investigation of 180° domain walls in $BaTiO_3$," *Phys. Rev. B* 53, R5969 (1996).

Parish M. M., Littlewood P. B., "Non-saturating magnetoresistance in heavily disordered semiconductors," *Nature* 426, 162–165 (2003).

Parkin S. S. P., Hayashi M., Thomas L., "Magnetic domain-wall racetrack memory," *Science* 320, 190 (2008).

Privratska J., "Possible appearance of spontaneous polarization and/or magnetization in domain walls associated with non-magnetic and non-ferroelectric domain pairs," *Ferroelectrics* 353, 116 (2007).

Privratska J., Janovec V., "Pyromagnetic domain walls connecting antiferromagnetic non-ferroelastic magnetoelectric domains," *Ferroelectrics* 204, 321 (1997).

Privratska J., Janovec V., "Spontaneous polarization and/or magnetization in non-ferroelastic domain walls: symmetry predictions," *Ferroelectrics* 222, 23 (1999).

Salje E. K. H., *Phase Transitions in Ferroelastic and Co-Elastic Materials* (Cambridge University Press, Cambridge, 1993).

Seidel J., Martin L. W., He Q., Zhan Q., Chu Y.-H., Rother A., Hawkridge M. E., Maksymovych P., Yu P., Gajek M., Balke N., Kalinin S. V., Gemming S., Wang F., Catalan G., Scott J. F., Spaldin N. A., Orenstein J., Ramesh R., "Conduction at domain walls in oxide multiferroics," *Nat. Mater.* 8, 229 (2009).

Shilo D., Ravichandran G., Bhattacharya K., "Investigation of twin-wall structure at the nanometre scale using atomic force microscopy," *Nat. Mater.* 3, 453–457 (2004).

Sluka T., Tagantsev A. K., Bednyakov P., Setter N., "Free electron gas at charged domain walls in insulating $BaTiO_3$," *Nat. Commun.* 4, 1808 (2013).

Stefani Ch., Ponet L., Shapovalov K., Chen P., Langenberg E., Schlom D. G., Artyurhin S., Stengel M., Domingo N., Catalan G., "Ferroelectric 180 degree walls are mechanically softer than the domains they separate," arXiv:2005.04249 (2020).

Stepkova V., Marton P., Hlinka J., "Stress-induced phase transition in ferroelectric domain walls of $BaTiO_3$," *J. Phys.: Condens. Mat.* 24, 212201 (2012).

Tagantsev A. K., Courtens E., Arzel L., "Prediction of a low-temperature ferroelectric instability in antiphase domain boundaries of strontium titanate," *Phys. Rev. B* 64, 224107 (2001).

Van Aert S., Turner S., Delville R., Schryvers D., Van Tendeloo G., Salje E. K. H., "Direct observation of ferrielectricity at ferroelastic domain boundaries in $CaTiO_3$ by electron microscopy," *Adv. Mater.* 24, 523 (2012).

Wada S., Yako K., Yokoo K., Kakemoto H., Tsurumi T., "Domain wall engineering in barium titanate single crystals for enhanced piezoelectric properties," *Ferroelectrics* 334, 17 (2006).

Wojdeł J. C., Íñiguez J., "Ferroelectric transitions at ferroelectric domain walls found from first principles," *Phys. Rev. Lett.* 112, 247603 (2014).

Xiao Y., Shenoy V. B., Bhattacharya K., "Depletion layers and domain walls in semiconducting ferroelectric thin films," *Phys. Rev. Lett.* 95, 247603 (2005).

Zhirnov V. A., "Contribution to the theory of domain walls in ferroelectrics," *Sov. Phys. JETP* 35, 822 (1959).

Zubko P., Catalan G., Buckley A., Welche P. R. L., Scott J. F., "Strain-gradient-induced polarization in $SrTiO_3$ single crystals," *Phys. Rev. Lett.* 99, 167601 (2007).

Zubko P., Catalan G., Tagantsev A. K., "Flexoelectric effect in solids," *Annu. Rev. Mater. Res.* 43, 387–421 (2013).

2

Novel Phases at Domain Walls

S. Farokhipoor[1], C. Magen[2], D. Rubi[3], and
B. Noheda

[1] *Zernike Institute for Advanced Materials, University of Groningen, Nijenborgh 4, 9747AG Groningen, The Netherlands*
[2] *Instituto de Ciencia de Materiales de Aragón, CSIC—Universidad de Zaragoza, Departamento de Física de la Materia Condensada, Pedro Cerbuna 12, 50009 Zaragoza, Spain*
[3] *Instituto de Nanociencia y Nanotecnología, Comisión Nacional de Energía Atómica and Consejo Nacional de Investigaciones Científicas y Técnicas, Gral. Paz 1499, San Martín, Argentina*

In addition to the distinct internal symmetry of domain walls discussed in Chapter 1, local stresses can induce new physical properties at domain walls. In this chapter, we discuss how the intense stress fields generated at the ferroelastic domain walls in some epitaxially strained oxide layers introduce selective chemical modifications, giving rise to novel 2D crystal structures at the walls. In the case of the strained $TbMnO_3$ films presented here, the substitution of Tb by Mn creates a net magnetic moment at each domain wall, thus making them distinct from the bulk-like antiferromagnetic domains. The possibility of tuning the domain wall density with the film thickness makes it possible to tailor the material's response at the nanoscale.

2.1 Introduction to TbMnO₃

The perovskite structure, with the general formula ABO_3, has cubic symmetry with the space group *Pm3m*. Oxygen anions are arranged in a corner-shared, face-centered fashion, forming a connected net of octahedral cages throughout the crystal structure in three dimensions. The smaller-size cation (B), in 6-fold coordination, is placed inside the octahedral cage, whereas A is a 12-fold coordinated cation located in the voids between the octahedra. At room temperature, most of the interesting perovskite-like materials are slightly distorted with respect to the structure described above, with one or both cations and/or the oxygens shifted along particular directions, lowering the symmetry. Among perovskites, manganites present a rich physics with interesting magnetic and

S. Farokhipoor, C. Magen, D. Rubi, and B. Noheda, *Novel Phases at Domain Walls* In: *Domain Walls: From Fundamental Properties to Nanotechnology Concepts.* Edited by: Dennis Meier, Jan Seidel, Marty Gregg, and Ramamoorthy Ramesh, Oxford University Press (2020).
© S. Farokhipoor, C. Magen, D. Rubi, and B. Noheda.
DOI: 10.1093/oso/9780198862499.003.0002

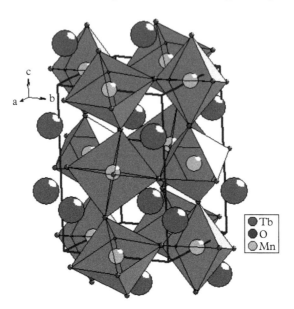

Figure 2.1 *Prototype orthorhombic crystal structure of TbMnO₃. The oxygen octahedra contained in one unit cell are shaded. Mn^{3+} cations are placed inside the octahedra, and the Tb^{3+} cations are located in the voids between the octahedra.*

electrical properties (Goodenough 1955; Coey et al. 1999; Salamon and Jaime 2001). Even though we focus on a perovskite manganite, TbMnO₃, the effects presented here may be generalized to different types of materials with similar constraints.

TbMnO₃ has a distorted perovskite structure with orthorhombic symmetry, as shown in Figure 2.1. In this particular atomic arrangement, the magnetic interactions between the 3d electrons of Mn^{3+} ions take place by superexchange mediated by the oxygen 2p orbitals, and TbMnO₃ shows magnetic frustration due to the competition between ferromagnetic and antiferromagnetic exchange interactions involving the magnetic moments at near neighbors' and next near neighbors' Mn^{3+} sites. This causes antiferromagnetic ordering below $T_N \approx 41K$, first in the form of a sinusoidal spin density wave and, upon further cooling, as a cycloidal spin modulation, below $T_{N'} \approx 25K$. The cycloid arises together with a relative displacement of the oxygens, due to the presence of the inverse Dzyaloshinskii–Moriya interaction, breaking the inversion symmetry and inducing ferroelectricity (Kimura et al. 2003; Kenzelmann et al. 2005).

Perovskite-based crystals are rarely homogeneous. Symmetry lowering from the high-temperature cubic (parent) phase gives rise to the formation of crystallographic domains or twins, that is, regions of identical crystal structure but different crystal orientation. Because different crystal orientations lead to different strain states, these domains are ferroelastic (stress can be used to tune the deformation between two or three different states) (Salje 1993). In bulk materials and in the absence of external forces, transitioning to the low-symmetry phase will take place by random nucleation, and subsequent growth, of domains at different parts of the crystal. The transition is completed when all domains meet, forming domain walls. In order to minimize the elastic energy, domain walls are formed along those atomic planes that are closely oriented in contiguous domains and can be brought to coincide by small adaptations (see also Chapters 1 and 4). If the crystal is free, this can take place by tilting of the twin domains, but if, on the other hand, the crystal is confined to a bulkier substrate, as in the case of epitaxial growth of thin films, twinning can be accompanied by very large strain gradients (Catalan et al. 2011). In this case of epitaxial growth, for an appropriate lattice mismatch between the film and the substrate, domain patterns can be formed by the alternation of two or more domains in a periodic fashion, rather than by random nucleation, in order to achieve lattice matching of the film with the substrate (Roitburd 1976). In this way, periodic arrays of domain walls are introduced during the transition to the distorted phase (Tagantsev et al. 2010).

Due to the crucial role that structural details play in determining the ferroic order parameters, it is naturally expected that the properties of multiferroics, and in particular manganites, are greatly modified (and possibly improved) via strain-engineered domain formation, as addressed in Chapter 5. Next to the structural and elastic degrees of freedom, in multiferroic materials, the magnetic and electric boundary conditions also play a crucial role in determining the formation of magnetic and ferroelectric domains, respectively. Depolarization or demagnetization fields that develop in the material as a result of the presence of a homogeneous polarization or magnetization, respectively, become stronger as the dimensions perpendicular to the net moment decrease, and thus there is a critical size below which a homogeneous magnetization or polarization cannot be stabilized. The material, then, will break into domains, the size of which will be determined by the balance between the dipolar/magnetic energy of the domains and the anisotropy energy, which determines the domain wall formation energy.

The simplest model that proposes this balance was reported for a slab of magnetic material by Landau and Lifshitz (Landau and Lifshitz 1935) and Kittel (Kittel 1946) by considering that the domain energy density is determined by the demagnetization field of the spins perpendicular to the surface. If the magnetic anisotropy is large enough, stripe domains, separated by 180° domain walls, form; while for lower anisotropy, closure domains and vortices are preferred, all of them causing the net magnetization to vanish. Added terms to the energy, such as the Dzyaloshinskii–Moriya (or asymmetric exchange) contribution, lead to other topological defects like skyrmions (Nagaosa and Tokura 2013). These considerations can be also applied to the case of ferroelectric domains (Mitsui and Furuichi 1953) and can be extended to magnetoelectric multiferroics (Daraktchiev et al. 2008). Ferroelectrics are largely anisotropic, compared to their magnetic counterparts, resulting in much narrower domain walls (often just one atom thick) and making it considerably more challenging to observe closure domains, vortices or skyrmions, which have only recently been reported in ferroelectrics (Nahas et al. 2015; Das et al. 2019).

As described in Schilling et al. (Schilling at al. 2006), the free energy of a ferroic material should contain a term that describes the energy density of the domains and another one that represents the domain wall formation energy. The former is proportional

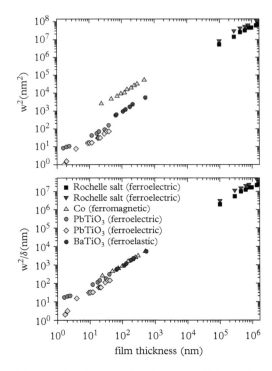

Figure 2.2 *(a) Square of the domain width as a function of the thickness, demonstrating that the quadratic dependence spans six orders of magnitude in thickness and that it is valid both for ferromagnets and for ferroelectrics. (b) If the square of the domain width is divided by the domain wall width, the quadratic law is independent of the nature of the dipolar interactions, and universality is revealed.*

Source: Taken from Catalan et al. 2012. Reproduced with permission from the American Physical Society.

to the domain size w, while the latter is proportional to both the number of domain walls (and thus proportional to $1/w$) and the domain wall area (and thus proportional to the film thickness, d). If other contributions, such as the wall–wall interactions, charge screening, etc. can be neglected, then the derivative of the free energy with these two terms leads to an equilibrium domain size of the form $w = Ad^{1/2}$, where A depends on the ratio between the anisotropy of the order parameter and the dipolar interaction or, in other words, on the domain wall width. This quadratic behavior can be shown in Figure 2.2(a) for different ferroelectric and ferromagnetic materials. It can also be observed in this figure that ferromagnetic systems give rise to a larger A pre-factor (≈ 10–100 nm$^{1/2}$) than that found for ferroelectric/ferroelastic systems (~ 1–5 nm$^{1/2}$), for which a strong anisotropy exists, reflecting the very different sizes of the domain walls in both cases. If the domain wall width is eliminated from the thickness scaling (see Figure 2.2(b)), then a universal behavior, independent of the physics of the interactions, is found (Catalan et al. 2009; Catalan et al. 2012).

In a similar manner, for ferroelastic domains, the domain width (w) is determined by the competition between the elastic energy in the domains and the formation energy of the domain walls and is a function of the thin film thickness (d). For the cases when $d > w$, the Roitburd scaling law applies (Roitburd 1976). Interestingly, this square root dependence holds for both epitaxial and freestanding layers. In the very thin film limit (for $d < w$), Roitburd's approximations are not valid, and a rigorous calculation of the elastic energy of the system leads to a nearly linear relation between the domain width and the film thickness (Pompe et al. 1993; Pertsev and Zembilgotov 1995). Thus, above the critical thickness for domain formation, the thinner the film, the larger the number of domain walls and the smaller the size of the domains.

The presence of a net magnetic moment has been detected in several antiferromagnetic manganites, such as $TbMnO_3$, $YbMnO_3$ and $YMnO_3$, when grown in thin film form (Marti et al. 2008, 2009, 2010; Rubi et al. 2008, 2009). It has been proposed that the ferromagnetism originates in the strain-induced modification of the balance between the different magnetic exchange interactions (Dong et al. 2009), which produces a canting of Mn spins (Marti et al. 2008, 2009, 2010). Interestingly, the magnetic moment seems to be induced independently of the ground state magnetic structure of the bulk material: $TbMnO_3$ displays a cycloidal spin structure, whereas $YbMnO_3$ and $YMnO_3$ have an E-type collinear spin structure. In both cases, the induced magnetization has been reported to follow a similar trend as the unit cell volume (Marti et al. 2010). This raises the question of whether the magnetic moment arises from strain effects linked to the modification of the bond lengths and angles, as proposed in the first instance, or whether it originates from a more general feature of epitaxial orthorhombic manganites (Fontcuberta 2015), such as the domain microstructure (Dong et al. 2009; Fontcuberta 2015).

In epitaxial orthorhombic $TbMnO_3$, the origin of the macroscopic net magnetic moment was reported to be the result of uncompensated Mn^{3+} spins upon the formation of a particular type of domain wall (Farokhipoor et al. 2014). Other examples of confinement of the magnetic moment in 2D can be found, such as the interface between $LaAlO_3$ and $SrTiO_3$ (Brinkman et al. 2007), or the ferromagnetically coupled interface between $CaMnO_3$ and $CaRuO_3$, due to the competition between antiferromagnetic superexchange and ferromagnetic double exchange between Mn_{4+} and Ru_{4+}. In none of these cases are the oxides ferromagnetic as single layer systems (Takahashi et al. 2001). The existence of a magnetic moment in a two-dimensional non-magnetic organic system (graphene) has also been reported (Ma et al. 2012; Yang et al. 2013).

In the following, we discuss the effects of epitaxy and the presence of domain walls on the magnetic properties of thin films of $TbMnO_3$ under a large strain induced by epitaxy on $SrTiO_3$ substrates. The presence of intense local stresses leads to selective atomic substitution, giving rise to the formation of a unique phase, which is responsible for the distinct magnetic properties at the domain walls of $TbMnO_3$ thin films, as mentioned above.

2.2   Formation of a Novel Phase at the Domain Walls of $TbMnO_3$

It has been shown that orthorhombic manganites grown on cubic $SrTiO_3$ substrates form crystallographic twins (Marti et al. 2008). Twinning is mainly determined by the symmetry relationships between the film and substrate materials, but it is affected by the growth kinetics and can differ depending on the growth conditions. $TbMnO_3$ thin films grown on (001)-$SrTiO_3$ with low deposition rates (approaching thermodynamic conditions) are reported to form four types of twin domains, all with the c-axis perpendicular to the substrate interface (Daumont et al. 2009). Despite the very large mismatch strain ($+4.1\%$ along $[100]_o$ and -5.7% along $[010]_o$), this microstructure allows the film to maintain partial coherence with the substrate, either along the $[100]_c$ or the $[010]_c$ (substrate) directions and, importantly, it determines the evolution of the lattice parameters with increasing thickness (Daumont et al. 2009): the partial coherence with the substrate and the crystal twinning are able to maintain the unit cell in-plane area constant. Thus, the out-of-plane lattice parameter and the unit cell volume remain basically unchanged for a large range of thicknesses, from 5 to 70 nm (Daumont et al. 2009). This domain/twin configuration prevents the accumulation of large shear stress with increasing thickness and allows slow relaxation from a fully coherent lattice toward the bulk orthorhombic structure by modifying a_o and b_o in opposite directions and by the same amounts. It can be rationalized that this is an efficient way of minimizing the elastic energy of the system in the presence of shear stress (Venkatesan et al. 2009). This phenomenon is very common as it takes place in the case of heteroepitaxy between cubic and orthorhombic lattices, the latter consisting of non-orthogonal pseudocubic units, but also between cubic and rhombohedral structures, where similar effects have been observed (Daumont et al. 2010).

The direct observation of the properties of these domain walls below the ordering temperature ($\sim$41 K) is very challenging due to their atomic-scale width. This is complicated by the fact that $TbMnO_3$ thin films are semiconducting, hampering nanoscale magnetic imaging using scanning tunneling microscopy (Bode et al. 2006). Instead, a combination of transmission electron microscopy (TEM), magnetometry, and first-principles calculations (see Chapter 3) was shown to be decisive for understanding the formation of two-dimensional magnetic sheets at the structural domain walls, as observed in the sketch shown in Figure 2.3.

In this way, the inverse dependence of the remanent magnetic moment with the film thickness (for thicknesses between 5 and 80 nm) could be explained. The evolution of the density of domain walls as a function of the thickness, evidenced by TEM, was consistent with the increased net magnetic moment for decreasing thickness; see Figure 2.4 (Daumont et al. 2009; Venkatesan et al. 2009). This phenomenon is attributed to the presence of chemical reconstruction at the structural domain walls. These domain walls are $(100)_o$ or $(010)_o$ planes through the A-cations, separating orthorhombic domains with opposite "zigzags," as shown in Figure 2.3. Interestingly, quantitative electron energy loss spectroscopy in scanning transmission electron microscopy mode (STEM-

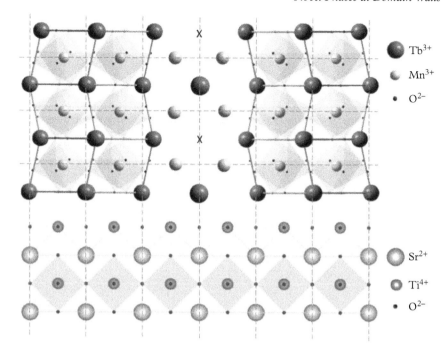

Figure 2.3 *A sketch of the atomic arrangement around a TbMnO₃ (110)ₒ domain wall between two orthorhombic domains (with the aₒ and bₒ directions interchanged with respect to each other). In the domains, the pseudocubic unit cells of orthorhombic TbMnO₃ form in zigzag fashion along the [001] direction. At the twin domain, the zigzag is mirrored, creating a short Tb-Tb bond distance at every other Tb atom of the domain wall. The experiments show that, at those sites with short bond distances, the Tb³⁺ cation is replaced with Mn³⁺. The cubic unit cells in the bottom row represent the SrTiO₃ underlying substrate.*

Source: Farokhipoor et al. (2014). Reproduced with permission from Springer Nature.

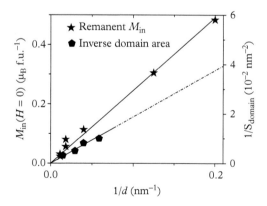

Figure 2.4 *Density of the domain walls and in-plane magnetic moment at H = 0 versus the inverse of the thickness.*

Source: Farokhipoor et al. (2014). Reproduced with permission from Springer Nature.

Figure 2.5 *High-angle annular dark field (HAADF) STEM image of region of the novel (encircled) phase, equivalent to ~12 consecutive domain walls.*

EELS) spectrum imaging shows that, at the domain walls, every other Tb^{3+} atomic column is substituted by Mn^{3+} ions along the growth direction. Indeed, as shown in the figure, the accommodation of opposite zigzag patterns at the domain walls gives rise to too short Tb^{3+}-Tb^{3+} bond distances for every other Tb atom, along the [001] direction. This unfavorable configuration induces the structure to rearrange itself by substituting Tb^{3+} by the smaller Mn^{3+} cation in order to locally release the strain at the boundaries of the twins (see also Chapter 3).

This novel phase is not only restricted to a single isolated 2D domain wall. In some regions, the strain conditions are such that this structure spreads over extended regions of the film, as though a succession of domain walls occurs. This is illustrated in Figure 2.5, in which the region separating two domains is not an atomic flat domain wall, but a volume equivalent to approximately 12 domain walls in width. The approximate stoichiometry of this new structure, assuming that the chemical substitution of alternate Tb columns along [001] direction is complete and there are no oxygen vacancies, would be $(Tb_xMn_{1-x})MnO_3$, with x ~ 0.5.

2.3   The Oxygen Stoichiometry of Off-Wall and On-Wall Areas in $TbMnO_3$ Thin Films Grown on $SrTiO_3$ Substrate

The chemical substitution at the domain walls is likely to affect the oxygen lattice, to accommodate to the different ionic radii and valence of the metal cations. Oxygen stoichiometry of the domain walls in $TbMnO_3$ can be explored by STEM-EELS, even with atomic resolution (Pennycook and Nellist 2011). Individual spectra of the specific EELS edge for a chemical element can be obtained by conventional background subtraction. Subsequent edge intensity integration and quantification in terms of the

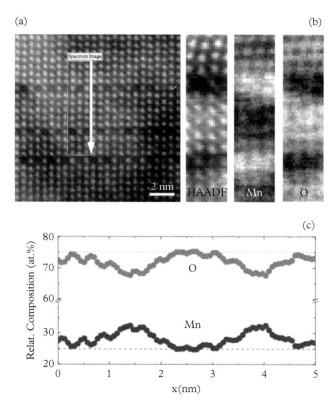

Figure 2.6 *STEM-EELS spectrum image of a region of 25-nm-thick TbMnO₃ thin film grown on SrTiO₃. (a) High-angle annular dark field (HAADF) reference image; (b) HAADF signal (in grey scale), Mn $L_{2,3}$ signal, and O K content; (c) profile of relative Mn/O ratio quantified from (b), determined along the white arrow in (a) by integrating 20 pixels horizontally. Dashed lines correspond to the nominal stoichiometry (Mn:O = 1:3).*

calculated inelastic scattering cross section of the different species can provide the compositional ratio between two or more chemical elements in a material (Egerton 2011). Figure 2.6 shows the STEM-EELS spectrum images collected in a region containing two domain walls in a 25-nm-thick TbMnO₃ film grown on SrTiO₃. While the estimated Mn:O ratio of the domains approaches the nominal 1:3 ratio (25% Mn at., 75% O at.) the oxygen content at the domain walls drops to approximately 70% O at. This non-stoichiometry of the wall is caused by Mn replacement of Tb columns, which is not compensated with higher O content and, therefore, increases the local Mn content at the domain wall.

The off-stoichiometry of the TbMnO₃ domain walls would impact the Mn valence, too. The Mn oxidation state has been frequently studied by analyzing the EELS fine structure of the O-K edge or the Mn $L_{2,3}$ edge, which are extremely dependent on the hybridization of the Mn-3d and O-2p orbitals, and thus on the oxygen content

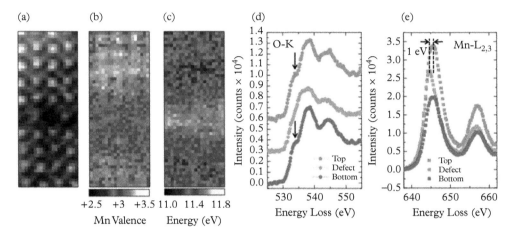

Figure 2.7 *Estimation of Mn nominal valence obtained from the EELS fine structure analysis of TbMnO₃ region containing a domain wall. (a) HAADF image of the region; (b) Mn valence determined from the shift of the O-K pre-peak; (c) relative energy shift of the L_3 and L_2 Mn edges; (d, e) examples of O-K and Mn $L_{2,3}$ edges in the different regions of the domains and the domain wall.*

Source: *Adapted from Farokhipoor et al. 2014. Reproduced with permission from Springer Nature.*

and chemical environment of Mn. The combined change in stoichiometry and novel crystal environment for Mn can be translated into a variation of the nominal Mn valence. This can be tracked by following the previous calibration performed in isoelectronic structures (lanthanum-calcium manganites as a function of the La/Ca content) in which the position of the O-K pre-peak relative to the main O peak is analyzed (Varela et al. 2006). By Gaussian fitting of both peaks, the energy difference between the maxima can be measured and correlated to the Mn oxidation state (Varela et al. 2006). Similarly, the energy difference between the L_2 and L_3 white lines of the EELS Mn edge also depends on the Mn valence. Figure 2.7 also shows a decrease in the Mn valence at the domain wall; while the domains average a nominal Mn oxidation state of approximately +3.2, the wall exhibits an average Mn valence of +3.0.

Even though a direct correlation between the stoichiometry and the nominal valence values obtained here cannot be made, both point to the same direction; i.e. it is plausible that Mn in the X columns would be less positively charged (that is, a +2 character), which would reduce the oxidation state of the defect to +3, assuming that the Mn in the B positions have the same valence as that measured in the domains. The evolution of the relative positions of the L_3 and L_2 white lines of Mn is consistent with this scenario; a shift of L_3 of about 1 eV to lower energy losses is observed, which has been reported in the literature and was associated with lower Mn valence (Schmid and Mader 2006).

All this points toward a quite unique driving mechanism behind the periodic structure that is observed in TbMnO₃/SrTiO₃, notably different from any other type of ordering reported in thin film heterostructures. Indeed, other studies present lattice-modulated patterns involving oxygen vacancies ordering, for instance, in cobaltites (Choi et al. 2012;

Gazquez et al. 2013; Biškup et al. 2014; Kwon et al. 2014). For $LaCoO_3$, the origin of magnetism is controversial, with one report ascribing it to the presence of oxygen vacancies (Gazquez et al. 2013; Biškup et al. 2014), whereas other reports correlate it with the high spin state of Co^{3+} (Choi et al. 2012; Kwon et al. 2014). In these works, the stripes look more like a periodically structured region with different chemical composition due to oxygen vacancies, and not necessarily located at the domain walls (Choi et al. 2012; Biškup et al. 2014; Kwon et al. 2014). Nevertheless, we postulate that this phenomenon could be more generally present in (001)-oriented $A^{3+}B^{3+}O_3$ orthorhombic perovskites, since they all show an A-cation zigzag configuration along the [001] direction and the same valence state for the small and large cations. In particular, multivalent B-cations, such as Mn, Ni, or Cr, should allow for more flexible coordination and formation of new phases at the domain walls. A similar magnetic trend with thickness has been observed in orthorhombic $ScMnO_3$ grown on $LaAlO_3$; a more detailed TEM and neutron diffraction studies is still pending (Wang et al. 2015).

2.4 Summary

In this chapter, we describe a route for the stabilization of novel two-dimensional phases at structural domain walls that self-assemble during growth, driven by the large stresses present at the ferroelastic twin boundaries. In the case of $TbMnO_3/SrTiO_3$ heterostructures, the change in chemistry takes place by selective substitution of half the Tb cations at the domain wall by Mn cations, with an accompanying decrease in Mn valence at the domain wall sites. This two-dimensional phase orders magnetically at low temperatures. We believe that this mechanism should also be active in other orthorhombic $A^{3+}B^{3+}O_3$ perovskites.

· ·

REFERENCES

Biškup N., Salafranca J., Mehta V., Oxley M. P., Suzuki Y., Pennycook S. J., Pantelides S. T., Varela M., "Insulating ferromagnetic $LaCoO_{3-\delta}$ films: a phase induced by ordering of oxygen vacancies," *Phys. Rev. Lett.* 112, 087202 (2014).

Bode M., Vedmedenko E. Y., Von Bergmann K., Kubetzka A., Ferriani P., Heinze S., Wiesendanger R., "Atomic spin structure of antiferromagnetic domain walls," *Nat. Mater.* 5, 477 (2006).

Brinkman A., Huijben M., van Zalk M., Huijben J., Zeitler U., Maan J. C., van de Wiel W. G., Rijnders G., Blank D. H. A., Hilgenkamp H., "Magnetic effects at the interface between non-magnetic oxides," *Nat. Mater.* 6, 493 (2007).

Catalan G., Lubk A., Vlooswijk A. H. G., Snoeck E., Magen C., Janssens A., Rispens G., Blank D. H. A., Noheda B., "Flexoelectric rotation of polarization in ferroelectric thin films," *Nat. Mater.* 10, 963 (2011).

Catalan G., Lukyanchuk I., Schilling A., Gregg J. M., Scott J. F., *J. Mater. Sci.* 44, 5307 (2009).

Catalan G., Seidel J., Ramesh R., Scott J. F., "Domain wall nanoelectronics," *Rev. Modern Phys.* 84, 119 (2012).

Choi W. S., Kwon J.-H., Jeen H, Hamann-Borrero J. E., Radi A., Macke S., Sutarto R., He F., Sawatzky G. A., Hinkov V., Kim M., Lee H. N., "Strain-induced spin states in atomically ordered cobaltites," *Nano. Lett.* 12, 4966 (2012).

Coey J. M. D., Viret M., von Molnar S., "Mixed-valence manganites," *Adv. Phys.* 48, 167 (1999).

Daraktchiev M., Catalan G., Scott J. F., "Landau theory of ferroelectric domain walls in magneto-electrics," *Ferroelectrics* 375, 122 (2008).

Das S., Tang Y. L., Hong Z., Gonçalves M. A. P., McCarter M. R., Klewe C., Nguyen K. X., Gómez-Ortiz F., Shafer P., Arenholz E., Stoica V. A., Hsu S.-L., Wang B., Ophus C., Liu J. F., Nelson C. T., Saremi S., Prasad B., Mei A. B., Schlom D. G., Íñiguez J., García-Fernández P., Muller D. A., Chen L. Q., Junquera J., Martin L. W., Ramesh R., "Observation of room-temperature polar skyrmions," *Nature* 568, 368 (2019).

Daumont C. J. M., Farokhipoor S., Ferri A., Wojdel J. C., Iniguez J., Kooi B. J., Noheda B., "Tuning the atomic and domain structure of epitaxial films of multiferroic $BiFeO_3$," *Phys. Rev. B* 81, 144115 (2010).

Daumont C. J. M., Mannix D., Venkatesan S., Rubi D., Catalan G., Kooi B. J., De Hosson J. Th. M., Noheda B., "Epitaxial $TbMnO_3$ thin films on $SrTiO_3$ substrates: a structural study," *J. Phys.: Condens. Mat.* 18, 182001 (2009).

Dong S., Yu R., Yunoki S., Liu J.-M., Dagotto E., "Double-exchange model study of multiferroic $RMnO_3$ perovskites," *Eur. Phys. J. B* 71, 339 (2009).

Egerton R. F., *Electron Energy-Loss Spectroscopy in the Electron Microscope*, 3rd ed. (Springer, Boston, 2011).

Farokhipoor S., Magén C., Venkatesan S., Iñiguez J., Daumont C. J. M., Rubi D., Snoeck E., Mostovoy M., de Graaf C., Müller A., Döblinger M., Scheu C., Noheda B., "Artificial chemical and magnetic structure at the domain walls of an epitaxial oxide," *Nature* 515, 379 (2014).

Fontcuberta J., "Multiferroic $RMnO_3$ thin films," *C. R. Phys.* 16, 204 (2015).

Gazquez J., Bose S., Sharma M., Torija M. A., Pennycock S. J., Leighton C., Varela M., "Lattice mismatch accommodation via oxygen vacancy ordering in epitaxial $La_{0.5}Sr_{0.5}CoO_{3-\delta}$ thin films," *APL Mater.* 1, 012105 (2013).

Goodenough J. B., "Theory of the role of covalence in the perovskite-type manganites [La, M(II)] MnO_3," *Phys. Rev.* 100, 564 (1955).

Kenzelmann M., Harris A. B., Jonas S., Broholm C., Schefer J., Kim S. B., Zhang C. L., Cheong S.-W., Vajk O. P., Lynn J. W., "Magnetic Inversion Symmetry Breaking and Ferroelectricity in $TbMnO_3$," *Phys. Rev. Lett.* 95, 087206 (2005).

Kimura Y., Goto T., Shintani H., Ishizaka K., Arima T., Tokura Y., "Magnetic control of ferroelectric polarization," *Nature* 426, 55 (2003).

Kittel C., "Theory of the structure of ferromagnetic domains in films and small particles," *Phys. Rev* 70, 965 (1946).

Kwon J.-H., Choi W. S., Kwon Y.-K., Jung R., Zuo J.-M., Lee H. N., Kim M., "Nanoscale spin-state ordering in $LaCoO_3$ epitaxial thin films," *Chem. Mater.* 26, 2496 (2014).

Landau L., Lifshitz E., "On the theory of the dispersion of magnetic permeability in ferromagnetic bodies," *Phys. Z. Sowjet.* 8, 153 (1935).

Ma Y., Dai Y., Guo M., Niu Ch., Zhu Y., Huang B., "Evidence of the existence of magnetism in pristine VX_2 monolayers (X = S, Se) and their strain-induced tunable magnetic properties," *ACS Nano.* 6, 1695 (2012).

Marti X., Sanchez F., Skumryev V., Laukhin V., Ferrater C., Garcia-Cuence M. V., Varela M., Fontcuberta J., "Crystal texture selection in epitaxies of orthorhombic antiferromagnetic $YMnO_3$ films," *Thin Solid Films* 516, 4899 (2008).

Marti X., Skumryev V., Cattoni A., Bertacco R., Laukhin V., Ferrater C., Garcia-Cuenca M. V., Varela M., Sanchez F., Fontcuberta J., "Ferromagnetism in epitaxial orthorhombic $YMnO_3$ thin films," *J. Magn. Magn.Mater* 321, 1719 (2009).

Marti X., Skumryev V., Ferrater C., Garcia-Cuence M. V., Varela M., Sanchez F., Fontcuberta J., "Emergence of ferromagnetism in antiferromagnetic $TbMnO_3$ by epitaxial strain," *Appl. Phys. Lett.* 96, 222505 (2010).

Mitsui T., Furuichi J., "Domain structure of rochelle salt and KH_2PO_4," *Phys. Rev.* 90, 193 (1953).

Nagaosa N., Tokura Y., "Topological properties and dynamics of magnetic skyrmions," *Nat. Nanotechnol.* 8, 899 (2013).

Nahas Y., Prokhorenko S., Louis L., Gui Z., Kornev I., Bellaiche L., "Discovery of stable skyrmionic state in ferroelectric nanocomposites," *Nat. Comm.* 6, 8542 (2015).

Pennycook S. J., Nellist P. D., *Scanning Transmission Electron Microscopy* (Springer, New York, 2011).

Pertsev N. A., Zembilgotov A. G., "Energetics and geometry of 90° domain structures in epitaxial ferroelectric and ferroelastic films," *J. Appl. Phys.* 78, 6170 (1995).

Pompe W., Gong X., Suo Z., Speck J. S., "Elastic energy release due to domain formation in the strained epitaxy of ferroelectric and ferroelastic films," *J. Appl. Phys.* 74, 6012 (1993).

Roitburd A. L., "Equilibrium structure of epitaxial layers," *Phys. Status Solidi A* 37, 329 (1976).

Rubi D., de Graaf C., Daumont C. J. M., Mannix D., Broer R., Noheda B., "Ferromagnetism and increased ionicity in epitaxially grown $TbMnO_3$ films," *Phys. Rev. B* 79, 014416 (2009).

Rubi D., Venkatesan S., Kooi B. J., De Hosson J. T. M., Palstra T. T. M., Noheda B., "Magnetic and dielectric properties of $YbMnO_3$ perovskite thin films," *Phys. Rev. B* 78, 020408 (2008).

Salamon M. B., Jaime M., "The physics of manganites: structure and transport," *Rev. Mod. Phys.* 73, 583 (2001).

Salje E. K., *Phase Transitions in Ferroelastic and Co-elastic Crystals* (Cambridge University Press, Cambridge, UK, p. 296, 1993).

Schilling A., Adams T. B., Bowman R. M., Gregg J. M., Catalan G., Scott J. F., "Scaling of domain periodicity with thickness measured in $BaTiO_3$ single crystal lamellae and comparison with other ferroics," *Phys. Rev. B* 74, 024115 (2006).

Schmid H. K., Mader W., "Oxidation states of Mn and Fe in various compound oxide systems," *Micron.* 37, 426 (2006).

Tagantsev A. K., Cross L. E., Fousek J., *Domains in Ferroic Crystals and Thin Films* (Springer, New York, 2010).

Takahashi K. S., Kawasaki M., Tokura Y., "Interface ferromagnetism in oxide superlattices of $CaMnO_3/CaRuO_3$," *Appl. Phys. Lett.* 79, 1324 (2001).

Varela M., Pennycook T. J., Tian W., Mandrus D., Pennycook S. J., Peña V., Sefrioui Z., Santamaria J., "Atomic scale characterization of complex oxide interfaces," *J. Mater. Sci.* 41, 4389 (2006).

Venkatesan S., Daumont C. J. M., Kooi B. J., Noheda B., de Hosson J. Th. M., "Nanoscale domain evolution in thin films of multiferroic $TbMnO_3$," *Phys. Rev. B* 80, 214111 (2009).

Wang F., Zhang Y. Q., Liu W., Ning X. K., Bai Y., Dai Z. M., Ma S., Zhao X. G., Li K., Zhang Z. D., "Abnormal magnetic ordering and ferromagnetism in perovskite $ScMnO_3$ film," *Appl. Phys. Lett.* 106, 232906 (2015).

Yang H. X., Hallal A., Terrade D., Waintal X., Roche S., Chshiev M., "Proximity effects induced in graphene by magnetic insulators: first-principles calculations on spin filtering and exchange-splitting gaps," *Phys. Rev. Lett.* 110, 046603 (2013).

3

First-Principles Studies of Structural Domain Walls

J. Íñiguez[1,2]

[1] *Materials Research and Technology Department, Luxembourg Institute of Science and Technology (LIST), 5 avenue des Hauts-Fourneaux, L-4362 Esch/Alzette, Luxemburg*
[2] *Department of Physics and Materials Science, University of Luxembourg, 41 rue du Brill, L-4422 Belvaux, Luxembourg*

3.1 Introduction

Structural domain walls are critical to the properties of multi-domain states in ferroelectric and ferroelastic materials. For one thing, the mobility of the walls, or the lack thereof, largely determines the response of a compound to external stimuli. For another, walls may have interesting functional properties of their own, either intrinsic or resulting from the various types of defects they usually attract (see also Chapter 1.4). While not new (Salje 2010), the latter concept has become particularly popular in recent times, together with the notion that suitably engineered walls may provide a path toward the production of novel densely packed nanodevices. This "the wall is the device" philosophy (Catalan et al. 2012) motivates most of the recent first-principles work on the topic, and is a focus of this chapter.

Why should walls present peculiar properties that set them apart from the domains? Why should they be the generally preferred location of point defects? The wall atomic structure is not a low-energy one; rather, it is one forced by the *structural boundary conditions* imposed by the surrounding domains that meet (clash) at the walls. Hence, the walls are at a relatively high-energy state in which the dominant interatomic couplings in the material are somewhat frustrated. Two expectations follow. First, the properties of the wall may depart from those of the domains, because their structures are different. Second, when it comes to accommodating defects, the walls seem to have a starting advantage as compared to the domains. In a domain, a defect creates a structural disruption that usually has a strong penalty associated with it, as it prevents the material from adopting its lowest-energy configuration. In contrast, the energy of the walls is not expected to increase as much in the presence of a defect[1]; in fact, it is even conceivable that certain defects might allow a wall to optimize its structure and energy (Zhao and Íñiguez 2019).

---

[1] This hand-waving argument is supported by a recent simulation study of $PbTiO_3$'s walls (Paillard et al. 2017); yet, exceptions to the rule are known too (Skjærvø et al. 2018).

J. Íñiguez, *First-Principles Studies of Structural Domain Walls* In: *Domain Walls: From Fundamental Properties to Nanotechnology Concepts.*
Edited by: Dennis Meier, Jan Seidel, Marty Gregg, and Ramamoorthy Ramesh, Oxford University Press (2020). © J. Íñiguez.
DOI: 10.1093/oso/9780198862499.003.0003

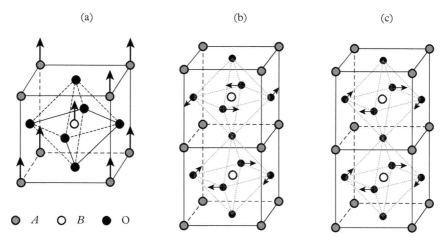

Figure 3.1 *Most common structural instabilities in perovskites. Structure of an ABO_3 cubic perovskite. A cations (shaded balls) occupy the cell corners, B cations (white balls) the cell center, and oxygen anions (black balls) the face centers. In panel (a), the arrows represent the local atomic displacements that are typical of a ferroelectric instability. Panels (b) and (c) show typical antiferrodistortive tiltings of the O_6 octahedral groups, in antiphase and in phase, respectively.*

What do we know about the structure of structural walls? Fortunately, more and more, thanks to the ongoing revolution in local-probe techniques. Yet, a detailed experimental characterization of the materials (devices) in the relevant conditions (of operation) remains a great challenge, and our detailed knowledge is still limited. Despite that, domain boundaries have long been discussed at a theoretical level, traditionally in the framework of Ginzburg–Landau models (Lajzerowicz and Niez 1979; Houchmandzadeh et al. 1991; Tagantsev et al. 2010). Within such approaches, the picture of the wall structure is straightforward: This is a region in which the ferroic order parameter (e.g. the electric polarization of Figure 3.1(a)) changes sign as we move from one domain to another; hence, being associated with a locally null order parameter, the wall looks like the high-symmetry phase of the compound (see also Chapter 4). A refinement of this picture is available for materials presenting two competing orders (e.g. the electric polarization of Figure 3.1(a) and the tilts of the O_6 octahedra of Figures 3.1(b) and 3.1(c)); in such cases, as already mentioned in Chapter 1.3, the weaker order parameter may appear at the wall region, where the strongest dominant order vanishes (see Figure 3.2(a)) (Houchmandzadeh et al. 1991; Tagantsev et al. 2001).

This basic understanding of structural walls has dominated our thinking for decades. Granted, the physical picture is beautiful and makes sense, and it should be essentially correct provided two basic conditions are satisfied. At a qualitative level, the Ginzburg–Landau models supporting it should contain all the order parameters (distortion fields) that can play a relevant role at the walls. At a quantitative level, the model coefficients—typically obtained from experimental information that essentially captures the behavior of the domains, not the walls—should give a fair description of the boundary region.

These are perfectly sound hypotheses that have allowed our community to work on this problem for decades (Chen 2008; Vasudevan et al. 2013). Nevertheless, today we are in the position to ask ourselves: Is this all we need, or are we missing important effects?

The question of how to treat structural domain walls theoretically is a no-brainer, at least in principle. We need methods that permit an unbiased description of unusual (typically, unknown) high-energy structures, with atomic resolution, and not relying on experimental information that we often lack. This is a perfect problem for modern first-principles simulation techniques (Martin 2004), one where they are likely to make a difference in terms of both gaining understanding and revealing unexpected behaviors. Indeed, while traditional model-Hamiltonian approaches are limited by our imagination (they are defined by a set of hand-picked variables and interactions among them), first-principles simulations allow us, like experiments, the luxury of serendipity, that is, of being surprised by results we would not have anticipated. Further, when it comes to problems involving defects or, more generally, differences in chemical bonding between domains and walls, a quantum-mechanical first-principles approach is all but irreplaceable.

Here I discuss representative first-principles studies of structural domain walls in ferroics, focusing on the compounds that have received the most attention by the simulations community so far: perovskite oxides. It is not my intention to be exhaustive. Rather, I describe in some detail a reduced number of case studies that come in handy to illustrate different effects and to highlight the added value of the first-principles investigations. As regards the simulation methods, I focus on applications of density functional theory (DFT) (Hohenberg and Kohn 1964; Kohn and Sham 1965), typically employing an approximation for an effective treatment of ionic cores (Pickett 1989; Blöchl 1994); since these are standard and abundantly reviewed methods, I do not discuss them here. I also include a section on the application to domain-wall problems of first-principles-based methods for large-scale simulations of ferroelectrics and ferroelastics, where our community has followed an original and productive path. Finally, I briefly comment on the opportunities and challenges for first-principles research in this field.

3.2 Basic Background

For the benefit of newcomers to the field, and to make this chapter reasonably self-contained, let me start by addressing some basic issues and terminology pertaining to structural walls and their physics. For the sake of concreteness, I consider the case of perovskite oxides, taking $PbTiO_3$ and $SrTiO_3$ as representative examples, noting that most of the ideas presented here are directly applicable to other materials families.

3.2.1 Structural Instabilities, Order Parameters, and Symmetry Breakings

Figure 3.1 shows sketches of the ABO_3 perovskite structure, and how the condensation of different structural instabilities breaks the symmetry of the prototype cubic phase to yield various low-symmetry phases. For example, Figure 3.1(a) displays a characteristic pattern of atomic displacements that, at a local level, yield an electric dipole. When repeated homogeneously in a large region of the lattice, this distortion results in sizable movement of bound charges and an associated electric polarization. Then, if such polar distortions are oriented differently in different regions, domains appear. Indeed, there are six symmetry-equivalent orientations of the local dipole shown in Figure 3.1(a), parallel or antiparallel to the three principal axes of the cubic lattice, implying that six equivalent ferroelectric domains may form in this case. The homogeneous replication of the distortion in Figure 3.1(a) reduces the symmetry of the perovskite lattice from cubic ($Pm\bar{3}m$ space group) to tetragonal ($P4mm$); this is exactly the case of $PbTiO_3$ (Lines and Glass 1977).

Figure 3.1(b) shows a second lattice instability that is ubiquitous in perovskites (Lines and Glass 1977). Locally it involves a rotation of the O_6 octahedron in the unit cell, and the tilts are modulated in anti-phase as one moves between neighboring lattice cells. This kind of distortion corresponds to an order parameter sometimes termed *antiferrodistortive*. There are six symmetry-equivalent ways to have it, differing in the axis and phase of the tilts, which yield six equivalent domains. Antiferrodistortive instabilities like that of Figure 3.1(b) occur spontaneously in materials as $SrTiO_3$, where the symmetry of the cubic lattice is reduced to $I4/mcm$. Also, first-principles simulations have revealed that they are *latent* in materials like $PbTiO_3$, where the antiferrodistortive mode competes (and losses) against the dominant ferroelectric order parameter (Ghosez et al. 1999; Wojdeł et al. 2013).

In the following, I will refer to the polarization of a domain by the three-dimensional vector P_α, where $\alpha = x, y, z$ labels the principal (pseudocubic) axes of the perovskite lattice. Similarly, I will denote the antiferrodistortive order parameter of anti-phase tilts by R_α, a three-dimensional (axial) vector whose direction and magnitude represent the axis and angle of the octahedral tilts, respectively. In the above examples, I have assumed that the symmetry-breaking distortions **P** and **R** are parallel to ⟨100⟩, but this does not need to be the case. Indeed, similar ferroelectric or antiferrodistortive distortions oriented along a ⟨110⟩ pseudocubic direction would yield orthorhombic low-symmetry phases, while rhombohedral ones would be obtained if the order parameters lay along a ⟨111⟩ axis. (In the following, all directions and planes are given in the pseudocubic setting.)

These are self-explanatory examples of symmetry-breaking structural transitions where the condensation of a (soft) phonon of a high-symmetry phase drives a transformation into a low-symmetry structure (Cowley 1980; Bruce 1980). However, structural domains can appear in cases that to do not match this picture exactly. An all-important example is that of multiferroic perovskite $BiFeO_3$ (Catalan and Scott 2009). In $BiFeO_3$ the ferroelectric transition does not involve the cubic perovskite phase. Instead, the role of the paraelectric phase is played by a strongly distorted structure (Arnold et al. 2010) featuring a complex—although extremely frequent (Lufaso and Woodward 2001; Chen et al. 2018b)—combination of octahedral tilts. This paraelectric phase is itself a low-symmetry state that holds no group–subgroup relationship with the ferroelectric phase; hence, it would be misleading (and incorrect) to talk about "symmetry breaking" in $BiFeO_3$'s ferroelectric transformation. Nevertheless, we can still postulate $BiFeO_3$'s cubic phase as a convenient high-symmetry reference to understand the ferroelectric domains occurring in the material (see Chapter 10 for details on the domains and domain walls in $BiFeO_3$). Hence, as regards the analysis of structural domains and walls, our symmetry arguments are a useful theoretical tool to interpret and predict behaviors, even if some of the ingredients (e.g. the high-symmetry reference) may be hypothetical in some cases.

3.2.2 The Role of Strain: Ferroelastic Features

As emphasized in Chapter 5, multi-domain structures may result from elastic boundary conditions acting on our materials (Salje 1993). Further, we know that strain effects are very important in perovskite oxides, "strain engineering" of thin films being one of the most powerful strategies to tune their properties (Schlom et al. 2007). In fact, many of the materials featured in this chapter (e.g. $SrTiO_3$ and $CaTiO_3$) are often described as being "ferroelastic." Thus, it may seem surprising that I have not included strain among the key structural variables in the previous section. This point warrants a clarification.

As regards domain walls, most of the materials of interest today—in particular, all compounds discussed in this chapter—are *not* proper ferroelastics. In none of these materials is strain the primary order parameter; further, strain by itself cannot drive a structural phase transition in any of them. (This statement relies on first-principles results showing that the high-symmetry reference structure is stable against pure homogeneous strains (King-Smith and Vanderbilt 1994; Chen et al. 2018b).) Rather, strain is a secondary distortion that *follows* the primary one. More precisely, in the ferroelectric (like $PbTiO_3$) and antiferrodistortive (like $SrTiO_3$) materials of interest here, the relationship between the symmetric strains $\{\eta_{\alpha\beta}\}$ and the **P** or **R** vectors has a simple analytical form. As shown by several authors (King-Smith and Vanderbilt 1994; Gu et al. 2012; Chen et al. 2018b), we have

$$\eta_{xx} = CQ_x^2 + C'\left(Q_y^2 + Q_z^2\right), \eta_{yy} = CQ_y^2 + C'\left(Q_z^2 + Q_x^2\right), \eta_{zz} = CQ_z^2 + C'\left(Q_x^2 + Q_y^2\right),$$
(3.1)

$$\eta_{xy} = C''Q_xQ_y, \eta_{yz} = C''Q_yQ_z, \eta_{zx} = C''Q_zQ_x,$$
(3.2)

where C, C', and C'' are material-dependent constants, and the vector **Q** can stand for either **P** or **R**. (The mathematical expressions for the strain are formally equivalent for both **P** or **R**, but the respective C coupling coefficients will, of course, differ.) Hence, the strain change across a ferroelectric or antiferrodistortive domain wall is fully determined by the change in the primary order parameter.

An important point to notice is that ferroelectric and antiferrodistortive domain walls may or may not involve a strain change. Consider for example the anti-phase antiferrodistortive wall separating domains with $\mathbf{R}^I \| [1,0,0]$ and $\mathbf{R}^{II} \| [-1,0,0]$; the strain components involved here are $\eta_{xx} \propto R_x^2$ and $\eta_{yy} = \eta_{zz} \propto R_x^2$, which do not depend on the sign of R_x; hence, this antiferrodistortive wall is not ferroelastic. In contrast, the 90° ferroelectric wall separating domains with $\mathbf{P}^I = (P,0,0)$ and $\mathbf{P}^{II} = (0,-P,0)$ does involve a change in strain: we have $\eta^{I}_{xx} = CP^2 \rightarrow \eta^{II}_{xx} = C'P^2$ and $\eta^{I}_{yy} = C'P^2 \rightarrow \eta^{II}_{yy} = CP^2$. Hence, this wall is ferroelastic.

A word of caution is in order here. We often refer to $PbTiO_3$ as a ferroelectric, and to $SrTiO_3$ as a ferroelastic, which seems to imply there is a fundamental difference between the two compounds in regard to their ferroelastic properties. As should be clear from the discussion above, this is not correct. $PbTiO_3$ and $SrTiO_3$ present essentially equivalent improper-ferroelastic behavior.

Finally, ferroelastic walls have an interesting property: as illustrated in Figure 3.2(b), they are polar by construction. Indeed, the symmetry breaking caused by the strain discontinuity at a ferroelastic wall necessarily yields a polar distortion. Since ferroelastic walls are obviously characterized by a strain gradient, it is natural to think of this effect as a flexoelectric one; indeed, the magnitude of the improper polarization occurring at ferroelastic walls will depend on the flexoelectric tensor of the material, which quantifies the polarization appearing in response to a strain gradient (Tagantsev and Yudin 2016). Yet, the emphasis on flexoelectricity may be misleading, as it seems to suggest that a polarization will occur in *any* wall having strain gradients associated with it, which is not correct. To understand this, first note that *any* wall (ferroelastic or not) associated with *any* order parameter presents strain gradients. Indeed, the local (isotropic) strain is always coupled to the (square of the) local order parameter; the volume of a cell at the wall, where $\mathbf{Q} = \mathbf{0}$, *cannot be* exactly the same as the volume of a cell within a domain, where $\mathbf{Q} \neq \mathbf{0}$, and thus, we always have strain gradients. The key is to realize that, in non-ferroelastic walls, such wall-related strain gradients integrate to zero, yielding no net electric polarization (instead, one has some sort of antipolar distortions originating from the corresponding flexoelectric couplings). What is specific to ferroelastic walls is that this integral is not zero, and a net polarization does appear.

3.2.3   Electric and Elastic Compatibility Conditions

Let us consider again the ferroelectric domains characterized by $\mathbf{P}^{I} = (P,0,0)$ and $\mathbf{P}^{II} = (0,-P,0)$, respectively. From the formulas above, we can deduce for the first domain $\eta^{I}_{xx} = CP^2$ and $\eta^{I}_{yy} = \eta^{I}_{zz} = C'P^2$, while for the second one $\eta^{II}_{xx} = \eta^{II}_{zz} = C'P^2$ and $\eta^{II}_{yy} = CP^2$. Thus, the wall separating these domains is both ferroelectric and ferroelastic. Can we say something a priori about the preferred orientation of such a wall? Yes indeed: Irrespective of the atomistic details of the ferroic distortions, walls will orient themselves so as to minimize two large contributions to the energy of any material, namely, electrostatic and elastic. (Electrostatic effects are all but restricted to insulating and semiconducting compounds, which are the cases of interest here.)

In order to minimize electrostatic energy, ferroelectric walls will tend to be *neutral*, implying that $\Delta\mathbf{P} \cdot \hat{n} = 0$, where $\Delta\mathbf{P} = \mathbf{P}^{II} - \mathbf{P}^{I}$ and $\hat{n}$ is the unit vector normal to the wall plane (see Figure 3.2(c)). Conversely, if $\Delta\mathbf{P} \cdot \hat{n} = \sigma_{\mathrm{wall}} \neq 0$, the wall is said to be *charged*, meaning that it would need to receive a σ_{wall} charge density in order to be electrically compliant. In this example, we have $\Delta\mathbf{P} = (-P,-P,0)$, which implies that walls in planes $(1,-1,0)$ or $(0,0,1)$ will be neutral; in contrast, walls in planes $(1,1,0)$ or $(1,0,0)$ would

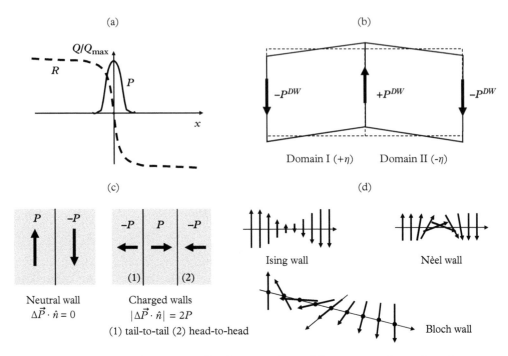

Figure 3.2 *Domain wall features. Panel (a) illustrates the possible occurrence of a weaker order parameter (P in this case) at the walls of a stronger order (R), as discussed by Houchmandzadeh et al. (1991). Panel (b) illustrates the symmetry breaking at a ferroelastic wall involving a shear strain η; at the walls, there is no symmetry between up and down, and an improper wall polarization (P^{wall}) is inevitable. Panel (c) shows two 180° ferroelectric walls, one neutral and the other charged. This charged wall configuration corresponds to the maximum $|\Delta\mathbf{P}\cdot\hat{n}|$ that is possible, and can be expected to be unstable against distortions that reduce the wall charge. Panel (d) shows sketches of walls with Ising, Nèel, and Bloch character, attending to the different ways in which the order parameter may evolve across the wall.*

be charged. A finite domain-wall charge might lead to a redistribution of mobile carriers and, hence, anomalous electronic transport as discussed, for example, in Chapter 6.

Additionally, in order to minimize the elastic energy, ferroelastic walls will tend to occupy planes that are *similarly strained* by both domains (Streiffer et al. 1998; Tagantsev et al. 2010). To understand this better, let us consider again the domains defined by $\mathbf{P}^{I} = (P,0,0)$ and $\mathbf{P}^{II} = (0,-P,0)$. These domains cause symmetry-equivalent, but different, spontaneous strains, and the dissimilarity can be captured by a *difference strain tensor* given by

$$\Delta\eta = \eta^{II} - \eta^{I} = \left(C - C'\right)P^2 \begin{pmatrix} -1 & 0 & 0 \\ 0 & 1 & 0 \\ 0 & 0 & 0 \end{pmatrix}. \tag{3.3}$$

To minimize its elastic energy, the wall (twin plane) between $\mathbf{P}^I$ and $\mathbf{P}^{II}$ must be such that both domains lead to the same spontaneous strains *within* that plane. Mathematically, this means that the application of the difference strain tensor to any vector $\mathbf{u}$ within the twin plane must render either a null result or, at most, a distortion normal to the wall. Hence, for example, for $\mathbf{u} = (0,0,1)$ we have $\Delta\boldsymbol{\eta}\,\mathbf{u} = \mathbf{0}$, implying that this direction is equally strained by both ferroelastic variants. Additionally, for $\mathbf{u} = (1,1,0)$ we get $\Delta\boldsymbol{\eta}\,\mathbf{u} = \left(C - C'\right) P^2 \left(-1,1,0\right)$; that is, the difference distortion is along a perpendicular direction. Combined, these two conditions imply that the elastic energy is minimized if the twin boundary lies in the $(1, -1, 0)$ plane. As mentioned above, this plane is also electrically neutral, which makes it a clear candidate to host the wall between $\mathbf{P}^I$ and $\mathbf{P}^{II}$. In contrast, for $\mathbf{u} = (0,1,0)$ we get $\Delta\boldsymbol{\eta}\,\mathbf{u} = \left(C - C'\right) P^2 \left(0,1,0\right)$, implying that planes containing this direction are not elastically compliant. For example, while electrically neutral, the $(0, 0, 1)$ plane is unfavorable from the point of view of the elastic energy; thus, we do not expect walls between $\mathbf{P}^I$ and $\mathbf{P}^{II}$ to be oriented in this way.

Finally, let me note that walls can further optimize their internal domain-wall structure by implementing the discontinuity of the order parameter in different ways. The most usual patterns are called Ising, Bloch and Nèel, respectively, as defined in Chapter 1. As it is essential for the following discussion, sketches of these different boundaries are also included in Figure 3.2(d). (For experimental studies on Ising, Bloch and Nèel walls, see Chapter 7.)

3.3 Case Studies of Ideal Domain Walls

In the following, I review a selection of representative works on structural walls in prototypic ferroic materials. I emphasize major milestones and discoveries, but do not attempt a detailed recollection of the works. The reader is strongly encouraged to cosult the original articles for more details.

3.3.1 Ferroelectric Materials: Perovskites $PbTiO_3$ and $BaTiO_3$

In my view, a review on this topic has to start with $PbTiO_3$, because this was the first material studied, remains the best-studied one and, as we will see in Section 3.5, its walls attract great attention today.

In 2002, Meyer and Vanderbilt published the article that set the standard for first-principles investigations of structural walls, tackling the ferroelectric boundaries of $PbTiO_3$ (Meyer and Vanderbilt 2002). These authors studied ideal (perfectly planar and neutral) walls associated with polarization changes from $\mathbf{P}^I = (0,0,P)$ to $\mathbf{P}^{II} = (0,0,-P)$ [180° wall within a $(1, 1, 0)$ plane] and to $\mathbf{P}^{II} = (0,P,0)$ [90° wall within a $(1, 1, 0)$ plane]. To do so, they made some clever choices of supercell (see Figure 3.3(a)) that allowed them to study the corresponding multi-domain structures with a minimal computational burden. (Finding such optimal supercells poses an excellent quiz for atomistic-simulation beginners!) They found that the walls are centered in PbO planes, are very narrow, and have an Ising character (see Figure 3.2(d)). They also computed activation energies for

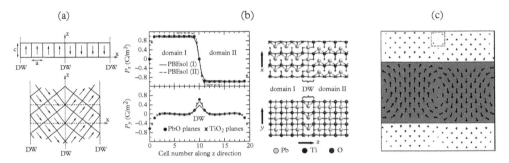

Figure 3.3 *Ferroelectric domain walls in PbTiO₃. Panel (a) shows periodically repeated supercells typically considered to investigate walls of 180° (top) and 90° (bottom) in PbTiO₃. In the former case, $\mathbf{P}^I$ and $\mathbf{P}^{II}$ are both along the [001] pseudocubic direction, and the wall is in the (1,0,0) plane. In the latter, always in the pseudocubic setting, we have $\mathbf{P}^I = (P,0,0)$ and $\mathbf{P}^{II} = (0,P,0)$, and the wall plane is (1, 1, 0). Note that we have symmetry-equivalent walls at the center and boundaries of the simulation supercells. Panel (b) summarizes the results by Wojdeł and Íñiguez (2014) for the internal structure of 180° walls: The domain polarizations are along x, and the wall is in a (001) plane. A wall polarization along y develops—as observed in first- (lines) and second- (symbols) principles simulations—which confers the wall a Bloch character. Details of the atomic distortions at domains and the wall are also shown. Panel (c) sketches the first-principles result for the multidomain structure of PbTiO₃ layers inserted between SrTiO₃ layers in a PbTiO₃/SrTiO₃ superlattice.*

Source: (a) Adapted from Meyer and Vanderbilt 2002. (c) Taken from Aguado-Puente and Junquera 2012.

wall motion, albeit within drastic approximations (the whole wall plane was assumed to move rigidly, and the transformation path was obtained from a simple interpolation) that yield a (probably much exaggerated) upper limit. They also discussed the electric potential associated with the wall discontinuity, revealing that even such idealized walls constitute shallow carrier traps. A brief summary of their results is included in Table 3.1.

While being the most important and influential, this study (Meyer and Vanderbilt 2002) was not the first first-principles investigation of the walls of PbTiO₃. Three years earlier, Pöykkö and Chadi (1999) had conducted a work restricted to the 180° boundaries (see Table 3.1), whose conclusions align with those of Meyer and Vanderbilt except in one aspect: the former authors report an internal polarization confined to the wall plane, while the latter did not find such a feature, in spite of explicitly looking for it. Thus, following the usual domain-wall classification depicted in Figure 3.2(d), Pöykkö and Chadi predict PbTiO₃'s walls to have a Bloch character, while Meyer and Vanderbilt's results (as well as others' (Behera et al. 2011)) suggest they have an Ising character.

More recent DFT investigations (Wojdeł and Íñiguez 2014; Wang et al. 2014; Liu and Cohen 2017b) have confirmed the Bloch polarization at the 180° boundaries (see Figure 3.3(b)). These works also discuss the energetics of the wall polar instability, showing that it has essentially the same nature as the polar distortion within the domains (i.e. it is dominated by the off-centering of the Pb cations). They also emphasize how the polar instability prevails in spite of the factors against it (Wojdeł and Íñiguez 2014); for example, the confinement to the wall plane implies the truncation of many

Table 3.1 *Summary of first-principles results on domain walls. Here we collect the basic features computed from first principles for representative materials. Domain-wall energies (E_{DW}) and barriers for wall motion (E_b) are in mJ/m^2, thickness (ξ) in nm, wall polarizations (P^{DW}) are in C/m^2. The employed density functional (LDA (Ceperley and Alder 1980), LDA+U (Anisimov et al. 1997), PBE (Perdew et al. 1996), or PBEsol (Perdew et al. 2008)) is indicated. The reported ξ corresponds to fitting the evolution of the order parameter across the wall to a hyperbolic tangent function of the form tanh(x/ξ), except in some cases in which the width was estimated by direct inspection of the data.*

PbTiO$_3$

Meyer and Vanderbilt (2002) (LDA)	180° (100)-wall, Ising: $E_{DW} = 132, \xi \approx 0.2$, $E_b = 37$
	90° (110)-wall, Ising: $E_{DW} = 35, \xi \approx 0.5, E_b \approx 1.4$
Pöykkö and Chadi (1999) (LDA)	180° (100)-wall, Bloch: $E_{DW} = 150, \xi \lesssim 0.4$, $P^{DW} \neq 0$
Wojdeł and Íñiguez (2014) (PBEsol)	180° (100)-wall, Bloch: $E_{DW} = 152, \xi \lesssim 0.4$, $P^{DW} \approx 0.3$
Wojdeł and Íñiguez (2014) (LDA)	180° (100)-wall, Bloch: $E_{DW} = 97, \xi \lesssim 0.4$, $P^{DW} \neq 0$
Wang et al. (2014) (LDA)	180° (100)-wall, Bloch: $E_{DW} = 118, \xi \lesssim 0.4$, $E_b = 38, P^{DW} \approx 0.3$
	180° (110)-wall: $E_{DW} = 117, \xi \lesssim 0.4, E_b = 21$, $P^{DW} \approx 0.4$

(All 180° (100)-DWs PbO centered; 90° DWs (Meyer and Vanderbilt (2002))

halfway between Pb-Ti-O planes)

BaTiO$_3$

Meyer and Vanderbilt (2002) (LDA)	Working with PbTiO$_3$-like tetragonal structure (not ground state)
	180° (100)-wall, BaO centered: $E_{DW} = 7.5, \xi$ not reported
Pöykkö and Chadi (1999) (LDA)	Working with rhombohedral ground state. For each **P**-rotation angle, lowest-energy DWs given
	180° Bloch wall: $E_{DW} = 24, \xi = 1.4, P^{DW} \approx 0.3$
	109° Ising wall: $E_{DW} = 11, \xi = 0.4$
	71° Ising wall: $E_{DW} = 4, \xi = 0.3$

Table 3.1 *Continued*

SrTiO₃	
Schiaffino and Stengel (2017) (LDA)	90° head-to-head: $E_{DW} \approx 0.35$, $\xi \approx 4$, $P_{imp}^{DW} \approx 1.5 \times 10^{-3}$
	90° head-to-tail: $E_{DW} \approx 0.4$, $\xi \approx 4$, $P_{imp}^{DW} \approx 2 \times 10^{-3}$
	Both walls are ferroelastic, polarization is improper
CaTiO₃	
Barone et al. (2014) (PBEsol)	$a^- a^- c^+$ structure, 90° rotation of axis for anti-phase tilts
	$E_{DW} \approx 16$, $\xi \lesssim 1$, $P_{imp}^{DW} \approx 0.3..$
	DW is ferroelastic, very large improper polarization
PbZrO₃	
Wei et al. (2014) (PBE)	Anti-phase wall for antiferroelectric order parameter
	(this kind of DW yields $P_{imp}^{DM} \neq 0$ even if it is not ferroelastic)
	$E_{DW} \approx 190$, $\xi \approx 1.5$, $P_{imp}^{DW} \approx 0.1$
BiFeO₃	
Lubk et al. (2009) (LDA+U)	180° (110)-wall: $E_{DW} = 829$, $\xi \lesssim 0.4$
	109° (100)-wall: $E_{DW} = 205$, $\xi \approx 0.4$
	71° (110)-wall: $E_{DW} = 363$, $\xi \approx 0.4$
Diéguez et al. (2013) (LDA+U)	180° (110)-wall: $E_{DW} = 82$, $\xi \approx 0.4$
	109° (100)-wall: $E_{DW} = 62$, $\xi \approx 0.4$
	71° (110)-wall: $E_{DW} = 167$, $\xi \approx 0.4$

(For each of the polarization-rotation angles indicated (180°, 109°, 71°), only the lowest-energy DWs obtained by Lubk et al. (2009) and Diéguez et al. (2013) are shown.)

Table 3.1 *Continued*

YMnO$_3$

Kumagai and Spaldin (2013) (LDA+U)	neutral wall containing [001] direction
	$\pi/3$ rotation of primary order, 180° switch of **P**
	$E_{\text{DW}} = 11, \xi \sim 0, E_b = 33$ (notably: $E_b \gg E_{\text{DW}}$!)
Småbråten et al. (2018) (LDA+U)	180° CDWs within (001) hexagonal plane
	$E_{\text{DW}} = 0.4 - 140, \xi \approx 0.2$ (CDWs with ultra-low energies!)

MAPbI$_3$

Liu et al. (2015) (PBE)	neutral (100)-wall, $E_{\text{DW}} = 3, \xi \lesssim 0.6$
	charged (100)-wall, $E_{\text{DW}} = 33, \xi \lesssim 0.6$, $\mid \Delta\mathbf{P} \cdot \hat{n} \mid = 0.08$ C/m^2
Chen et al. (2018a) (PBEsol)	charged (100)-wall, $E_{\text{DW}} = 9 - 46$ depends on domain width
	$\xi \lesssim 0.6$ for all domain widths; DWs metallic for widths > 13 nm

CsPbI$_3$

Warwick et al. (2019) (PBEsol)	Working with SrTiO$_3$-like antiferrodistortive structure that mimics
	MAPbI$_3$ room-temperature phase (not ground state of CsPbI$_3$)
	90° head-to-head: $E_{\text{DW}} \approx -1.2, \xi \approx 0.5$, $P_{\text{imp}}^{\text{DW}} \approx 0.06$
	90° head-to-tail: $E_{\text{DW}} \approx 4.4, \xi \approx 0.5$, $P_{\text{imp}}^{\text{DW}} \approx 0.02$
	$E_{\text{DW}} < 0$ denotes local wall relaxation toward ground state

ferroelectricity-favoring (dipole–dipole) interactions; also, the strain state imposed by the domains goes against the development of a polarization perpendicular to $\mathbf{P}^I$ or $\mathbf{P}^{II}$, which nevertheless occurs. By now, the Bloch character of $PbTiO_3$'s 180° walls has been shown to be robust against the most critical approximations in the DFT simulations; most notably, the same qualitative wall structure is obtained for all typical exchange-correlation potentials and various treatments of the core electrons. Hence, this is a solid first-principles prediction. Further, today we count with convincing direct and indirect experimental evidence backing it (Shafer et al. 2018; Das et al. 2019).

In hindsight, I think this difficulty of determining the character of $PbTiO_3$'s walls reflects a complication that pertains to this material and probably many other ferroics. Considered in their idealized high-symmetry (Ising-like) structure, $PbTiO_3$'s walls present a very flat energy landscape for the polar distortions, and the development (or not) of a spontaneous wall polarization may well depend on usually unimportant details of the calculations, including the tolerance for residual forces and stresses used in a structural relaxation. We experienced these issues ourselves when performing the calculations for the study by Wojdeł and Íñiguez (2014); in fact, it was thanks to the "second-principles" simulations described below (which are much more effective for exploring the energy landscape) that we discovered the Bloch-like lowest-energy structure, which we later ratified from first principles. Thus, the lesson here is: When running a simulation to identify the lowest-energy configuration of a wall, perform some sort of simulated temperature annealing that maximizes the chances of escaping from shallow local minima or very flat saddle points.

Ferroelectric walls with Bloch character have also been predicted in $BaTiO_3$ (Taher-inejad et al. 2012; Li et al. 2014). The mechanism behind $BaTiO_3$'s wall polarization seems very similar to that of $PbTiO_3$—simply, the native ferroelectric instability of the material develops along a different direction at the wall—suggesting that this behavior may be quite general. Interesting properties of these wall polarizations have been highlighted by several authors; for example, it has been predicted that temperature (in $PbTiO_3$ (Wojdeł and Íñiguez 2014)) or pressure (in $BaTiO_3$ (Marton et al. 2013)) can be used to eliminate it. In particular, by heating $PbTiO_3$, it is possible to induce a transition of its 180° walls from Bloch-like to Ising-like (Wojdeł and Íñiguez 2014).

I should also note that $PbTiO_3$'s walls (Lee et al. 2009; Behera et al. 2011), as well as similar 180° boundaries in $LiNbO_3$ (Lee et al. 2009) and $BaTiO_3$ (Li et al. 2014), have been claimed to have a Nèel-like character (Figure 3.2(d)). I believe this conclusion may not be warranted. As mentioned in Section 3.2.2, *any* wall (ferroelectric or not, ferroelastic or not) presents strain gradients along its normal direction. Except in ferroelasic walls, these strain gradients always integrate to zero, and so do the local flexoelectric polarizations that they cause. As far as I can tell, all existing claims for Nèel-like ferroelectric walls follow this pattern: 180° non-ferroelastic boundaries displaying small polarizations on each side of the wall, normal to the wall plane, and opposing each other. Given that this is a feature of *any* structural boundary, and that it does not really match the picture of a Nèel wall in Figure 3.2(d) (which involves breaking the symmetry of the Ising-like configuration), I doubt it deserves emphasis.

The walls of $PbTiO_3$ and $BaTiO_3$ have also been investigated in larger-scale first-principles simulations. For example, Aguado-Puente and Junquera (2008) considered ultra-thin $BaTiO_3$ films (forced by epitaxial strain to present a tetragonal structure with **P** normal to the film plane) in a DFT simulation in which the electrodes ($SrRuO_3$), and thus the electrode–ferroelectric interfaces, were treated explicitly. Interestingly, they observed that the stray fields enter the interfacial layers of $SrRuO_3$, so that the ferroelectric domains close within the electrode. Similar issues have later been investigated in free-standing $PbTiO_3$ films (Shimada et al. 2010) and, most remarkably, in $PbTiO_3/SrTiO_3$ superlattices (Aguado-Puente and Junquera 2012). In particular, the work on superlattices revealed vortex-like domain structures occurring in the $PbTiO_3$ layers (Figure 3.3(c)), which were hinted at (Zubko et al. 2010) and later explicitly observed (Yadav et al. 2016) experimentally, and which are one of the main reasons for today's focus on topological skyrmion-like features in strong ferroelectrics (Section 3.5). (On a related note, a DFT discussion of nontrivial dipole topologies in $BaTiO_3$ nanowires can be found in the work by Hong et al. (2010).) Finally, let me mention the work of Jiang et al., who combined DFT and non-equilibrium Green's function methods to understand how walls affect electronic transport across ultra-thin $PbTiO_3$ films (Jiang et al. 2016); their beautiful simulations, with explicit consideration of Pt electrodes, reveal a large enhancement of the leakage current owing to a greater number of conduction channels when the walls are present. While challenging, the application of such advanced and detailed simulation techniques should become more frequent in the future, as they are required to obtain a deeper understanding of wall functionality.

3.3.2 Antiferrodistortive Materials: Perovskites $SrTiO_3$ and $CaTiO_3$

Now we turn our attention to materials where the primary order parameter is an antiferrodistortive one, involving concerted anti-phase or in-phase rotations of the oxygen octahedra that constitute the backbone of the perovskite structure (Figures 3.1(b) and 3.1(c)).

Historically, the best-investigated compound in this category may be $SrTiO_3$. The interest in $SrTiO_3$ is twofold. On one hand, it is known to be a quantum paraelectric, where the development of a ferroelectric polarization is prevented by zero-point quantum fluctuations (Müller and Burkard 1979; Zhong and Vanderbilt 1996). On the other hand, it is a model compound for competing instabilities in perovskite oxides (Zhong and Vanderbilt 1995; Wojdeł et al. 2013); indeed, first-principles theory indicates that, in the absence of antiferrodistortive order, $SrTiO_3$ would be a regular ferroelectric. Naturally, this rich behavior suggests that very interesting effects are likely to occur in the antiferrodistortive domain walls of $SrTiO_3$.

When studying $SrTiO_3$'s walls, and in particular their possible polarity, an important distinction must be made. On one hand, as explained in Section 3.2.2, one expects $SrTiO_3$'s *ferroelastic* walls to be polar. This polarity constitutes an improper ferroelectric order that is not switchable. On the other hand, much has been discussed about the possible presence of switchable ferroelectricity at $SrTiO_3$'s walls, ferroelastic or not

(Tagantsev et al. 2001; Scott et al. 2012). Because the dominant antiferrodistortive order diminishes in the wall region, one might expect the weaker ferroelectric order to develop spontaneously there. That would be a switchable proper ferroelectricity, confined to the antiferrodistortive wall.

A first-principles investigation of (some of) the domain walls of $SrTiO_3$ has been accomplished only very recently, for two reasons. On the one hand, $SrTiO_3$ is a tricky material to treat from first principles, as the traditional DFT approximation for the exchange-correlation functional (i.e. the so-called local-density approximation or LDA (Ceperley and Alder 1980)) predicts that this compound is a very soft dielectric, but not ferroelectric (King-Smith and Vanderbilt 1994). In other words, according to the LDA, bulk $SrTiO_3$ would never develop ferroelectricity, even in absence of antiferrodistortive order or quantum fluctuations. More recent and accurate density functionals (e.g. the so-called PBEsol (Perdew et al. 2008)) correct this result and give a qualitatively correct description of the antiferrodistortive–ferroelectric competition in $SrTiO_3$, predicting the material to present a weak ferroelectric instability (Wahl et al. 2008). On the other hand, at variance with the cases of the ferroelectric walls of $PbTiO_3$ and $BaTiO_3$, the most interesting antiferrodistortive walls in $SrTiO_3$ can be expected to be very wide, thus requiring large simulation supercells for a realistic first-principles treatment. This difference is related to the relatively high energy cost associated with having gradients of the antiferrodistortive order parameter along certain directions (i.e. perpendicular to the rotation axis), while similar gradients of the ferroelectric polarization have a relatively small energy penalty associated to them. (This is related to the fact that the bands related with $SrTiO_3$'s antiferrodistortive instability are more dispersive than those associated with $PbTiO_3$'s ferroelectric instability (Wojdeł et al. 2013).) This is a major obstacle, and it has been necessary to wait for the development of powerful enough computers and DFT codes to tackle $SrTiO_3$'s walls from first principles.

It has been worth the wait. In 2017, Schiaffino and Stengel (2017) published the first DFT investigation of ferroelastic twin boundaries in $SrTiO_3$, which yielded a number of interesting discoveries. These authors considered two types of walls: "head-to-head" boundaries separating $\mathbf{Q}^I = (0, Q, 0)$ and $\mathbf{Q}^{II} = (-Q, 0, 0)$ domains, and "head-to-tail" walls between $\mathbf{Q}^I = (-Q, 0, 0)$ and $\mathbf{Q}^{II} = (0, -Q, 0)$ domains. In both cases, the walls lay in the $(1, 1, 0)$ crystallographic plane. Beyond the quantification of wall energies, widths (about 8 nm at 0 K), and electric polarization (see Table 3.1), this work revealed the key role of structural details that might have seemed unimportant and would never be (had never been) considered in effective descriptions of $SrTiO_3$. Most remarkably, it was shown that antipolar distortions of the Ti cations become active at the walls (due to the symmetry breaking intrinsic to the boundaries) and play a key role in stabilizing the wall structure; this happens despite the fact that such distortions are high-energy ones, essentially negligible as regards the behavior of (mono-domain) $SrTiO_3$. Further, this work provides us with a detailed discussion of the Ginzburg–Landau model that would be required to obtain an accurate description of the DFT results. In doing so, the authors reveal (and show how to compute from first principles) a number of original couplings that, despite their now obvious importance, would have been all but impossible to identify in a purely phenomenological approach. Most remarkable is the "rotopolar" interaction between polarization and antiferrodistortive tilt gradients, closely related to the flexoelectric effect and key to quantifying the development of the improper wall polarization.

An all-important family of antiferrodistortive perovskites (which gathers about half of all perovskite compounds (Lufaso and Woodward 2001; Chen et al. 2018b)) is formed by orthorhombic materials featuring a combination of anti-phase ($\mathbf{R}$) and in-phase ($\mathbf{M}$) O_6 tilts. A representative example is $CaTiO_3$, where a given domain would display anti-phase tilts about the [1,1,0] pseudocubic direction ($\mathbf{R} = (R,R,0)$) and in-phase tilts about [0,0,1] ($\mathbf{M} = (0,0,M)$), a pattern labeled $a^-a^-c^+$ in the self-explanatory notation introduced by Glazer (1972). This complex tilt pattern gives rise to many symmetry-equivalent domain variants (24), and a very large number of possible domain walls. In 2014, Barone et al. (2014) used DFT to investigate one specific wall that is commonly observed experimentally in $CaTiO_3$, namely, a ferroelastic boundary between variants $\mathbf{R}^I = (R,R,0)$ / $\mathbf{M}^I = (0,0,M)$ and $\mathbf{R}^{II} = (-R,R,0)$ /$\mathbf{M}^{II} = (0,0,M)$, lying within the (1, 0, 0) plane. The simulations yielded a relatively thin wall (well below 1 nm at 0 K) whose details are summarized in Table 3.1; most importantly, these authors quantified, for the first time from first principles, the improper polarization associated with one such ferroelastic boundary. Further, Barone et al. also investigated the same wall in multiferroic (ferroelastic and antiferromagnetic) compound $CaMnO_3$, obtaining similar results. Interestingly, the walls of $CaMnO_3$ are (improper) ferroelectric and magnetic, which hints at a novel strategy to stabilize magnetoelectric multiferroic states.

As compared to $SrTiO_3$'s broad twin boundaries, those of $CaTiO_3$ and $CaMnO_3$ are predicted to be much thinner. Interestingly, recent DFT calculations of non-ferroelastic anti-phase boundaries in $SrTiO_3$—e.g. the (0, 0, 1) wall between $\mathbf{R}^I = (0,0,R)$ and $\mathbf{R}^{II} = (0,0,-R)$ variants—yield similarly wide boundaries (Bristowe 2019). In contrast, very thin walls have been predicted (and observed experimentally) for other orthorhombic perovskites like $TbMnO_3$ (Farokhipoor et al. 2014) (see also Chapter 2), aligning with the results of Barone et al. (2014) for $CaTiO_3$ and $CaMnO_3$. Moreover, Zhao and Íñiguez (2020) have recently investigated all possible twin boundaries in {100} planes for a set of representative $a^-a^-c^+$ compounds ($CaTiO_3$, $LaGaO_3$, $YAlO_3$, $GdAlO_3$, $NdAlO_3$): very thin walls are generally obtained. All in all, these results challenge the common notion that antiferrodistortive boundaries are expected to be wide, and point to the importance of the specific material and/or tilt structure. Interpretation of the increasingly available DFT data, via suitable effective models, will be needed to elucidate this issue and reveal the physical parameters controlling the width and related behavior (e.g. mobility) of these walls.

Finally, let me mention a provocative case of an antiferrodistortive compound: $PbZrO_3$. Because of the dominant role that O_6 tilts play in its phase transitions (Íñiguez et al. 2014), $PbZrO_3$ rightly belongs among the antiferrodistortive materials mentioned above. Yet, the discussion of this material is usually focused on its striking antiferroelectric behavior (Lines and Glass 1977). Interestingly, such behavior makes the walls special: as shown in Wei et al. 2014 by DFT simulation and experiment, $PbZrO_3$'s anti-phase boundaries present a large (improper) electric polarization arising from its Pb-driven antipolar distortion, which is not compensated at the walls.

3.3.3 Materials with Multiple Primary Order Parameters: Perovskite $BiFeO_3$

Ferroelectric and antiferrodistortive instabilities are typically observed to compete, with one of them prevailing over (and suppressing) the other (Zhong and Vanderbilt 1995; Wojdeł et al. 2013). Nevertheless, there are materials in which both instabilities are similarly (and very) strong, and their competition relatively weak, which renders stable structures featuring both (Gu et al. 2018). The most notorious example is room-temperature multiferroic $BiFeO_3$ (Catalan and Scott 2009; Diéguez et al. 2011), which is arguably the main compound responsible for the recent revival of the domain-wall field.

The best-studied phase of $BiFeO_3$, stable over a wide range of temperatures and pressures including ambient conditions, is characterized by P and R vectors aligned along the same $\langle 111 \rangle$ direction. This rhombohedral ($R3c$) structure displays a modulated antiferromagnetic order, which makes $BiFeO_3$ a magnetoelectric multiferroic. Being one of the few (maybe the only) room-temperature magnetoelectric multiferroic compound that is easy to synthesize, $BiFeO_3$ has been a focus of attention for the past 15 years. In particular, a landmark investigation of its walls by Seidel et al. (2009) revealed that they are much more conductive than the domains themselves, hinting at all-new possibilities for wall-based electronics employing circuits that one could write/erase at will, by simply writing/erasing the corresponding ferroelectric domains (Sharma et al. 2017). Naturally, these results brought $BiFeO_3$'s walls into the spotlight, and motivated several DFT works on them. (See Chapters 10 and 15 for further experimental studies on domains and domain walls in $BiFeO_3$.)

To date, the most exhaustive first-principles investigation of $BiFeO_3$'s walls may be due to Diéguez et al. (2013), who studied all possible neutral walls that can exist within {100} and {110} planes. (In passing, note that this work provides useful guidelines for carrying out a systematic study of this type.) One major difficulty we had to face in that work traces back to $BiFeO_3$'s polymorphic nature (Diéguez et al. 2011; Singh et al. 2018): because this compound can adopt many different structures of similar energy (in particular, Bi^{3+} cations are comfortable in very diverse oxygen environments), a default relaxation of the walls can easily fall into a local minimum of the potential energy surface. Hence, it proved necessary to run repeated simulated-annealing simulations in order to be reasonably confident that the lowest-energy wall structures had been identified. This difficulty is apparent in the discrepancies among published wall energies, and is not a minor matter: If our DFT relaxation of a given wall gets stuck in a high-energy local minimum, we will predict wall properties that may differ from those of the *correct* (i.e. more likely to occur) lowest-energy wall configuration. As shown in Table 3.1, the wall energies obtained by Diéguez et al. are considerably lower than others published, and have been ratified by other authors (Ren et al. 2013); hence, they probably correspond to more relevant wall structures, and here we will focus on their results.

Diéguez et al. (2013) suggest that the neutral walls of $BiFeO_3$ do not involve any significant change in the electronic structure of the compound. Together with additional observations (e.g. the walls do not significantly affect the superexchange interactions of the compound, so we cannot expect large changes in the spin order that might in turn affect the band gap (Diéguez and Íñiguez 2019)), these results suggest that the experimentally measured enhanced conductivity is not an intrinsic property of the walls; rather, it should be linked to extrinsic factors, such as defects and/or peculiarities of the wall–electrode contact. Indeed, that is today the most widely accepted interpretation of the experimental results for $BiFeO_3$ (Farokhipoor and Noheda 2011). Further, similar effects have been measured at the walls of other ferroelectric compounds (Guyonnet et al. 2011), questioning a cause specific to $BiFeO_3$. (An explicit conductivity calculation as that by Jiang et al. (2016) has not been attempted in $BiFeO_3$. Yet, we must note that, to be realistic, such a $BiFeO_3$ study should probably go beyond the ultra-thin limit that is currently treatable in such DFT-based transport simulations.)

Admittedly, at the time we were disappointed by our conclusion that no intrinsic mechanism causes a high conductivity in $BiFeO_3$ walls. Fortunately, our $BiFeO_3$ study also brought about some unexpected and quite interesting discoveries. For example, having results for a long list of possible walls allowed us to tackle the following question: which factors determine the domain-wall energy, making some boundaries intrinsically more likely to occur than others? Since $BiFeO_3$'s walls are usually called "ferroelectric," our initial focus was to understand the correlation between domain-wall energy and the corresponding change in **P**. That did not take us far, and it quickly became clear that, instead, we should pay attention to the changes in the antiferrodistortive tilts. Indeed, as shown in Figure 3.4, the energy of the walls correlates strongly with the deformation

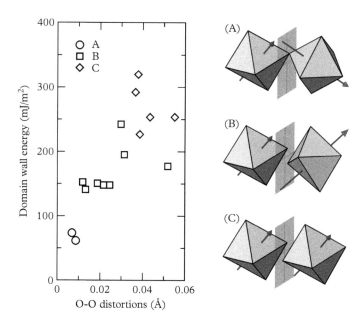

Figure 3.4 *Explaining domain-wall energies in $BiFeO_3$. The energies of ferroelectric walls in $BiFeO_3$, as obtained from first principles, correlate strongly with the degree of distortion of the O_6 octahedral groups at the wall, which in turn depends on the antiferrodistortive discontinuity. Three different levels of matching—i.e. A: undistorted O_6 groups; B: slightly distorted (one tilt mismatch); C: strongly distorted (double tilt mismatch)—can be identified and used to rationalize the results.*
Source: Taken from Diéguez et al. 2013.

of the O_6 groups at the boundary (see Figure 3.4); this is in turn determined by $\Delta \mathbf{R}$ across the wall and the wall orientation. Hence, the antiferrodistortive order parameter controls the energy of the "ferroelectric walls" of $BiFeO_3$.

Our results also revealed that the most stable walls have a structure that strongly resembles a low-energy polymorph of the material. Thus, for example, for the (100) (equivalently, *(1,0,0)*) boundary between $\mathbf{P}^I = (P, P, P)/\mathbf{R}^I = (R, R, R)$ and $\mathbf{P}^{II} = (P, -P, -P)/\mathbf{R}^{II} = (-R, R, R)$, the structure at the wall can be seen as a $a^+ b^- b^-$ tilt pattern, that is, the one corresponding to the orthorhombic phase of $CaTiO_3$ discussed above. This phase is also a low-lying polymorph of $BiFeO_3$ (Diéguez et al. 2011), which can be stabilized at high temperature (Arnold et al. 2010) or high pressure (Guennou et al. 2011), or by doping with lanthanides (Kan et al. 2010; González-Vázquez et al. 2012); hence, this wall looks very much like bulk $BiFeO_3$ under slightly different circumstances! The polytypic nature of $BiFeO_3$'s domain walls has been further evidenced in first-principles studies discussing so-called *nano-twined* structures (Prosandeev et al. 2011; Bellaiche and Íñiguez 2013) that $BiFeO_3$-related compounds are believed to form and which can be thought of either as the regular rhombohedral phase of $BiFeO_3$ with a massive density of walls, or as coexisting rhombohedral and orthorhombic $BiFeO_3$ structures.

The significance of these first-principles results, which highlight the all-important role that the antiferrodistortive tilts play in $BiFeO_3$, should not be overlooked. For example, they strongly suggest we should be cautious about effective theories of $BiFeO_3$ that do *not* include the tilts explicitly; the corresponding results, particularly those involving multi-domain states and wall properties, might well be qualitatively incorrect.[2]

Finally, Ren et al. (2013) used DFT to monitor the behavior of a few representative (low-energy) $BiFeO_3$ walls as a function of epitaxial strain. The authors observed changes in the relative stability of the different boundaries, particularly as the material approaches the regime (at about 6% epitaxial compression) where it transforms into a different (super-tetragonal) polymorph with a very large cell aspect ratio (Béa et al. 2009). This first-principles investigation provides us with an intriguing—and unique to this date, I believe—example of how walls can be modified by external perturbations, like epitaxial strain in this case.

3.3.4   Improper Ferroelectrics: Hexagonal Manganites

Improper ferroelectrics are materials where one or more nonpolar (typically, antiferrodistortive) order parameters drive a structural phase transition, as a result of which an electric polarization appears as a secondary order (Dvořák 1974). Hence, whenever the primary order parameter changes across a wall, the secondary polarization will (typically) change as well, making the boundary a ferroelectric one. All-important families of improper ferroelectrics are those formed by Ruddlesden–Popper layered

---

[2] It is difficult to exaggerate the importance of antiferrodistortive tilts in $BiFeO_3$. For example, we have good reasons to believe that their evolution during ferroelectric switching plays a critical role in (permits) the electric-field-driven reversal of the weak magnetization of this compound (Heron et al. 2014), an effect most likely missed if we treat $BiFeO_3$ as a ferroelectric-only material.

perovskites like $Ca_3Mn_2O_7$ or $Ca_3Ti_2O_7$ (Benedek and Fennie 2011; Oh et al. 2015), and hexagonal manganites like $YMnO_3$ or $ErMnO_3$ (Fennie and Rabe 2005; Choi et al. 2010; Meier et al. 2012). Because of their peculiar lattice topologies—which impose geometric constraints absent in the perovskite structure—the multi-domain structures of these materials display fascinating topological properties, which has attracted much attention in recent years.

The rich physics of domains in improper ferroelectrics is discussed in Chapter 6 of this book. Here let me just note that, because of their structural complexity, they pose a great challenge to first-principles simulation, as large simulation boxes would be required to treat their most appealing topological properties. To the best of my knowledge, walls in layered perovskites remain unstudied from first principles, and there are only a handful of works tackling walls in hexagonal manganites.

The first works on the manganites are due to Kumagai and Spaldin (Meier et al. 2012; Kumagai and Spaldin 2013), who in 2013 published a landmark study of neutral walls in $YMnO_3$ (Kumagai and Spaldin 2013). These authors considered representative boundaries separating two variants of the antiferrodistortive order parameter (the so-called α^+ and β^-; see Figure 3.5), oriented in planes with low Miller indexes and

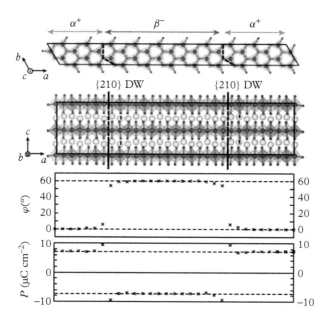

Figure 3.5 *Ferroelectric domain walls in $YMnO_3$. The figure shows different views of a multi-domain configuration that presents two walls. The primary order parameter involves tilts of the MnO_5 trigonal bipyramids (shaded polyhedra), and switches between 0° and 60° at the walls. The secondary polarization is mainly associated with the off-centering of the Y cations (light-colored balls), and switches at the walls following the primary order.*
Source: Taken from Kumagai and Spaldin 2013.

containing the hexagonal direction (because the polarization lies along the hexagonal axis, such walls are neutral). Among many interesting findings, this work shows that neutral walls in $YMnO_3$ are atomically sharp, with a nominal zero width. Further, the computed wall energies are relatively low as compared to those of perovskite oxides with similarly strong structural instabilities (e.g. $PbTiO_3$ and $BiFeO_3$; see Table 3.1). To explain these results, the authors make an interesting point: In the perovskites, the wall structure deviates from the low-energy ground state, imposing a frustration of sorts; in contrast, in $YMnO_3$ the walls preserve the basic structural features that characterize the domains (tilting of MnO_5 groups, off-centering of the Y atoms), being reminiscent of a stacking fault in a close-packed structure. Interestingly, Kumagai and Spaldin also showed that these sharp wall structures are quite stable, as they estimated activation barriers for wall motion that are actually higher than the wall energy. This is in striking contrast with the case of perovskite oxides like $PbTiO_3$, where comparatively less stable (higher-energy) walls are relatively easy to move (see Table 3.1).

Topology being sovereign in the physics of these manganites, Kumagai and Spaldin stretched their DFT simulations to discuss these matters. To that end, they ran a remarkable simulation of *winding* walls, with the wall plane changing as one moves along the boundary. They considered a periodically repeated supercell of 300 atoms that, while small from an experimental perspective, constitutes (even today) the upper limit for a very accurate first-principles simulation of a multiferroic like $YMnO_3$. The authors observed that the walls remain ultra-thin in spite of being curved, and were able to connect their results with the tendency of the material to form robustly stable (topologically protected) stripe-domain phases, as experimentally observed. With further calculations, they were also able to explicitly verify the theoretical expectation (Artyukhin et al. 2014) that $YMnO_3$ does not present purely antiferrodistortive or purely ferroelectric walls; instead, its boundaries always present a mixed antiferrodistortive/ferroelectric character. Finally, Kumagai and Spaldin (2013) also discussed the holy grail of topological defects in $YMnO_3$, namely, the experimentally observed regions at which the six possible domain variants meet. Based on their results for simpler domains, they suggested that these regions could actually be formed by a single point (i.e. a line along the hexagonal axis) with Y atoms sitting in their high-symmetry paraelectric positions. Today this is the generally accepted picture of these mesmerizing sixfold defects in hexagonal manganites.

Beyond the obvious interest of their results for $YMnO_3$, this study (Kumagai and Spaldin 2013) constitutes a great example of how one can make the most of (size-limited) DFT simulations to investigate seemingly beyond-scope problems and deliver a relevant discussion of the important physics, with atomistic insight and quantitative detail.

3.3.5 Perovskite Halides

It was recently noticed that organic-inorganic perovskite halides may present very high photovoltaic power conversion efficiencies (over 23% today and reaching 28% in perovskite/Si tandems (NREL 2019)), which has made them a central focus of attention (Park et al. 2016). The origin of such high efficiencies remains unclear, and some have proposed that part of the explanation may be linked to the domain walls that these ferroic materials present (Huang et al. 2017). (For a discussion of photovoltaic effects at domain walls, see Chapter 9.)

Indeed, perovskite halides display a rich variety of structural phases and transitions akin to those occurring in their oxide counterparts, featuring octahedral-tilting antiferrodistortive modes and ferroelectric distortions. Interestingly, the most promising halides are materials like methylammonium lead triiodide, $CH_3NH_3PbI_3$ or $MAPbI_3$, where the A-site of the perovskite lattice is occupied by the polar molecule CH_3NH_3 (Figure 3.6(a)). The presence of such molecular groups sets these materials apart from the usual perovskites; in particular, it makes (anti)ferroelectric behavior almost inevitable, and can be expected to strongly affect the structural, dynamic, and response properties of these compounds.

Once the existence of ferroelectric order in perovskite halides was established, the focus turned to possible photoferroic effects and, eventually, to the question of whether the corresponding domain boundaries might affect the photovoltaic behavior.

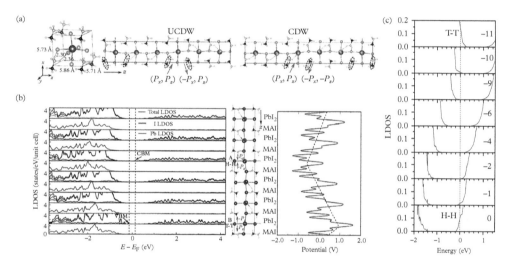

Figure 3.6 *Ferroelectric domain walls in $CH_3NH_3PbI_3$. Panel (a) shows the structure of this perovskite halide, as well as representative examples of the neutral (or uncharged, UCDW) and charged (CDW) domain walls it may present. Panel (b) shows the spatially resolved density of states and potential obtained from first principles for a configuration with charged walls. Note that we have both head-to-head (H-H) and tail-to-tail (T-T) walls, as in the sketch of Figure 3.2(c). The potential slope corresponds to depolarizing fields acting inside the domains, and causing the band bending that is visible in the density of states. Yet, the material remains insulating. Panel (c) shows the density of states of a similar charged-wall configuration, but for thicker domains. In that limit, the conduction band minimum (CBM) moves below the Fermi energy at the H-H wall, while the valence band maximum (VBM) goes above the Fermi level at the T-T wall. Hence, the walls become metallic, with an excess of electrons and holes, respectively. Yet, the depolarizing fields are not screened and remain present within the domains, as can be noticed from the persisting band bending.*

Source: Panels (a) and (b) are taken from Liu et al. 2015, and panel (c) from Chen et al. 2018a.

In particular, it was suggested by Frost et al. (2014) that the depolarizing fields could aid the electron-hole separation. This is a subtle point, because such fields are expected only in domain configurations where the walls are nominally charged and not perfectly screened. As we will see below, such configurations are hard to obtain in classic ferroelectrics like $PbTiO_3$ or $BiFeO_3$; yet, here the assumption was that the hybrid perovskite halides present them spontaneously. This possibility was confirmed in a landmark DFT study by Liu et al., who investigated a collection of representative neutral and charged boundaries in $MAPbX_3$ compounds ($X = Cl, Br, I$) (Liu et al. 2015). These authors found that, indeed, these materials can form low-energy, ultra-thin charged walls while retaining an insulating character, i.e. without any electronic compensation of the $\Delta \mathbf{P} \cdot \hat{n} \neq 0$ discontinuity at the walls (see Figures 3.6(a) and 3.6(b)). In such conditions, internal (depolarizing) fields appear within the domains, which in turn results in a band bending and a reduction of the electronic gap (although not necessarily of the optical gap). In a later first-principles study, Chen et al. (2018a) discussed the stability of such charged walls in $MAPbI_3$ as a function of the width of the domains, showing that the band gap closes, and the walls become metallic, for sufficiently wide domains of about 13 nm (Figure 3.6(c)). The results of Chen et al. show that, in these materials, one can simultaneously have metallic walls and carrier-separating electric fields within the domains; hence, the multi-domain configurations truly resemble nano p-n junctions with built-in electrodes!

The mentioned works suggest that the spontaneous occurrence of charged domain walls in perovskite halides is closely related to the relatively small values of the electric polarization, which have similarly small polar discontinuities associated with them. Nevertheless, I think much remains to be learned about this issue, maybe by comparing the hybrid perovskites with their oxide counterparts. One should also bear in mind that, beyond these charged ferroelectric walls, these compounds present other structural boundaries that may be relevant to their photovoltaic performance. Especially interesting are the domain walls occurring in nonpolar phases. Along these lines, Warwick et al. (2019) have recently investigated from first principles the ferroelastic boundaries of $CsPbI_3$ (a simpler and relevant analog of $MAPbI_3$), observing relatively large wall polarizations (see Table 3.1). Further, these authors found structural and electronic features suggesting that different walls may act as channels for carriers of different charge.

In summary, domain walls in perovskite halides might play a role in their superior photovoltaic properties. Remarkably, because this is a relatively new topic, and because today we count with simulation tools and strategies to study structural walls, the first-principles community is making relevant contributions to the research effort from the start.

3.4 Studies of Non-ideal Domain Walls

The examples above show that DFT investigations of walls have led to many discoveries and insights, pushing the field beyond what was known from experiment or phenomenological theory. Yet, these were ideal, relatively simple situations in which the walls were not very different from the domains. Whenever that is not the case—because the walls have a dissimilar composition, a markedly different electronic structure, etc.—first-principles methods become all but mandatory for a relevant theoretical investigation.

3.4.1 Defects (and Virtues) at Domain Walls

Defects in ferroics are turning from villains to heroes, as the focus shifts from avoiding them (because they cause fatigue, leakage, etc.) to engineering them (so that they result in enhanced properties). Despite some notable works, the investigation of defects in ferroic materials is underdeveloped when compared to other fields (chiefly, semiconductors), especially in what concerns first-principles work (Freysoldt et al. 2014). In particular, studies of defects in structural walls are in their infancy. Indeed, one has to have a clear motivation to tackle a problem that is technically so complex and computationally so demanding. (Large simulation supercells are typically required for reliable simulations of isolated defects; in the presence of domain walls, such a requirement makes simulations all but unfeasible!)

Nevertheless, there has been progress. A first pioneering study was due to Pöykkö and Chadi (2000), who in 2000 published a basic first-principles investigation of vacancy–impurity complexes at the 180° domain walls of $PbTiO_3$. These authors showed that oxygen vacancies tend to bind to Pt substitutional impurities (which replace Ti) and pin domain walls. While suffering from many shortcomings (this simulation was run in a periodically repeated cell of only 80 atoms!), this early study predicted the basic features of the interaction between defects and domain boundaries. More recently, applying well-known techniques to simulate charged defects, Paillard et al. investigated cation and oxygen vacancies in $PbTiO_3$ (Paillard et al. 2017), considering both domains and 180° ferroelectric walls. Most interestingly, they found that all vacancies are more easily formed at the walls, thus confirming the conclusions of previous DFT investigations (Pöykkö and Chadi 2000; He and Vanderbilt 2003; Tomoda et al. 2015). (Pinning of walls by oxygen vacancies has also been predicted from DFT in $LiNbO_3$ (Lee 2016).) Further, Paillard et al. proposed a physical picture (and phenomenological model) to explain this: in essence, the walls can be more easily disrupted by vacancies because their structure is not energy-optimized to begin with. In contrast, the work of Skjærvø et al. (2018) on neutral walls of $YMnO_3$ showed that, in this case, some oxygen vacancies tend to move away from the walls (see Figure 3.7(a)), which is explained in terms of local strain and chemical bond effects. Beyond their many merits, these two studies provide excellent guidelines on how to conduct a thorough investigation of point defects interacting with walls.

From the perspective of computational materials design, defects at walls offer many interesting possibilities. Since the domain-wall revival started, experimental groups have tried to control the concentration and nature of wall defects, in order to enhance the domain-wall properties (chiefly, electric conductivity) or make them more clearly distinct from those of the domains. Going one step further, several works combining experiment and DFT simulations (Biškup et al. 2014; Farokhipoor et al. 2014; Becher et al. 2015)

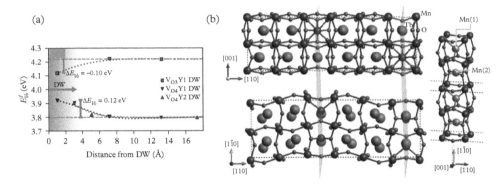

Figure 3.7 *Defects and emerging structures at domain walls. Panel (a) shows the defect formation energy of two types of oxygen vacancies in YMnO₃, as computed from first principles and as a function of the distance to a wall. According to these results, the most likely oxygen vacancy occurs within the domains. Panel (b) shows several atomistic views of ferroelastic domain walls (marked by shaded planes) in perovskite TbMnO₃, as obtained from DFT simulations. Half of the Tb³⁺ cations (biggest balls) are replaced by the smaller Mn³⁺ cations (medium-sized balls) at the wall plane, adopting the peculiar oxygen-coordination environments (square planar, tetrahedral) shown in the sketch on the right. Hence, these walls constitute a two-dimensional crystal with a composition that differs from that of the domains. Source: Panel (a) is taken from Skjærvø et al. 2018 and (b) from Farokhipoor et al. 2014.*

show that it is possible to obtain walls with long-range-ordered defect structures within them. In other words, we have the opportunity to create new two-dimensional crystals, with novel or optimized properties, at structural walls. Experimentally verified examples include TbMnO₃ (Farokhipoor et al. 2014)—where long-range substitution of Tb by Mn at its ferroelastic walls yields 2D crystals with peculiar magnetic properties (see Figure 3.7(b) and Chapter 2)—and CaMnO₃ (Becher et al. 2015)—where ordered arrays of oxygen vacancies render insulating walls within conductive domains. (Ordered vacancy planes observed in LaCoO₃₋δ films, while not identified with walls, have been proposed to be responsible for the insulating ferromagnetic character of the materials (Biškup et al. 2014).) DFT simulations have also hinted at the formation of ordered oxygen-vacancy structures at the walls of PbTiO₃ (He and Vanderbilt 2003; Xu et al. 2016), emphasizing the interesting magnetic properties they would present. Finally, motivated by Farokhipoor et al. (2014), a recent DFT work (Zhao and Íñiguez 2019) predicts that it is possible to start from well-known rather inert (insulating, non-magnetic, nonpolar) perovskite oxides (e.g. LaGaO₃, CaTiO₃) and dope their ferroelastic walls to make them strongly polar (even with a switchable ferroelectric polarization), metallic, and/or magnetic. Hence, according to an increasing number of first-principles predictions and experimental evidence, the possibility of engineering walls, via chemical doping or oxidation/reduction, appears to be very promising. The challenge for experimentalists is to actually produce these dream-like domain-wall structures in a controlled and reproducible way.

3.4.2 Charged Ferroelectric Walls

In *strong* ferroelectrics with large polarizations—like $PbTiO_3$, $BaTiO_3$, or $BiFeO_3$— charged domain walls have traditionally been regarded as an oddity. We knew that they can appear as a transient state during ferroelectric switching (Shur 2008; Maksymovych et al. 2011), or locally in non-planar walls; yet, they were all but impossible to manipulate or investigate in detail. The situation has recently changed. Driven by the prospect of having conductive boundaries in an insulating matrix, the efforts to stabilize charged walls in strong ferroelectrics have multiplied, notably led by Setter and collaborators (Sluka et al. 2012, 2013; Bednyakov et al. 2018), and more recently by the team of Seidel (Sharma et al. 2017) and others. Progress has been enormous, and today we can say that the possibility of dynamically creating and manipulating charged walls in materials like $BaTiO_3$ and $BiFeO_3$ is a reality.

A first-principles investigation of such charged walls in strong ferroelectric materials remains a great challenge, though. Let us imagine we run a first-principles simulation of a (100)-wall in $PbTiO_3$, separating domains with $\mathbf{P}^I = (P,0,0)$ and $\mathbf{P}^{II} = (-P,0,0)$, on the left and right sides of the wall plane, respectively (Figure 3.2(c)). Let us assume the lattice is perfect (no defects) and the system globally neutral (no doping). Because $PbTiO_3$'s polarization is so large, the discontinuity $\Delta \mathbf{P} \cdot \hat{n} \approx -1.6$ C/m^2 associated with this head-to-head wall is huge, and generates similarly large depolarizing fields inside the domains; the situation is unstable, and the material will evolve toward a different configuration in our simulation, typically mono-domain. Now, if our first-principles simulation supercell is periodically repeated (as they usually are), this head-to-head domain wall has a tail-to-tail counterpart, with a $\Delta \mathbf{P} \cdot \hat{n}$ of opposite sign. For sufficiently large domain sizes, because the depolarizing fields bend the electronic bands, the walls will eventually become metallic, with an excess of (conduction-band) electrons at the head-to-head boundary and a related excess of (valence-band) holes at the tail-to-tail wall (as shown in Figure 3.6(b) for charged walls in $MAPbI_3$). At this point, the polarization discontinuity is (partly) screened, the depolarizing fields reduced, and the possibility of stabilizing the charged walls appears. Using the second-principles methods described below (García-Fernández et al. 2016), we estimate that, for $PbTiO_3$, the minimum domain size yielding (meta-)stable charged-wall solutions is about 10 nm (García-Fernández et al. 2019), which is in the limit of what is feasible with DFT simulations today. (Our explicit DFT simulations indicate that domains of 6.5 nm remain unstable (García-Fernández et al. 2019).) Thus, the obvious fact: we still lack the most elemental first-principles characterization of charged domain walls in strong ferroelectrics.

Having said this, let me note a few DFT studies of charged walls in strong ferroelectrics, albeit in peculiar conditions. For example, Wu and Vanderbilt (2006) considered heterovalent substitutions in planes of the $PbTiO_3$ lattice (δ-doping with Sc^{3+} and Nb^{5+}, both substituting for Ti^{4+}), thus forcing the occurrence of head-to-head and tail-to-tail walls to screen the ensuing internal electric fields. A similar approach was used by Rahmanizadeh et al. (2014): oxygen-related defects were introduced in the $PbTiO_3$ lattice to force a tail-to-tail wall, inevitably leading to the formation of a head-to-head wall at a different location of the periodically repeated (relatively small) supercell; the authors could thus investigate the properties of the head-to-head boundary, observing charge and orbital orders of the screening electrons occupying Ti-3d levels. More interestingly, Jiang et al. studied the motion of neutral 180° walls in PbTiO$_3$, considering deviations from a perfectly planar structure and, thus, locally charged regions (Jiang et al. 2017). This notable investigation provides many valuable insights; in particular, it reveals that the steps are relatively smooth, which allows the wall to reduce the energy associated with the $\Delta \mathbf{P} \cdot \hat{n} \neq 0$ discontinuity.

The situation is different in hybrid perovskite halides, where charged walls occur spontaneously and have been studied from first principles (Liu et al. 2015; Chen et al. 2018a). Aside from specific features of these compounds that may make it easier to obtain such meta-stable states (difficulty of polar molecules like CH$_3$NH$_3$ to rotate freely, relatively large dielectric permittivity ε_∞), the main difference with strong ferroelectric compounds is the size of the electric polarization: it is relatively small in the halides, so that charged walls can even be obtained as stable insulating solutions where there is no electronic screening of the polar discontinuity.

A similar situation occurs in improper ferroelectrics like hexagonal manganites and layered perovskites, where spontaneous formation of charged walls has been experimentally observed (Meier et al. 2012; Oh et al. 2015). The electric polarization is relatively small in these materials as well, which yields narrow walls amenable to DFT investigations (Meier et al. 2012; Småbråten et al. 2018). In fact, because walls in improper ferroelectrics are defined by the change of the primary nonpolar order parameter, these materials provide us with an excellent playground for forming and tuning charged walls, as the contribution of the $\Delta \mathbf{P} \cdot \hat{n} \neq 0$ discontinuity to the total wall energy may be relatively small (Småbråten et al. 2018). Thus, by construction, improper ferroelectrics are particularly well suited for charged-wall-related studies or applications (see Chapter 6).

Finally, it is worth noting the case of hyperferroelectric compounds (Garrity et al. 2014). These are materials with an unstable band of polar longitudinal phonons, which amounts to saying that the occurrence of charged walls constitutes an instability of their high-symmetry (paraelectric) structure. Examples of hyperferroelectrics, as predicted from first principles, include some exotic ferroelectric materials (e.g. recently discovered hexagonal *ABC* compounds like LiBeBi and NaZnSb (Garrity et al. 2014)), but also more familiar ones (e.g. LiNbO$_3$ (Li et al. 2016), La$_2$Ti$_2$O$_7$ or Ca$_3$Ti$_2$O$_7$ (Zhao et al. 2018)). Their peculiar hyperferroelectric behavior can be associated with a variety of factors, e.g. relatively large ε_∞ values (i.e. good electronic screening) and ferroelectric instabilities that do not rely on strong dipole–dipole interactions (e.g. with a marked chemical or steric origin). These compounds are likely to exhibit charged walls spontaneously, as has indeed been observed in (Ca,Sr)$_3$Ti$_2$O$_7$ (Oh et al. 2015). Liu and Cohen (2017a) provide an interesting first-principles investigation of charged walls in hexagonal *ABC* ferroelectrics and LiNbO$_3$, discussing the basic structural and electronic features of such peculiar boundaries, such as the possibility of confining two-dimensional electron/hole gases.

3.5 Large-Scale Simulations of Structural Domain Walls Based on First-Principles Models

Most of the first-principles simulations cited in this chapter involve periodically repeated supercells of about 100 atoms, which is a size that I think can be considered standard and tractable within modern DFT studies of transition-metal oxides with a high-accuracy requirement (i.e. resolved to within 1 meV per atom). Further, the vast majority of these works involve relaxations of (meta)stable walls, considered in the limit of 0 K and assuming a classical (particle-like) behavior of the atoms. While enough to learn many interesting lessons about walls, this is hardly satisfactory. Using these methods, an investigation of realistic wall configurations as a function of temperature is all but impossible, let alone the study of field-driven thermally activated processes like wall motion and ferroic switching.

Over the years, the first-principles community has produced many ideas and methods to run simulations in larger length and time scales while retaining the accuracy and predictive power of the DFT calculations (Elstner et al. 1998; Cui et al. 2001; Makov et al. 2009; Gaus et al. 2011; Behler 2014). Truth be told, not many of the general methods have been applied to the investigation of the ferroelectric and ferroelastic materials of interest here, as our community has somehow followed its own path, strongly conditioned by previous phenomenological works in the field. Indeed, the techniques best developed and most applied to structural ferroics—namely, shell models, bond-valence models, and effective Hamiltonians—build upon a large body of previous literature on those very materials. The strategy is simple: you borrow an existing phenomenological model potential, extend it a bit if needed, and compute the corresponding parameters by fitting to selected results obtained from first principles. This has allowed several groups to run molecular-dynamics and Monte Carlo simulations of 1,000–100,000 atoms, at finite temperatures, with applied electric and mechanical fields. Progress—in terms of insights and better understanding, and also mesmerizing predictions later confirmed—has been, and continues to be, enormous. It would be preposterous to attempt here a summary of first-principles-based model potential works, even if I restrict myself to the ambit of domain walls. I will just mention a few illustrative works that I find particularly important and inspiring.

First-principles-based shell models—as those initially introduced, e.g. to interpret neutron and Raman measurements of phonons (Cowley 1964; Migoni et al. 1976)— have been constructed for many key perovskite oxides, including $BaTiO_3$ (Tinte et al. 1999), $PbTiO_3$ and $PbTiO_3$-based relaxors (Sepliarsky et al. 2004), and $BiFeO_3$ (Graf et al. 2014). Applications to wall-related problems include the investigation of topological multi-domain toroidal structures in $PbTiO_3$ nanoparticles (Stachiotti and Sepliarsky 2011), or the study of wall-specific response properties in $PbTiO_3$ stripe-domain structures (Liu and Cohen 2017b).

More original from a methodological perspective was the transmutation of the bond-valence method—an empirical and incredibly useful way to measure the chemical *satisfaction* of ions in a structure (Brown 2009), so that predictions of equilibrium configurations can be made—into a model potential from which energies, forces, and stresses can be computed (Etxebarria et al. 2005; Shin et al. 2005). This smart idea has been much developed and exploited by Rappe and collaborators (see Chapter 14), who have applied the thus constructed potentials for $PbTiO_3$ (fitted to a combination of DFT and experimental data (Shin et al. 2005)) in very-large-scale investigations of wall motion (Shin et al. 2007) and ferroelectric switching (Liu et al. 2016). Beyond their obvious importance, these works constitute beautiful examples of how to connect atomistic simulation and existing empirical theories addressing these critical problems; in my view, they are a must-read for anyone interested in this field.

Finally, the most developed and applied strategy is the so-called "effective Hamiltonian" approach, which can be viewed as an extension of the classic "ϕ^4" Hamiltonians (Giddy et al. 1989; Radescu et al. 1995) long employed in the investigation of phase transitions driven by soft modes. The method first developed by Rabe, Vanderbilt, and others—who worked with relatively simple materials like GeTe (Rabe and Joannopoulos 1987a, 1987b), $BaTiO_3$ (Zhong et al. 1994, 1995), $SrTiO_3$ (Zhong and Vanderbilt 1995), $PbTiO_3$ (Waghmare and Rabe 1997), or $KNbO_3$ (Krakauer et al. 1999)—was later much extended by Bellaiche and collaborators to treat chemically disordered compounds (e.g. $PbZr_{1-x}Ti_xO_3$ (Bellaiche et al. 2000)), finite systems (Fu and Bellaiche 2003; Naumov et al. 2004), magnetoelectric multiferroics accounting for spins and spin–lattice interactions (from the initial application to $BiFeO_3$ (Kornev et al. 2007; Prosandeev et al. 2013) to recent models of $YMnO_3$ (Wu et al. 2017)), etc. The effective Hamiltonian approach has been applied to wall-related problems since its early days (Padilla et al. 1996). Among many interesting results, let me highlight what may be one of the most exciting discoveries in ferroelectrics in the past 10 years: The prediction of exotic dipole arrangements with nontrivial topologies in ferroelectric nanostructures (Figure 3.8(a)) (Fu and Bellaiche 2003; Naumov et al. 2004; Louis et al. 2012; Nahas et al. 2015). Many deemed these landmark results as an oddity at the time; yet, they were the avant-garde of one of the hottest areas of activity in ferroelectrics today!

All these effective methods rely on approximations that of course limit them. In this context, we have recently introduced what we call "second-principles" methods (Wojdeł et al. 2013; García-Fernández et al. 2016; Escorihuela-Sayalero et al. 2017), in an attempt to extend the scope of effective models as regards both accuracy and physical phenomena that can be treated. Our second-principles lattice potentials can be seen as an extension of the effective Hamiltonian approach, such that all atoms are considered in the model (thus removing the coarse graining associated with the use of local modes in effective Hamiltonians (Rabe and Waghmare 1995; Íñiguez et al. 2000)); they employ a generic polynomial expression for the energy, whose precision can be systematically improved in a trivial way. In this way, the lattice potentials are designed to reach first-principles accuracy, if required.[3]

Yet, this is not the most interesting feature of the new scheme. Indeed, unlike the effective schemes presented above, the second-principles approach is designed to include

---

[3] Because they use of a polynomic Taylor-like expansion around a reference point, our current second-principles methods assume the existence of an underlying lattice. Hence, they are not ideally suited for dealing with strong changes in chemical bonds, which may break that underlying lattice topology.

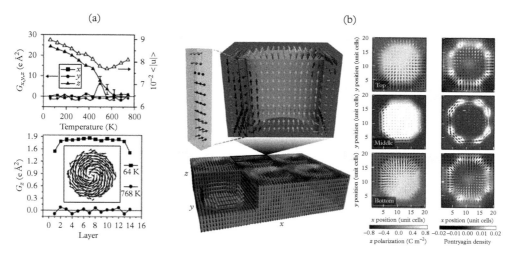

Figure 3.8 *Large-scale simulation of multi-domain states. Panel (a) shows results from effective Hamiltonian simulations predicting the development of a toroidal dipole structure upon cooling a disk-shaped nanostructure of $PbZr_{0.5}Ti_{0.5}O_3$. The different components of the toroidal moment G_α ($\alpha = x, y, z$) are shown as a function of temperature, as is the average absolute value of the local dipoles ($\langle|u|\rangle$); also shown is the evolution of G_z along the vertical of the disk at two different temperatures. Panel (b) show sketches of the bubble domains forming in a $PbTiO_3$ layer stacked between $SrTiO_3$ layers, as obtained from second-principles simulations. The cuts show the local dipoles (colors for the out-of-plane component, drawn arrows for the in-plane components) as well as the corresponding Pontryagrin density obtained from them. At their middle plane, the bubbles are Bloch skyrmions, but they turn into Nèel skyrmions close to the $PbTiO_3/SrTiO_3$ interfaces.*
Source: Panel (a) is taken from Naumov et al. 2004 and (b) from Das et al. 2019.

a description of the relevant electronic bands, treating lattice and electrons on the same footing. This methodology, which we call "second-principles DFT," was introduced by García-Fernández et al. (2016) and is implemented in the code scale-up. While some important aspects are still being developed (easier construction of electron–lattice models from DFT data, efficient dynamical simulations of both electrons and lattice), these techniques will soon allow us to run large-scale simulations (10,000–100,000 atoms) of charged walls, optical excitations, electronic conductivity, magnetism and magnetoelectric effects, etc.

So far, the most relevant application of second-principles methods concerns a purely structural problem, namely, the occurrence of nontrivial wall-related topologies in ferroelectrics. In a nutshell: The lattice model developed by Wojdeł and Íñiguez for $PbTiO_3$ (Wojdeł et al. 2013) was key to establish the Bloch character of the 180° walls of this material (Wojdeł and Íñiguez 2014). It was then found that this very feature is responsible for the chirality observed experimentally in $PbTiO_3/SrTiO_3$ superlattices (Shafer et al. 2018). Further, second-principles simulations predicted that the Bloch wall polarization can be harnessed to create skyrmion-like arrangements of electric dipoles in $PbTiO_3$ films, simply by writing columnar nanodomains (Pereira Gonçalves et al. 2019);

simultaneously, such mesmerizing dipole textures have been predicted theoretically, and corroborated experimentally, in $PbTiO_3/SrTiO_3$ superlattices in which column-like domains occur spontaneously (Das et al. 2019) (see Figure 3.8(b)). Together with other works reviewed elsewhere in this book, these results confirm the advent of topology and skyrmion physics in strong proper ferroelectrics!

Beyond these examples, second-principles simulations have also revealed the key role that walls play in the negative-capacitance response of the $PbTiO_3$ layers in $PbTiO_3/SrTiO_3$ superlattices (Zubko et al. 2016; Yadav et al. 2016). They have also shown the possibility of using domain walls to create resistance against thermal transport (Seijas-Bellido et al. 2017) and even as phonon filters (Royo et al. 2017). Finally, while out of the scope of this chapter, let me mention that there have been valuable applications of empirical atomistic models to wall-related problems (Calleja et al. 2003; Gonçalves-Ferreira et al. 2008), which the reader may want to consult.

3.6 Outlook

Domain walls are, and will continue to be, an excellent playground for the application first-principles methods, and a great motivation for their further development.

Existing first-principles techniques still have a lot to offer. For example, to understand better the dynamical and response properties of domain walls, and the optimization possibilities they may provide; to investigate the novel cross-couplings, electronic or magnetic behaviors they may present; to explore engineering possibilities by controlling wall composition or defect concentration and arrangement; to elucidate wall physics in materials displaying richer phenomena, such as polar metals, nickelates with metal-insulator transitions, iridates with exotic phases driven by strong spin-orbit interactions, etc. Developing our fundamental understanding of questions like these and moving toward a simulation-driven engineering of the intrinsic properties of domain walls require predictive quantum-mechanical atomistic simulations. Thus, I am convinced that the application of DFT and other first-principles methods to wall-related problems will continue to grow and inspire us in the coming years.

Then, there are problems beyond the scope of current first-principles methods, because they involve too large length and time scales, or too complex physical ingredients. The challenge here is to develop techniques to tackle such difficult problems, a slow process that usually requires specialists. While recent developments—e.g. the mentioned second-principles methods combining lattice and electrons (García-Fernández et al. 2016), an eventual systematic construction of Ginzburg–Landau potentials from first principles (Schiaffino and Stengel 2017), etc.—offer some hope to continue advancing, the longer-term evolution of the field is in my opinion not so clear. Looking back at the early 1990s, one notes that most of the pioneers in the application of DFT methods to ferroelectric and ferroelastic oxides—not to mention those working on strongly correlated oxides—had a very strong electronic-structure background; indeed, many of those researchers made relevant contributions to first-principles theory and methods themselves. The situation has clearly changed: In the early 2000s, DFT software became widely available and enabled a black-box approach to simulations, solid technical expertise being no longer required. Consequently, the capacity for methodological innovation has decreased in our community (in all communities), and it will continue to do so. Hence, I think that in the midterm, if we want to see progress in walls and related challenging problems, we will have to publicize better the opportunities in our field and engage researchers from other areas, particularly experts in the development of advanced simulation methods, to import their tools or the people themselves. The challenge from our experimental colleagues is quite obvious: light-triggered functionalities and photovoltaic effects, ultra-fast phenomena, transport and non-equilibrium properties, correlated quantum phases, integration with 2D materials, hybrid and organic materials, flexible functional materials, etc. The structural walls in ferroics are likely to play an important role in most of those fronts, but we still lack the methods to handle them. Exciting physics and relevant research are guaranteed. It is a challenge worth taking!

· ·

REFERENCES

Aguado-Puente P., Junquera J., "Ferromagneticlike closure domains in ferroelectric ultrathin films: first-principles simulations," *Phys. Rev. Lett.* 100, 177601 (2008).

Aguado-Puente P., Junquera J., "Structural and energetic properties of domains in $PbTiO_3/SrTiO_3$ superlattices from first principles," *Phys. Rev. B* 85, 184105 (2012).

Anisimov V. I., Aryasetiawan F., Lichtenstein A. I., "First-principles calculations of the electronic structure and spectra of strongly correlated systems: the LDA+U method," *J. Phys.: Condens. Mat.* 9(4), 767 (1997).

Arnold D. C., Knight K. S., Catalan G., Redfern S. A. T., Scott J. F., Lightfoot P., Morrison F. D. "The β-to-γ transition in $BiFeO_3$: a powder neutron diffraction study," *Adv. Funct. Mater.* 20(13), 2116–2123 (2010).

Artyukhin S., Delaney K. T., Spaldin N. A., Mostovoy M., "Landau theory of topological defects in multiferroic hexagonal manganites," *Nat. Mater.* 13, 42–49 (2014).

Barone P., Di Sante D., Picozzi S., "Improper origin of polar displacements at $CaTiO_3$ and $CaMnO_3$ twin walls," *Phys. Rev. B* 89, 144104 (2014).

Béa H., Dupe B., Fusil S., Mattana R., Jacquet E., Warot-Fonrose B., Wilhelm F., Rogalev A., Petit S., Cros V., Anane A., Petroff F., Bouzehouane K., Geneste G., Dkhil B., Lisenkov S., Ponomareva I., Bellaiche L., Bibes M., Barthélémy A., "Evidence for Room-Temperature multiferroicity in a compound with a giant axial ratio," *Phys. Rev. Lett.* 102, 217603–217605 (2009).

Becher C., Maurel L., Aschauer U., Lilienblum M., Magén C., Meier D., Langenberg E., Trassin M., Blasco J., Krug I. P., Algarabel P. A., Spaldin N. A., Pardo J. A., Fiebig M., "Strain-induced coupling of electrical polarization and structural defects in $SrMnO_3$ films," *Nat. Nanotechnol.* 10, 661–665 (2015).

Bednyakov P. S., Sturman B. I., Sluka T., Tagantsev A. K., Yudin P. V., "Physics and applications of charged domain walls," *NPJ Comput. Mater.* 4, 65 (2018).

Behera R. K., Lee C.-W., Lee D., Morozovska A. N., Sinnott S. B., Asthagiri A., Gopalan V., Phillpot S. R., "Structure and energetics of 180° domain walls in $PbTiO_3$ by density functional theory," *J. Phys.: Condens. Mat* 23(17), 175902 (2011).

Behler J., "Representing potential energy surfaces by high-dimensional neural network potentials," *J. Phys.: Condens. Mat* 26(18), 183001 (2014).

Bellaiche L., García A., anderbilt D. "Finite-temperature properties of $Pb(Zr_{1-x}Ti_x)O_3$ alloys from first principles," *Phys. Rev. Lett.* 84(23), 5427 (2000).

Bellaiche L., Íñiguez J., "Universal collaborative couplings between oxygen-octahedral rotations and antiferroelectric distortions in perovskites," *Phys. Rev. B* 88, 014104 (2013).

Benedek N. A., Fennie C. J., "Hybrid improper ferroelectricity: a mechanism for controllable polarization-magnetization coupling," *Phys. Rev. Lett.* 106, 107204 (2011).

Biškup N., Salafranca J., Mehta V., Oxley M. P., Suzuki Y., Pennycook S. J., Pantelides S. T., Varela M., "Insulating ferromagnetic $LaCoO_{3-\delta}$ films: a phase induced by ordering of oxygen vacancies," *Phys. Rev. Lett.* 112, 087202 (2014).

Blöchl P. E., "Projector augmented-wave method," *Phys. Rev. B* 50(24), 17953–17979 (1994).

Bristowe N. C., *Private Communication* (2019).

Brown I. D., "Recent developments in the methods and applications of the bond valence model," *Chem. Rev.* 109(12), 6858–6919 (2009).

Bruce A. D., "Structural phase transitions. ii. static critical behaviour," *Adv. Phys.* 29(1), 111–217 (1980).

Calleja M., Dove M. T., Salje E. K. H. "Trapping of oxygen vacancies on twin walls of $CaTiO_3$: a computer simulation study," *J. Phys.: Condens. Mat.* 15(14), 2301–2307 (2003).

Catalan G., Seidel J., Ramesh R., Scott J. F., "Domain wall nanoelectronics," *Rev. Mod. Phys.* 84(1), 119–156 (2012).

Catalan G., Scott J. F., "Physics and applications of bismuth ferrite," *Adv. Mater.* 21(24), 2463–2485 (2009).

Ceperley D. M., Alder B. J., "Ground state of the electron gas by a stochastic method," *Phys. Rev. Lett.* 45, 566–569 (1980).

Chen L., Paillard C., Zhao H. J., Íñiguez J., Yang Y., Bellaiche L., "Tailoring properties of hybrid perovskites by domain-width engineering with charged walls," *NPJ Comput. Mater.* 4, 75 (2018a).

Chen L.-Q., "Phase-field method of phase transitions/domain structures in ferroelectric thin films: a review," *J. Am. Ceram. Soc.* 91(6), 1835–1844 (2008).

Chen P., Grisolia M. N., Zhao H. J., González-Vázquez O. E., Bellaiche L., Bibes M., Liu B.-G., Íñiguez J., "Energetics of oxygen-octahedra rotations in perovskite oxides from first principles," *Phys. Rev. B* 97, 024113 (2018b).

Choi T., Horibe Y., Yi H. T., Choi Y. J., Wu W., Cheong S.-W. "Insulating interlocked ferroelectric and structural antiphase domain walls in multi-ferroic $YMnO_3$," *Nat. Mater.* 9, 253–258 (2010).

Cowley R. A., "Lattice dynamics and phase transitions of strontium titanate," *Phys. Rev.* 134, A981–A997 (1964).

Cowley R. A. "Structural phase transitions. i. landau theory," *Adv. Phys.* 29(1), 1–10 (1980).

Cui Q., Elstner M., Kaxiras E., Frauenheim T., Karplus M. "A qm/mm implementation of the self-consistent charge density functional tight binding (scc-dftb) method," *J. Phys. Chem. B* 105(2), 569–585 (2001).

Das S., Tang Y. L., Hong Z., Gonalves M. A. P., McCarter M. R., Klewe C., Nguyen K. X., Gómez-Ortiz F., Shafer P., Arenholz E., Stoica V. A., Hsu S.-L., Wang B., Ophus C., Liu J. F., Nelson C. T., Saremi S., Prasad B., Mei A. B., Schlom D. G., Íñiguez J., García-Fernández P., Muller D. A., Chen L. Q., Junquera J., Martin L. W., Ramesh R., "Observation of room-temperature polar skyrmions," *Nature* 568, 368–372 (2019).

Diéguez O., Aguado-Puente P., Junquera J., Íñiguez J., "Domain walls in a perovskite oxide with two primary structural order parameters: first-principles study of BiFeO$_3$," *Phys. Rev. B* 87(2), 024102 (2013).

Diéguez O., González-Vázquez O. E., Wojdeł J. C., Íñiguez J., "First-principles predictions of low-energy phases of multiferroic BiFeO$_3$," *Phys. Rev. B* 83, 094105 (2011).

Diéguez O., Íñiguez J., *Unpublished* (2019).

Dvořák V., "Improper ferroelectrics," *Ferroelectrics* 7(1), 1–9 (1974).

Elstner M., Porezag D., Jungnickel G., Elsner J., Haugk M., Frauenheim Th., Suhai S., Seifert G. "Self-consistent-charge density-functional tight-binding method for simulations of complex materials properties," *Phys. Rev. B* 58, 7260–7268 (1998).

Escorihuela-Sayalero C., Wojdeł J. C., Íñiguez J., "Efficient systematic scheme to construct second-principles lattice dynamical models," *Phys. Rev. B* 95, 094115 (2017).

Etxebarria I., Perez-Mato J. M., García A., Blaha P., Schwarz K., Rodriguez-Carvajal J., "Comparison of empirical bond-valence and first-principles energy calculations for a complex structural instability," *Phys. Rev. B* 72, 174108 (2005).

Farokhipoor S., Magén C., Venkatesan S., Íñiguez J., Daumont C. J. M., Rubi D., Snoeck E., Mostovoy M., de Graaf C., Müller A., Döblinger M., Scheu C., Noheda B. "Artificial chemical and magnetic structure at the domain walls of an epitaxial oxide," *Nature* 515, 379–383 (2014).

Farokhipoor S., Noheda B. "Conduction through 71° domain walls in BiFeO$_3$ thin films," *Phys. Rev. Lett.* 107(12), 127601 (2011).

Fennie C. J., Rabe K. M. "Ferroelectric transition in YMnO$_3$ from first principles," *Phys. Rev. B* 72, 100103 (2005).

Freysoldt C., Grabowski B., Hickel T., Neugebauer J., Kresse G., Janotti A., Van de Walle C. G., "First-principles calculations for point defects in solids," *Rev. Mod. Phys.* 86, 253–305 (2014).

Frost J. M., Butler K. T., Brivio F., Hendon C. H., van Schilfgaarde M., Walsh A., "Atomistic origins of high-performance in hybrid halide perovskite solar cells," *Nano Lett.* 14(5), 2584–2590 (2014).

Fu H., Bellaiche L., "Ferroelectricity in barium titanate quantum dots and wires," *Phys. Rev. Lett.* 91, 257601 (2003).

García-Fernández P., Íñiguez J., Junquera J., *Unpublished* (2019).

García-Fernández P., Wojdeł J. C., Íñiguez J., Junquera J., "Second-principles method for materials simulations including electron and lattice degrees of freedom," *Phys. Rev. B* 93, 195137 (2016).

Garrity K. F., Rabe K. M., Vanderbilt D., "Hyperferroelectrics: proper ferroelectrics with persistent polarization," *Phys. Rev. Lett.* 112, 127601 (2014).

Gaus M., Cui Q., Elstner M., "Dftb3: extension of the self-consistent-charge density-functional tight-binding method (scc-dftb)," *J. Chem. Theory Comput.* 7(4), 931–948 (2011).

Ghosez Ph., Cockayne E., Waghmare U. V., Rabe K. M., "Lattice dynamics of BaTiO$_3$, PbTiO$_3$, and PbZrO$_3$: a comparative first-principles study," *Phys. Rev. B* 60(2), 836 (1999).

Giddy A. P., Dove M. T., Heine V. "What do landau free energies really look like for structural phase transitions?," *J. Phys.: Condens. Mat.* 1(44), 8327 (1989).

Glazer A. M., "The classification of tilted octahedra in perovskites," *Acta Crystall. B* 28, 3384–3392 (1972).

Goncalves-Ferreira L., Redfern S. A. T., Artacho E., Salje E. K. H. "Ferrielectric twin walls in CaTiO$_3$," *Phys. Rev. Lett.* 101(9), 097602 (2008).

González-Vázquez O. E., Wojdeł J. C., Diéguez O., Íñiguez J., "First-principles investigation of the structural phases and enhanced response properties of the BiFeO$_3$-LaFeO$_3$ multiferroic solid solution," *Phys. Rev. B* 85, 064119 (2012).

Graf M., Sepliarsky M., Tinte S., Stachiotti M. G., "Phase transitions and antiferroelectricity in BiFeO$_3$ from atomic-level simulations," *Phys. Rev. B* 90, 184108 (2014).

Gu T., Scarbrough T., Yang Y., Íñiguez J., Bellaiche L., Xiang H. J., "Cooperative couplings between octahedral rotations and ferroelectricity in perovskites and related materials," *Phys. Rev. Lett.* 120, 197602 (2018).

Gu Y., Rabe K., Bousquet E., Gopalan V., Chen L.-Q. "Phenomenological thermodynamic potential for CaTiO$_3$ single crystals," *Phys. Rev. B* 85, 064117 (2012).

Guennou M., Bouvier P., Chen G. S., Dkhil B., Haumont R., Garbarino G., Kreisel J., "Multiple high-pressure phase transitions in BiFeO$_3$," *Phys. Rev. B* 84, 174107 (2011).

Guyonnet J., Gaponenko I., Gariglio S., Paruch P., "Conduction at domain walls in insulating Pb(Zr$_{0.2}$Ti$_{0.8}$)O$_3$ thin films," *Adv. Mater.* 23(45), 5377–5382 (2011).

He L., Vanderbilt D., "First-principles study of oxygen-vacancy pinning of domain walls in PbTiO$_3$," *Phys. Rev. B* 68, 134103 (2003).

Heron J. T., Bosse J. L., He Q., Gao Y., Trassin M., Ye L., Clarkson J. D., Wang C., Liu J., Salahuddin S., Ralph D. C., Schlom D. G., Íñiguez J., Huey B. D., Ramesh R., "Deterministic switching of ferromagnetism at room temperature using an electric field," *Nature* 516, 370–373 (2014).

Hohenberg P., Kohn W., "Inhomogeneous electron gas," *Phys. Rev.* 136(3B), B864 (1964).

Hong J., Catalan G., Fang D. N., Artacho E., Scott J. F., "Topology of the polarization field in ferroelectric nanowires from first principles," *Phys. Rev. B* 81, 172101 (2010).

Houchmandzadeh B., Lajzerowicz J., Salje E., "Order parameter coupling and chirality of domain walls," *J. Phys.: Condens. Mat.* 3(27), 5163 (1991).

Huang J., Yuan Y., Shao Y., Yan Y., "Understanding the physical properties of hybrid perovskites for photovoltaic applications," *Nat. Rev. Mater.* 2, 17042 (2017).

Íñiguez J., García A., Pérez-Mato J. M. "Optimized local modes for lattice-dynamical applications," *Phys. Rev. B* 61(5), 3127 (2000).

Íñiguez J., Stengel M., Prosandeev S., Bellaiche L., "First-principles study of the multimode antiferroelectric transition in PbZrO$_3$," *Phys. Rev. B* 90, 220103 (2014).

Jiang X. W., Yang Q., Cao J. X., "The effect of domain walls on leakage current in PbTiO$_3$ thin films," *Phys. Lett. A* 380(9), 1071–1074 (2016).

Jiang Y. X., Wang Y. J., Chen D., Zhu Y. L., Ma X. L., "First-principles study of charged steps on 180° domain walls in ferroelectric PbTiO$_3$," *J. Appl. Phys.* 122(5), 054101 (2017).

Kan D., Pálová L., Anbusathaiah V., Cheng C. J., Fujino S., Nagarajan V., Rabe K. M., Takeuchi I., "Universal behavior and electric-field-induced structural transition in rare-earth-substituted BiFeO$_3$," *Adv. Funct. Mater.* 20(7), 1108–1115 (2010).

King-Smith R. D., Vanderbilt D., "First-principles investigation of ferroelectricity in perovskite compounds," *Phys. Rev. B* 49(9), 5828 (1994).

Kohn W., Sham L. J., "Self-consistent equations including exchange and correlation effects," *Phys. Rev.* 140(4A), A1133 (1965).

Kornev I. A., Lisenkov S., Haumont R., Dkhil B., Bellaiche L., "Finite-temperature properties of multiferroic BiFeO$_3$," *Phys. Rev. Lett.* 99(22), 227602–227604 (2007).

Krakauer H., Yu R., Wang C.-Z., Rabe K. M., Waghmare U. V., "Dynamic local distortions in KNbO$_3$," *J. Phys.: Condens. Mat.* 11(18), 3779–3787 (1999).

Kumagai Y., Spaldin N., "Structural domain walls in polar hexagonal manganites," *Nat. Commun.* 4, 1540 (2013).

Lajzerowicz J. J., Niez J. J., "Phase transition in a domain wall," *J. Phys. Lett. Paris* 40, L165–L169 (1979).

Lee D., "The effect of domain wall on defect energetics in ferroelectric LiNbO$_3$ from density functional theory calculations," *J. Korean Ceram. Soc.* 53(3), 312–316 (2016).

Lee D., Behera R. K., Wu P., Xu H., Li Y. L., Sinnott S. B., Phillpot S. R., Chen L. Q., Gopalan V., "Mixed Bloch-Néel-Ising character of 180° ferroelectric domain walls," *Phys. Rev. B* **80**, 060102 (2009).

Li M., Gu Y., Wang Y., Chen L.-Q., Duan W., "First-principles study of 180° domain walls in BaTiO$_3$: mixed Bloch-Néel-Ising character," *Phys. Rev. B* 90, 054106 (2014).

Li P., Ren X., Guo G.-C., He L., "The origin of hyperferroelectricity in LiBO$_3$ (b = v, nb, ta, os)," *Sci. Rep.* 6, 34085 (2016).

Lines M. E., Glass A. M., *Principles and Applications of Ferroelectrics and Related Materials.* Oxford Classic Texts in the Physical Sciences (Clarendon Press, Oxford, 1977).

Liu S., Cohen R. E., Stable charged antiparallel domain walls in hyper-ferroelectrics. *J. Phys.: Condens. Mat.* 29(24), 244003 (2017a).

Liu S., Cohen R. E., "Origin of stationary domain wall enhanced ferroelectric susceptibility," *Phys. Rev. B* 95, 094102 (2017b).

Liu S., Grinberg I., Rappe A. M., "Intrinsic ferroelectric switching from first principles," *Nature* 534, 360–363 (2016).

Liu S., Zheng F., Koocher N. Z., Takenaka H., Wang F., Rappe A. M., "Ferroelectric domain wall induced band gap reduction and charge separation in organometal halide perovskites," *J. Phys. Chem. Lett.* 6(4), 693–699 (2015).

Louis L., Kornev I., Geneste G., Dkhil B., Bellaiche L., "Novel complex phenomena in ferroelectric nanocomposites," *J. Phys.: Condens. Mat.* 24(40), 402201 (2012).

Lubk A., Gemming S., Spaldin N. A., "First-principles study of ferroelectric domain walls in multiferroic bismuth ferrite," *Phys. Rev. B* 80, 104110 (2009).

Lufaso M. W., Woodward P. M., "Prediction of the crystal structures of perovskites using the software program *SPuDS*," *Acta Crystall. B* 57(6), 725–738 (2001).

Makov G., Gattinoni C., Vita De A., "*Ab initio* based multiscale modelling for materials science," *Model. Simul. Mater. Sci. Eng.* 17(8), 084008 (2009).

Maksymovych P., Seidel J., Chu Y. H., Wu P., Baddorf A. P., Chen L.-Q., Kalinin S. V., Ramesh R., "Dynamic conductivity of ferroelectric domain walls in BiFeO$_3$," *Nano Lett.* 11(5), 1906–1912 (2011).

Martin R. M., *Electronic Structure: Basic Theory and Practical Methods* (Cambridge University Press, Cambridge, UK, 2004).

Marton P., Stepkova V., Hlinka J., "Divergence of dielectric permittivity near phase transition within ferroelectric domain boundaries," *Phase Transit.* 86, 103–108 (2013).

Meier D., Seidel J., Cano A., Delaney K., Kumagai Y., Mostovoy M., Spaldin N. A., Ramesh R., Fiebig M., "Anisotropic conductance at improper ferroelectric domain walls," *Nat. Mater.* 11, 284–288 (2012).

Meyer B., Vanderbilt D., "*Ab initio* study of ferroelectric domain walls in PbTiO$_3$," *Phys. Rev. B* 65(10), 104111 (2002).

Migoni R., Bilz H., Bäuerle D., "Origin of Raman scattering and ferroelectricity in oxidic perovskites," *Phys. Rev. Lett.* 37, 1155–1158 (1976).

Müller K. A., Burkard H., "SrTiO$_3$: an intrinsic quantum paraelectric below 4 k," *Phys. Rev. B* 19, 3593–3602 (1979).

Nahas Y., Prokhorenko S., Louis L., Gui Z., Kornev I., Bellaiche L., "Discovery of stable skyrmionic state in ferroelectric nanocomposites," *Nat. Commun.* 6, 8542 (2015).

Naumov I. I., Bellaiche L., Fu H., "Unusual phase transitions in ferroelectric nanodisks and nanorods," *Nature* 432(7018), 737–740 (2004).

NREL. *Photovoltaic efficiency chart* (2019). https://www.nrel.gov/pv/assets/pdfs/pv-efficiency-chart.20181221.pdf.

Oh Y. S., Luo X., Huang F.-T., Wang Y., Cheong S.-W., "Experimental demonstration of hybrid improper ferroelectricity and the presence of abundant charged walls in $(Ca,Sr)_3Ti_2O_7$ crystals," *Nat. Mater.* 14, 407–413 (2015).

Padilla J., Zhong W., Vanderbilt D., "First-principles investigation of 180° domain walls in $BaTiO_3$," *Phys. Rev. B* 53, R5969–R5973 (1996).

Paillard C., Geneste G., Bellaiche L., Dkhil B., "Vacancies and holes in bulk and at 180° domain walls in lead titanate," *J. Phys.: Condens. Mat.* 29(48), 485707 (2017).

Park N.-G., Gärtzel M., Miyasaka T., Zhu K., Emery K., "Towards stable and commercially available perovskite solar cells," *Nat. Energy* 1, 16152 (2016).

Perdew J., Ruzsinszky A., Csonka G., Vydrov O., Scuseria G., Constantin L., Zhou X., Burke K., "Restoring the density-gradient expansion for exchange in solids and surfaces," *Phys. Rev. Lett.* 100(13), 136406 (2008).

Perdew J. P., Burke K., Ernzerhof M., "Generalized gradient approximation made simple," *Phys. Rev. Lett.* 77(18), 3865 (1996).

Pereira Gonçalves M. A., Escorihuela-Sayalero C., García-Fernández P., Junquera J., Íñiguez J., "Theoretical guidelines to create and tune electric skyrmion bubbles," *Sci. Adv.* 5(2), eaau7023 (2019).

Pickett W. E., "Pseudopotential methods in condensed matter applications," *Comput. Phys. Rep.* 9(3), 115–197 (1989).

Pöykkö S., Chadi D. J. "*Ab initio* study of 180° domain wall energy and structure in $PbTiO_3$," *Appl. Phys. Lett.* 75(18), 2830–2832 (1999).

Pöykkö S., Chadi D. J., "*Ab initio* study of dipolar defects and 180° domain walls in $PbTiO_3$," *J. Phys. Chem. Solids* 61(2), 291–294 (2000).

Prosandeev S., Kornev I. A., Bellaiche L., "Magnetoelectricity in $BiFeO_3$ films: first-principles-based computations and phenomenology," *Phys. Rev. B* 83(2), 020102 (2011).

Prosandeev S., Wang D., Ren W., Íñiguez J., Bellaiche L., "Novel nanoscale twinned phases in perovskite oxides," *Adv. Funct. Mater.* 23, 234–240 (2013).

Rabe K. M., Joannopoulos J. D., "*Ab initio* determination of a structural phase transition temperature," *Phys. Rev. Lett.* 59, 570–573 (1987a).

Rabe K. M., Joannopoulos J. D., "Theory of the structural phase transition of gete," *Phys. Rev. B* 36, 6631–6639 (1987b).

Rabe K. M., Waghmare U. V., "Localized basis for effective lattice hamiltonians: lattice Wannier functions," *Phys. Rev. B* 52(18), 13236 (1995).

Radescu S., Etxebarria I., Perez-Mato J. M., "The landau free energy of the three-dimensional Φ^4 model in wide temperature intervals," *J. Phys.: Condens. Mat.* 7(3), 585 (1995).

Rahmanizadeh K., Wortmann D., Bihlmayer G., Blügel S., "Charge and orbital order at head-to-head domain walls in $PbTiO_3$," *Phys. Rev. B* 90, 115104 (2014).

Ren W., Yang Y., Diéguez O., Íñiguez J., Choudhury N., Bellaiche L., "Ferroelectric domains in multiferroic $BiFeO_3$ films under epitaxial strains," *Phys. Rev. Lett.* 110, 187601 (2013).

Royo M., Escorihuela-Sayalero C., Íñiguez J., Rurali R., "Ferroelectric domain wall phonon polarizer," *Phys. Rev. Mater.* 1, 051402 (2017).

Salje E. K. H., *Phase Transitions in Ferroelastic and Co-elastic Crystals* (Cambridge University Press, Cambridge, 1993).

Salje E. K. H., "Multiferroic domain boundaries as active memory devices: trajectories towards domain boundary engineering," *ChemPhysChem* 11, 940–950 (2010).

Schiaffino A., Stengel M., "Macroscopic polarization from antiferrodistortive cycloids in ferroe-lastic $SrTiO_3$," *Phys. Rev. Lett.* 119, 137601 (2017).

Schlom D. G., Chen L.-Q., Eom C.-B., Rabe K. M., Streiffer S. K., Triscone J.-M., "Strain tuning of ferroelectric thin films," *Ann. Rev. Mater. Res.* 37, 589 (2007).

Scott J. F., Salje E. K. H., Carpenter M. A., "Domain wall damping and elastic softening in $SrTiO_3$: evidence for polar twin walls," *Phys. Rev. Lett.* 109(18), 187601 (2012).

Seidel J., Martin L. W., He Q., Zhan Q., Chu Y.-H., Rother A., Hawkridge M. E., Maksymovych P., Yu P., Gajek M., Balke N., Kalinin S. V., Gemming S., Wang F., Catalan G., Scott J. F., Spaldin N. A., Orenstein J., Ramesh R., "Conduction at domain walls in oxide multiferroics," *Nat. Mater.* 8(3), 229–234 (2009).

Seijas-Bellido J. A., Escorihuela-Sayalero C., Royo M., Ljungberg M. P., Wojdeł J. C., Íñiguez J., Rurali R., "A phononic switch based on ferroelectric domain walls," *Phys. Rev. B* **96**, 140101 (2017).

Sepliarsky M., Wu Z., Asthagiri A., Cohen R. E., "Atomistic model potential for $PbTiO_3$ and pmn by fitting first principles results," *Ferroelectrics* 301(1), 55–59 (2004).

Shafer P., García-Fernández P., Aguado-Puente P., Damodaran A. R., Yadav A. K., Nelson C. T., Hsu S.-L., Wojdeł J. C., Íñiguez J., Martin L. W., Arenholz E., Junquera J., Ramesh R., "Emergent chirality in the electric polarization texture of titanate superlattices," *Proc. Natl. Acad. Sci. U. S. A.* 115(5), 915–920 (2018).

Sharma P., Zhang Q., Sando D., Lei C. H., Liu Y., Li J., Nagarajan V., Seidel J., "Nonvolatile ferroelectric domain wall memory," *Sci. Adv.* 3(6), e1700512 (2017).

Shimada T., Tomoda S., Kitamura T., "*Ab initio* study of ferroelectric closure domains in ultrathin $PbTiO_3$ films," *Phys. Rev. B* 81, 144116 (2010).

Shin Y.-H., Cooper V. R., Grinberg I., Rappe A. M., "Development of a bond-valence molecular-dynamics model for complex oxides," *Phys. Rev. B* 71(5), 054104 (2005).

Shin Y.-H., Grinberg I., Chen I.-W., Rappe A. M., "Nucleation and growth mechanism of ferroelectric domain-wall motion," *Nature* 449, 881–884 (2007).

Shur V. Ya., "Domain nanotechnology in ferroelectrics: nano-domain engineering in lithium niobate crystals," *Ferroelectrics* 373(1), 1–10 (2008).

Singh A., Singh V. N., Canadell E., Íñiguez J., Diéguez O., "Polymorphism in bi-based perovskite oxides: a first-principles study," *Phys. Rev. Mater.* 2, 104417 (2018).

Skjærvø S. H., Småbråten D. R., Spaldin N. A., Tybell T., Selbach S. M., "Oxygen vacancies in the bulk and at neutral domain walls in hexagonal $YMnO_3$," *Phys. Rev. B* 98, 184102 (2018).

Sluka T., Tagantsev A. K., Bednyakov P., Setter N., "Free-electron gas at charged domain walls in insulating $BaTiO_3$," *Nat. Commun.* 4, 1808 (2013).

Sluka T., Tagantsev A. K., Damjanovic D., Gureev M., Setter N., "Enhanced electromechanical response of ferroelectrics due to charged domain walls," *Nat. Commun.* 3, 748 (2012).

Småbråten D. R., Meier Q. N., Skjærvø S. H., Inzani K., Meier D., Selbach S. M., "Charged domain walls in improper ferroelectric hexagonal manganites and gallates," *Phys. Rev. Mater.* 2, 114405 (2018).

Stachiotti M. G., Sepliarsky M., "Toroidal ferroelectricity in $PbTiO_3$ nanoparticles," *Phys. Rev. Lett.* 106, 137601 (2011).

Streiffer S. K., Parker C. B., Romanov A. E., Lefevre M. J., Zhao L., Speck J. S., Pompe W., Foster C. M., Bai G. R., "Domain patterns in epitaxial rhombohedral ferroelectric films. i. geometry and experiments," *J. Appl. Phys.* 83(5), 2742–2753 (1998).

Tagantsev A., Cross L. E., Fousek J., *Domains in Ferroic Crystals and Thin Films* (Springer, New York, 2010).

Tagantsev A. K., Courtens E., Arzel L., "Prediction of a low-temperature ferroelectric instability in antiphase domain boundaries of strontium titanate," *Phys. Rev. B* 64, 224107 (2001).

Tagantsev A. K., Yudin P. V., *Flexoelectricity in Solids* (World Scientific, Singapore, 2016).

Taherinejad M., Vanderbilt D., Marton P., Stepkova V., Hlinka J., "Bloch-type domain walls in rhombohedral BaTiO$_3$," *Phys. Rev. B* 86, 155138 (2012).

Tinte S., Stachiotti M. G., Sepliarsky M., Migoni R. L., Rodriguez C. O., "Atomistic modelling of BaTiO$_3$ based on first-principles calculations," *J. Phys.: Condens. Mat.* 11(48), 9679 (1999).

Tomoda S., Shimada T., Ueda T., Wang J., Kitamura T., "Hybrid functional study on the ferroelectricity of domain walls with o-vacancies in PbTiO$_3$," *Mech. Eng. J.* 2(3), 15–00037–15–00037 (2015).

Vasudevan R. K., Wu W., Guest J. R., Baddorf A. P., Morozovska A. N., Eliseev E. A., Balke N., Nagarajan V., Maksymovych P., Kalinin S. V., "Domain wall conduction and polarization-mediated transport in ferroelectrics," *Adv. Funct. Mater.* 23(20), 2592–2616 (2013).

Waghmare U. V., Rabe K. M., "*Ab initio* statistical mechanics of the ferroelectric phase transition in PbTiO$_3$," *Phys. Rev. B* 55(10), 6161 (1997).

Wahl R., Vogtenhuber D., Kresse G., "SrTiO$_3$ and BaTiO$_3$ revisited using the projector augmented wave method: performance of hybrid and semilocal functionals," *Phys. Rev. B* 78(10), 104116 (2008).

Wang Y. J., Chen D., Tang Y. L., Zhu Y. L., Ma X. L., "Origin of the bloch-type polarization components at the 180° domain walls in ferroelectric PbTiO$_3$," *J. Appl. Phys.* 116(22), 224105 (2014).

Warwick A. R., Íñiguez J., Haynes P. D., Bristowe N. C., "First-principles study of ferroelastic twins in halide perovskites," *J. Phys. Chem. Lett.* 10(6), 1416–1421 (2019).

Wei X.-K., Tagantsev A. K., Kvasov A., Roleder K., Jia C.-L., Setter N., "Ferroelectric translational antiphase boundaries in nonpolar materials," *Nat. Commun.* 5(3031), (2014).

Wojdeł J. C., Hermet P., Ljungberg M. P., Ghosez P., Íñiguez J., "First-principles model potentials for lattice-dynamical studies: general methodology and example of application to ferroic perovskite oxides," *J. Phys.: Condens. Mat.* 25(30), 305401 (2013).

Wojdeł J. C., Íñiguez J., "Ferroelectric transitions at ferroelectric domain walls found from first principles," *Phys. Rev. Lett.* 112(24), 247603 (2014).

Wu X., Petralanda U., Zheng L., Ren Y., Hu R., Cheong S.-W., Artyukhin S., Lai K., "Low-energy structural dynamics of ferroelectric domain walls in hexagonal rare-earth manganites," *Sci. Adv.* 3(5), e1602371 (2017).

Wu X., Vanderbilt D., "Theory of hypothetical ferroelectric superlattices incorporating head-to-head and tail-to-tail 180° domain walls," *Phys. Rev. B* 73, 020103 (2006).

Xu T., Shimada T., Araki Y., Wang J., Kitamura T., "Multiferroic domain walls in ferroelectric PbTiO$_3$ with oxygen deficiency," *Nano Lett.* 16(1), 454–458 (2016).

Yadav A. K., Nelson C. T., Hsu S. L., Hong Z., Clarkson J. D., Schlepuëtz C. M., Damodaran A. R., Shafer P., Arenholz E., Dedon L. R., Chen D., Vishwanath A., Minor A. M., Chen L. Q., Scott J. F., Martin L. W., Ramesh R., "Observation of polar vortices in oxide superlattices," *Nature* 530(7589), 198–201 (2016).

Zhao H. J., Filippetti A., Escorihuela-Sayalero C., Delugas P., Canadell E., Bellaiche L., Fiorentini V., Íñiguez J., "Metascreening and permanence of polar distortion in metallized ferroelectrics," *Phys. Rev. B* 97, 054107 (2018).

Zhao H. J., Íñiguez J., "Creating multiferroic and conductive domain walls in common ferroelastic compounds," *NPJ Comput. Mater.* 5, 92 (2019).

Zhao H. J., Íñiguez J., "Domain-wall polarization in CaTiO$_3$-type orthorhombic perovskites," *In preparation* (2020).

Zhong W., Vanderbilt D., "Competing structural instabilities in cubic perovskites," *Phys. Rev. Lett.* 74(13), 2587 (1995).

Zhong W., Vanderbilt D., "Effect of quantum fluctuations on structural phase transitions in SrTiO$_3$ and BaTiO$_3$," *Phys. Rev. B* 53(9), 5047 (1996).

Zhong W., Vanderbilt D., Rabe K. M., "Phase transitions in BaTiO$_3$ from first principles," *Phys. Rev. Lett.* 73(13), 1861 (1994).

Zhong W., Vanderbilt D., Rabe K. M., "First-principles theory of ferroelectric phase transitions for perovskites: the case of BaTiO$_3$," *Phys. Rev. B* 52(9), 6301 (1995).

Zubko P., Stucki N., Lichtensteiger C., Triscone J. M., "X-ray diffraction studies of 180° ferroelectric domains in PbTiO$_3$/SrTiO$_3$ superlattices under an applied electric field," *Phys. Rev. Lett.* 104(18), 187601 (2010).

Zubko P., Wojdeł J. C., Hadjimichael M., Fernandez-Pena S., Sené A., Luk'yanchuk I., Triscone J.-M., Íñiguez J., "Negative capacitance in multidomain ferroelectric superlattices," *Nature* 534(7608), 524–528 (2016).

4

Fundamental Properties of Ferroelectric Domain Walls from Ginzburg–Landau Models

P. Ondrejkovic[1], P. Marton[1,2], V. Stepkova[1], and J. Hlinka[1]

[1]*Institute of Physics, The Czech Academy of Sciences, Na Slovance 2, 182 00 Prague 8, Czech Republic*
[2]*Institute of Mechatronics and Computer Engineering, Technical University of Liberec, Studentská 2, 461 17 Liberec, Czech Republic*

Chapter 1, Section 1.2.2 has already briefly mentioned the phenomenological approach to ferroelectric domain walls. In this chapter, contemporary possibilities, prospects, and limitations of phase-field simulations and Ginzburg–Landau–Devonshire models of domain walls are discussed in detail, focusing on the most-studied ferroelectric oxides $BaTiO_3$, $KNbO_3$, $PbTiO_3$, as well as on various complex perovskite oxides like PZT and lead-based relaxor ferroelectrics. It reviews their available model potentials and the key predictions reported in the last decade. The overview is complemented by original data allowing comparisons and an assessment of future prospects to be made.

4.1 Introduction

One of the interesting aspects of the perovskite ferroelectric oxides is that their spontaneous polarization can choose among multiple equivalent crystallographic directions (typically, 6 directions in tetragonal ferroelectric phases, 8 directions in rhombohedral phases, 12 directions in orthorhombic phases, and 24 in the case of monoclinic ones). The richness of the available domain states is reflected in the multiple kinds of ferroelectric domain walls. Since the atomic displacements related to the ferroelectricity of perovskite oxides are rather small and leave the chemical bonding almost intact, the ferroelectric domain boundaries can be, at least in principle, formed reversibly and as fully coherent interfaces. Consequently, such domain walls have a well-defined intrinsic atomic structure, and they also can easily slide from place to place under suitable external stimuli, as mentioned in Chapters 10 to 12. The mobility of the domain

P. Ondrejkovic, P. Marton, V. Stepkova, and J. Hlinka, *Fundamental Properties of Ferroelectric Domain Walls from Ginzburg–Landau Models*
In: *Domain Walls: From Fundamental Properties to Nanotechnology Concepts.* Edited by: Dennis Meier, Jan Seidel, Marty Gregg, and Ramamoorthy Ramesh, Oxford University Press (2020). © P. Ondrejkovic, P. Marton, V. Stepkova, and J. Hlinka.
DOI: 10.1093/oso/9780198862499.003.0004

walls is often either exploited or needs to be circumvented when considering the actual functional properties of these materials. Similarly, the well-defined atomic structure at the nanometer scale of the domain wall is favorable for exploiting its intrinsic nanoscale electronic, optical, or transport properties (Catalan et al. 2012; Meier 2015; Bednyakov et al. 2018).

In the absence of mobile charge carriers and inhomogeneous built-in fields, the orientation of domain walls in insulating ferroelectric media is naturally selected in a way to minimize the bound polarization charge accumulation. Therefore, only the head-to-tail (equivalent to tail-to-head) domain wall arrangements, minimizing the discontinuity of the normal component of the spontaneous polarization, are realized. Similarly, in order to minimize the macroscopic stresses in the multi-domain ferroelectric medium, all coherent domain walls with macroscopic lateral spread have to choose orientations in which the tangential components of the spontaneous strain are equal. The implications of these basic requirements in terms of classification of plausible domain types are well understood, verified by many observations of macroscopic domains in real materials and conveniently summarized, for example, by Fousek and Janovec (1969), Tagantsev et al. (2010), and Marton et al. (2010).

Obviously, the concept of idealized mechanically compatible and formally charge-neutral domain walls considered in this chapter does not apply strictly to the full spectrum of domain wall types discussed in this book. In fact, whenever the domain wall properties are significantly influenced by the nearby surfaces or interfaces, dislocations, or ionic defect segregation as mentioned in Chapter 3, one should consider going beyond the simple pristine bulk domain wall considerations. Nevertheless, the concept of idealized mechanically compatible domain walls serves as a very useful reference and classification tool, and anyway it remains the most obvious starting point for a more thorough analysis of any other more subtle case. For the reader's convenience, the summary of permissible orientations of mechanically compatible charge-neutral domain walls in the tetragonal, orthorhombic, and rhombohedral proper ferroelectric phases of perovskite oxides are displayed in Figure 4.1 (reproduced with permission from the work of Marton et al. 2010).

Among the most-studied perovskite oxide ferroelectrics, there are quite a few systems which can be driven to distinct ferroelectric phases by a slight variation of temperature, chemical composition, pressure, or electric field. Roughly speaking, there are two groups of these compounds. The first group includes compounds in which the competing ferroelectric phases are most easily switched by changing the temperature; the second one contains solid solutions in which the ferroelectric–ferroelectric transformation is very sensitive to the exact chemical composition. Well-known representatives of the first group are $BaTiO_3$ and $KNbO_3$, which both possess a rhombohedral ferroelectric ground state, but the orthorhombic and tetragonal ferroelectric phases become more stable as the temperature is increased. Prototypical examples of the second group are the morphotropic phase boundary (MPB) materials such as $PbZr_{1-x}Ti_xO_3$ (PZT100x) and solid solutions of relaxor perovskites with tetragonal $PbTiO_3$, like $Pb(Mg_{1/3}Nb_{2/3})_{1-x}Ti_xO_3$ (PMN-100xPT) or its very close analog solid solution, $Pb(Zn_{1/3}Nb_{2/3})_{1-x}Ti_xO_3$ (PZN-100xPT) (Lines and Glass 1977).

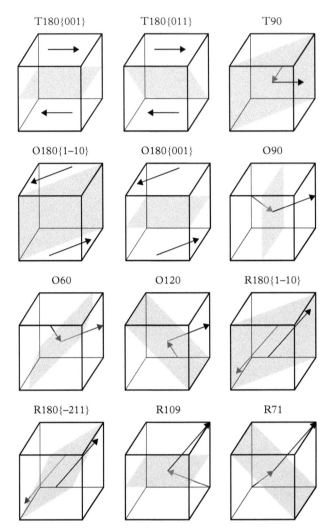

Figure 4.1 *Set of mechanically compatible and electrically neutral domain walls in the three ferroelectric phases of BaTiO₃ within the minimal KGLD model introduced in Section 4.2. In the case of 180° domain walls, where the orientation is not determined by symmetry, walls with the most important crystallographic orientations are displayed. Domain walls are labelled by a symbol composed of the letter specifying the ferroelectric phase T, O, or R, staying for the tetragonal, orthorhombic or rhombohedral, respectively, number indicating the polarization rotation angle 180°, 120°, 109°, 90°, 71°, or 60° and, if needed, the orientation of the domain wall normal with respect to the parent pseudocubic reference structure (after Marton et al. 2010).*

Source: Reprinted figure with permission from Marton et al. 2010. Copyright (2010) by the American Physical Society.

In spite of the tremendous increased interest in ferroelectric domain walls, up to now we do not have readily available standard experimental and computational tools allowing us to determine the atomic structure of ferroelectric domain walls with the same precision as current knowledge of the structure of the bulk crystals. Therefore, very often, it is convenient to consider domain wall properties within the framework of Landau-type phenomenological models. This approach consists in a phenomenological description of the expressions of the parametric dependence of thermodynamical potentials on the essential degrees of freedom, called order parameters. Clearly, the key order parameters of the ferroelectric perovskite oxides are the vector of local spontaneous polarization and strain. For most purposes, it is sufficient to keep the few lowest symmetry-allowed terms in the free-energy expansion, as in the original Devonshire model for $BaTiO_3$ (Devonshire 1949). Obviously, an additional order parameter might be unavoidable when a further symmetry lowering is present, such as in phases with frozen oxygen octahedra tilts (Haun et al. 1989a; Marton et al. 2017).

In addition, to treat properly the energy costs of spatially inhomogeneous polarization at the domain wall, the terms involving gradients of polarization and electrostatic energy, related to the locally uncompensated bound charge (Semenovskaya and Khachaturyan 1998), need to be included. The latter term can be neglected when fully charge-neutral domain walls are calculated. Historically, the intrinsic properties of ferroelectric domain walls in perovskite ferroelectrics were first considered in this way by Zhirnov (1958). The corresponding theoretical framework is usually denoted as the Ginzburg–Landau–Devonshire (GLD) theory. When the electric dipole–dipole interaction is included, the theory contains an explicit penalty for the bound charge accumulation, and slightly charged configurations can be studied. This Khachaturyan–Ginzburg–Landau–Devonshire minimal phenomenological framework, denoted here as the KGLD model, is particularly suitable for phase-field simulations since it does not require the imposition of any other *a priori* geometric limitations on the trial domain wall arrangements. Even further extensions of the phenomenology are needed when strongly charged domain walls are considered (Sluka et al. 2016), but this already requires specific knowledge about the nature of the available charge carriers, and therefore we shall not address it here. All these approaches relate domain wall properties to a small number of bulk material tensor properties. Therefore, they can be used to study general trends even if the structures of the domain walls of interest are not well known, or to guess natural domain states in thin films, to simulate the properties of some composite ferroelectric devices, etc. (Pertsev et al. 1998, 2003; Jin et al. 2003; Su and Landis 2007; Slutsker et al. 2008; Chen 2008; Eliseev et al. 2009, 2013; Hlinka et al. 2009; Tagantsev et al. 2010; Sluka et al. 2012; Stepkova et al. 2012, 2014; Yudin et al. 2012; Ondrejkovic et al. 2013; Gu et al. 2014; Hong et al. 2017).

The purpose of this chapter is to map out the recent progress in GLD modeling of perovskite oxide ferroelectrics and to review the possibility of predicting the basic characteristics of domain walls in the most-studied systems within the GLD and KGLD approaches. The chapter is organized as follows. The mathematical framework of the GLD and KGLD theories is briefly introduced in Section 4.2. Then, Section 4.3 compares the available material-specific numerical coefficients of the models used in recent publications. Using them, we have assembled several complete sets that are helpful for grasping the essence of the properties of a few representative oxide ferroelectrics. These are presented and discussed in Section 4.4. Finally, in Sections 4.5 and 4.6, these complete parameter sets are used to calculate and compare the desired domain wall characteristics as well as to discuss some interesting anomalies, such as chiral polarization profiles in so-called Bloch domain walls (see Chapter 3, Figure 3.2(d)). The general discussion of the results in Section 4.7 allows us to draw a picture of future prospects for phase-field modeling.

4.2   Phenomenological Model for Perovskite Ferroelectrics

The central mathematical objects in phenomenological theories of domain walls are the GLD and KGLD functionals. GLD functionals are simply defined as an integral over the GLD density, which is usually constructed as a systematic expansion of the selected thermodynamical potential in polynomials of the order parameter components and their spatial gradients. There are different options for the choice of the thermodynamical potential and independent thermodynamical variables, different number of terms and their combinations included in the expansion, and also different choices of the basic independent polynomial terms. All these differences result in somewhat different flavors of GLD modeling. Rather than argue about the advantages and disadvantages of different options, here we shall briefly introduce the model that we have most frequently used when dealing with $BaTiO_3$ (Hlinka and Marton 2006; Hlinka et al. 2009; Marton et al. 2010).

4.2.1   Minimal KGLD Model

The excess Gibbs free-energy functional F of a ferroelectric phase is expressed in terms of lowest-order polynomials of the ferroelectric polarization P_i, its spatial derivatives $P_{i,j} = \partial P_i / \partial x_j$ and strain components $e_{ij} = (\partial u_i / \partial x_j + \partial u_j / \partial x_i)/2$ with $i,j = 1-3$ as

$$F = \int f_{GLD}\left[\{P_i, P_{i,j}, e_{ij}\}\right] d\mathbf{r} + F_{dip}\left[\{P_i\}\right], \tag{4.1}$$

where the first term f_{GLD} consists of Landau, Ginzburg (gradient), elastic, and electrostriction free-energy densities

$$f_{GLD} = f_L^{(e)}\{P_i\} + f_G\{P_{i,j}\} + f_C\{e_{ij}\} + f_q\{P_i, e_{ij}\} \tag{4.2}$$

and the other term in Equation (4.1) describes the electrostatic energy (see Equation (4.7)).

The Landau potential is usually expanded up to the sixth order in components of polarization for cubic symmetry $(m\overline{3}m)$:

$$
\begin{aligned}
f_{\mathrm{L}}^{(e)} = {} & \alpha_1 \left(P_1^2 + P_2^2 + P_3^2\right) \\
& + \alpha_{11}^{(e)} \left(P_1^4 + P_2^4 + P_3^4\right) \\
& + \alpha_{12}^{(e)} \left(P_1^2 P_2^2 + P_2^2 P_3^2 + P_1^2 P_3^2\right) \\
& + \alpha_{111} \left(P_1^6 + P_2^6 + P_3^6\right) \\
& + \alpha_{112} \left[P_1^4 \left(P_2^2 + P_3^2\right) + P_2^4 \left(P_1^2 + P_3^2\right) + P_3^4 \left(P_1^2 + P_2^2\right)\right] \\
& + \alpha_{123} P_1^2 P_2^2 P_3^2,
\end{aligned}
\tag{4.3}
$$

where P_i are the components of the polarization vector.
The gradient term f_G is considered in the form

$$
\begin{aligned}
f_G = {} & \tfrac{1}{2} G_{11} \left(P_{1,1}^2 + P_{2,2}^2 + P_{3,3}^2\right) \\
& + G_{12} \left(P_{1,1} P_{2,2} + P_{2,2} P_{3,3} + P_{1,1} P_{3,3}\right) \\
& + \tfrac{1}{2} G_{44} \left[\left(P_{1,2} + P_{2,1}\right)^2 + \left(P_{2,3} + P_{3,2}\right)^2 \right. \\
& \left. \qquad\quad + \left(P_{3,1} + P_{1,3}\right)^2\right].
\end{aligned}
\tag{4.4}
$$

The free-energy density of the elastic part is given by

$$
\begin{aligned}
f_C = {} & \tfrac{1}{2} C_{ijkl} e_{ij} e_{kl} = \tfrac{1}{2} C_{\alpha\beta} e_\alpha e_\beta \\
= {} & \tfrac{1}{2} C_{11} \left[e_1^2 + e_2^2 + e_3^2\right] \\
& + C_{12} \left[e_2 e_3 + e_1 e_3 + e_1 e_2\right] \\
& + \tfrac{1}{2} C_{44} \left[e_4^2 + e_5^2 + e_6^2\right],
\end{aligned}
\tag{4.5}
$$

where $C_{\alpha\beta}$ are the components of the elastic (stiffness) tensor in Voigt notation ($C_{11} = C_{1111}$, $C_{12} = C_{1122}$, and $C_{44} = C_{1212}$), and e_α are the components of the deformation tensor in Voigt notation ($e_1 = e_{11}$, $e_2 = e_{22}$, $e_3 = e_{33}$, $e_4 = 2e_{23}$, $e_5 = 2e_{13}$, and $e_6 = 2e_{12}$).

The electrostriction part is defined by

$$
\begin{aligned}
f_q = {} & -q_{ijkl} e_{ij} P_k P_l \\
= {} & -q_{11} \left[e_1 P_1^2 + e_2 P_2^2 + e_3 P_3^2\right] \\
& -q_{12} \left[e_1 \left(P_2^2 + P_3^2\right) + e_2 \left(P_1^2 + P_3^2\right) + e_3 \left(P_1^2 + P_2^2\right)\right] \\
& -q_{44} \left[e_4 P_2 P_3 + e_5 P_1 P_3 + e_6 P_1 P_2\right],
\end{aligned}
\tag{4.6}
$$

where $q_{\alpha\beta}$ are the components of the electrostriction tensor in Voigt notation ($q_{11} = q_{1111}$, $q_{12} = q_{1122}$, and $q_{44} = 2q_{1212}$).

The electrostatic energy in Equation (4.1) is defined by

$$F_{\text{dip}}\left[\{P_i\}\right] = -\frac{1}{2} \int \left[\mathbf{E}_{\text{dip}}\left(\mathbf{r}\right) \cdot \mathbf{P}\left(\mathbf{r}\right)\right] d\mathbf{r}. \tag{4.7}$$

This contribution is associated with the long-range interaction of individual electric dipoles with the electric fields of all other dipoles, expressed via the inhomogeneous depolarization field $\mathbf{E}_{\text{dip}}$ (Hu and Chen 1998; Semenovskaya and Khachaturyan 1998). This inhomogeneous part of the polarization field is given by

$$\mathbf{E}_{\text{dip}}\left(\mathbf{r}\right) = -\frac{1}{4\pi\,\epsilon_0\epsilon_B} \int \left[\frac{\mathbf{P}\left(\mathbf{r}'\right)}{|\mathbf{R}|^3} - \frac{3\left(\mathbf{P}\left(\mathbf{r}'\right) \cdot \mathbf{R}\right)\mathbf{R}}{|\mathbf{R}|^5}\right] d\mathbf{r}', \tag{4.8}$$

where $\mathbf{R} = \mathbf{r} - \mathbf{r}'$, ϵ_B is the relative background permittivity of the medium (without the primary order parameter contribution), and ϵ_0 is the permittivity of vacuum.

As a matter of fact, the above defined model allows to address the full set of mechanically compatible domain walls in all ferroelectric phases of BaTiO$_3$ (see Figure 4.1). It includes all the relevant terms necessary to define the thickness and orientation of electrically and mechanically compatible domain walls, and still, all terms are taken in the lowest order. In this sense, it could be considered as a minimal model for perovskite ferroelectric materials.

4.2.2 Alternative Expressions and Other Useful Relationships

As already mentioned, the potential terms introduced above are sometimes written in a slightly different form. For example, the f_G potential sometimes is written as

$$\begin{aligned}
f_G = \tfrac{1}{2}M_{11}\left(P_{1,1}^2 + P_{2,2}^2 + P_{3,3}^2\right) \\
+ M_{14}\left(P_{1,1}P_{2,2} + P_{2,2}P_{3,3} + P_{1,1}P_{3,3}\right) \\
+ \tfrac{1}{2}M_{44}\left(P_{1,2}^2 + P_{2,1}^2 + P_{2,3}^2 + P_{3,2}^2 + P_{3,1}^2 + P_{1,3}^2\right)
\end{aligned} \tag{4.9}$$

or as

$$\begin{aligned}
f_G = \tfrac{1}{2}H_{11}\left(P_{1,1}^2 + P_{2,2}^2 + P_{3,3}^2\right) \\
+ H_{12}\left(P_{1,1}P_{2,2} + P_{2,2}P_{3,3} + P_{1,1}P_{3,3}\right) \\
+ \tfrac{1}{2}H_{44}\left[\left(P_{1,2} + P_{2,1}\right)^2 + \left(P_{2,3} + P_{3,2}\right)^2 + \left(P_{3,1} + P_{1,3}\right)^2\right] \\
+ \tfrac{1}{2}H'_{44}\left[\left(P_{1,2} - P_{2,1}\right)^2 + \left(P_{2,3} - P_{3,2}\right)^2 + \left(P_{3,1} - P_{1,3}\right)^2\right],
\end{aligned} \tag{4.10}$$

Table 4.1 *Relations between gradient coefficients of the Ginzburg potential expressed in different conventions.*

f_G in Equation (4.4)	f_G in Equation (4.9)	f_G in Equation (4.10)
G_{11}	$= M_{11}$	$= H_{11}$
G_{12}	$= M_{14} - M_{44}$	$= H_{12} - 2H'_{44}$
G_{44}	$= M_{44}$	$= H_{44} + H'_{44}$

where the $M_{\alpha\beta}$ and $H_{\alpha\beta}$ coefficients are related to the $G_{\alpha\beta}$ coefficients in Equation (4.4) as presented in Table 4.1. The form of the gradient potential in Equation (4.9) is convenient for the isotropic potential ($M_{11} = M_{44}$ and $M_{14} = 0$). Both forms are in the continuum (long wavelength) limit equivalent to our standard form.

The elastic tensor components relate deformation to stress by Hooke's law: $\sigma_\alpha = C_{\alpha\beta}e_\beta$ (in matrix form), where the stress components are denoted by $\sigma_1 = \sigma_{11}, \sigma_2 = \sigma_{22}, \sigma_3 = \sigma_{33}, \sigma_4 = \sigma_{23}, \sigma_5 = \sigma_{13}$, and $\sigma_6 = \sigma_{12}$. In the case of the inverse relationship $e_\alpha = S_{\alpha\beta}\sigma_\beta$, the deformation and the stress components are related via the elastic compliance tensor components $S_{\alpha\beta}$. We use the following notation: $S_{11} = S_{1111}, S_{12} = 2S_{1122}$, and $S_{44} = 4S_{1212}$. From the above notations, the stiffness tensor in matrix form is the simple inverse of the compliance tensor in matrix form. Explicitly, their components are related as follows:

$$C_{11} = \frac{S_{11}+S_{12}}{(S_{11}-S_{12})(S_{11}+2S_{12})},$$
$$C_{12} = \frac{-S_{12}}{(S_{11}-S_{12})(S_{11}+2S_{12})}, \qquad (4.11)$$
$$C_{44} = \frac{1}{S_{44}}.$$

In the case of a stress-free homogeneous sample, the free energy can be derived in a simple Landau expansion $f_L = f_L^{(e)} + f_C + f_q$. Thus, the elastic and electrostriction terms result in re-normalization of the bar expansion coefficients $\alpha_{11}^{(e)}$ and $\alpha_{12}^{(e)}$ when minimizing the free energy with respect to the strain components (in the homogeneous sample). Then, the α_{11}, α_{12} coefficients of the relaxed f_L are related to bar $\alpha_{11}^{(e)}, \alpha_{12}^{(e)}$ as follows:

$$\alpha_{11} = \alpha_{11}^{(e)} - \frac{1}{6}\left[\frac{\hat{q}_{11}^2}{\hat{C}_{11}} + 2\frac{\hat{q}_{22}^2}{\hat{C}_{22}}\right],$$
$$\alpha_{12} = \alpha_{12}^{(e)} - \frac{1}{6}\left[2\frac{\hat{q}_{11}^2}{\hat{C}_{11}} - 2\frac{\hat{q}_{22}^2}{\hat{C}_{22}} + 3\frac{q_{44}^2}{C_{44}}\right] \qquad (4.12)$$

with

$$\hat{C}_{11} = C_{11} + 2C_{12},$$
$$\hat{C}_{12} = C_{11} - C_{12},$$
$$\hat{q}_{11} = q_{11} + 2q_{12}, \qquad (4.13)$$
$$\hat{q}_{12} = q_{11} - q_{12}.$$

The equilibrium spontaneous strain of the stress-free homogeneous sample is given by

$$
\begin{aligned}
e_1 &= Q_{11}P_1^2 + Q_{12}P_2^2 + Q_{12}P_3^2, \\
e_2 &= Q_{12}P_1^2 + Q_{11}P_2^2 + Q_{12}P_3^2, \\
e_3 &= Q_{12}P_1^2 + Q_{12}P_2^2 + Q_{11}P_3^2, \\
e_4 &= Q_{44}P_2P_3, \\
e_5 &= Q_{44}P_3P_1, \\
e_6 &= Q_{44}P_1P_2,
\end{aligned}
\tag{4.14}
$$

where $Q_{\alpha\beta}$ are the bulk electrostriction coefficients expressed as follows:

$$
\begin{aligned}
Q_{11} &= Q_{1111} = \tfrac{1}{3}\left[\frac{\hat{q}_{11}}{\hat{C}_{11}} + 2\frac{\hat{q}_{22}}{\hat{C}_{22}}\right], \\
Q_{12} &= Q_{1122} = \tfrac{1}{3}\left[\frac{\hat{q}_{11}}{\hat{C}_{11}} - \frac{\hat{q}_{22}}{\hat{C}_{22}}\right], \\
Q_{44} &= 4Q_{1212} = \frac{q_{44}}{C_{44}}.
\end{aligned}
\tag{4.15}
$$

4.3 Numerical Values of Free-Energy Coefficients of Selected Perovskite Ferroelectrics

Material-specific numerical coefficients of the above phenomenological model have been estimated by many authors in the past, using either experimental results (e.g. Berlincourt and Jaffe 1958; Haun et al. 1987, 1989b, 1989d; Li et al. 2005; Hlinka and Marton 2006; Hlinka 2008) or microscopic calculations (e.g. Behera et al. 2011; Taherinejad et al. 2012; Marton et al. 2017). The models were typically derived for particular individual compounds or pseudo-binary systems. Typically, some of the parameters are taken from earlier measurements or calculations. As far as we know, numerical coefficients derived for different perovskite oxides were more systematically compiled only by Rabe et al. (2007); otherwise, these important model parameters are dispersed within the original journal publications. Here we try to map out the current status of KGLD model development for the most important perovskite oxide ferroelectrics.

In order to facilitate the comparisons, we focus on the parameters of the above minimal model. When a different form of the potential is used, for example, when different additional numerical pre-factors are introduced for convenience, we try our best to recalculate values corresponding to the model defined above. For a more straight-forward comparison, we have summarized ambient-temperature numerical coefficients of the potential. Data of BaTiO$_3$ and KNbO$_3$ are summarized in Table 4.2 and for PbTiO$_3$ in Table 4.3, while Tables 4.4 and 4.5 show similar data for PZT, PMN-PT, and PZN-PT.

Table 4.2 *List of parameters used for BaTiO$_3$ (A–H) and KNbO$_3$ (I) at 298 K in the literature.*

	A	B	C	D	E	F	G	H	I	Unit
α_1	(−2.772)	(−3.712)	—	—	—	—	—	—	(−15.047)	10^7 J m C^{-2}
α_{11}	(−6.476)	−2.097	—	—	—	—	—	—	−6.360	10^8 J m^5 C^{-4}
α_{12}	3.230	7.974	—	—	—	—	—	—	9.660	10^8 J m^5 C^{-4}
α_{111}	(8.004)	1.294	—	—	—	—	—	—	2.810	10^9 J m^9 C^{-6}
α_{112}	4.470	−1.950	—	—	—	—	—	—	−1.990	10^9 J m^9 C^{-6}
α_{123}	4.910	−2.500	—	—	—	—	—	—	6.030	10^9 J m^9 C^{-6}
α_{1111}	—	3.863	—	—	—	—	—	—	1.740	10^{10} J m^{13} C^{-8}
α_{1112}	—	2.529	—	—	—	—	—	—	0.599	10^{10} J m^{13} C^{-8}
α_{1122}	—	1.637	—	—	—	—	—	—	2.500	10^{10} J m^{13} C^{-8}
α_{1123}	—	1.367	—	—	—	—	—	—	−1.170	10^{10} J m^{13} C^{-8}
G_{11}	—	—	—	—	—	5.1	0.138	0.093	—	10^{-10} J m^3 C^{-2}
G_{12}	—	—	—	—	—	−0.2	0	−0.093	—	10^{-10} J m^3 C^{-2}
G_{44}	—	—	—	—	—	0.2	0.138	0.093	—	10^{-10} J m^3 C^{-2}
q_{11}	—	—	—	14.20	14.20	—	—	—	—	10^9 J m C^{-2}
q_{12}	—	—	—	−0.74	−0.74	—	—	—	—	10^9 J m C^{-2}
q_{44}	—	—	—	1.57	3.14	—	—	—	—	10^9 J m C^{-2}
Q_{11}	—	—	11.0	—	—	—	—	10.0	12.0	10^{-2} m^4 C^{-2}
Q_{12}	—	—	−4.5	—	—	—	—	−3.4	−5.3	10^{-2} m^4 C^{-2}
Q_{44}	—	—	2.9	—	—	—	—	2.9	5.2	10^{-2} m^4 C^{-2}
C_{11}	—	—	—	2.750	—	—	—	1.780	‡2.559	10^{11} J m^{-3}
C_{12}	—	—	—	1.790	—	—	—	0.964	‡0.804	10^{11} J m^{-3}
C_{44}	—	—	—	0.543	—	—	—	1.220	‡0.901	10^{11} J m^{-3}
ε_B	—	—	—	—	—	7.35	—	—	—	

Temperature-dependent parameters are in round brackets. Components of the stiffness elastic tensor $C_{\alpha\beta}$ marked with ‡ were calculated from the compliance elastic tensor in the literature using Equation (4.11). Data in columns after: Bell (2001) (A), Li et al. (2005) (B), Yamada (1972) (C), Hu and Chen (1998) (D), Hlinka et al. (2009) (E), Hlinka and Marton (2006) (F), Ahluwalia et al. (2005) (G), Li and Chen (2006) (H), Liang et al. (2009) (I)

Table 4.3 *List of parameters used for PbTiO$_3$ at 298 K in the literature.*

	A	B	C	D	E	F	G	H	I	Unit
α_1	(−17.090)	(−17.887)	—	—	—	—	—	—	—	10^7 J m C^{-2}
α_{11}	−0.725	(−0.550)	—	—	—	—	—	—	—	10^8 J m^5 C^{-4}
α_{12}	7.500	3.390	—	—	—	—	—	—	—	10^8 J m^5 C^{-4}
α_{111}	0.261	0.257	—	—	—	—	—	—	—	10^9 J m^9 C^{-6}
α_{112}	0.610	0.695	—	—	—	—	—	—	—	10^9 J m^9 C^{-6}
α_{123}	−3.660	1.313	—	—	—	—	—	—	—	10^9 J m^9 C^{-6}
α_{1111}	—	—	—	—	—	—	—	—	—	10^{10} J m^{13} C^{-8}
α_{1112}	—	—	—	—	—	—	—	—	—	10^{10} J m^{13} C^{-8}
α_{1122}	—	—	—	—	—	—	—	—	—	10^{10} J m^{13} C^{-8}
α_{1123}	—	—	—	—	—	—	—	—	—	10^{10} J m^{13} C^{-8}
G_{11}	—	—	1.000	1.038	0.455	1.260	0.214	2.769	—	10^{-10} J m^3 C^{-2}
G_{12}	—	—	−1.000	−1.038	−0.559	−0.879	−0.126	−2.769	—	10^{-10} J m^3 C^{-2}
G_{44}	—	—	1.000	1.038	0.559	0.879	0.126	2.769	—	10^{-10} J m^3 C^{-2}
Q_{11}	8.90	8.40	—	—	—	—	—	—	—	10^{-2} m^4 C^{-2}
Q_{12}	−2.60	−2.50	—	—	—	—	—	—	—	10^{-2} m^4 C^{-2}
Q_{44}	6.75	3.50	—	—	—	—	—	—	—	10^{-2} m^4 C^{-2}
C_{11}	—	—	—	—	—	—	—	—	‡1.746	10^{11} J m^{-3}
C_{12}	—	—	—	—	—	—	—	—	‡0.794	10^{11} J m^{-3}
C_{44}	—	—	—	—	—	—	—	—	‡1.111	10^{11} J m^{-3}
ε_B	—	—	—	—	—	—	—	—	—	

Temperature-dependent parameters are in round brackets. Components of the stiffness elastic tensor $C_{\alpha\beta}$ marked with ‡ were calculated from the compliance elastic tensor in the literature using Equation (4.11). Data in columns after: Haun et al. 1987 (A), Heitmann and Rossetti 2014 (B), Hlinka 2008 (C), Li et al. 2002 (D), Behera et al. 2011(E,F), Shin et al. 2007 (G), Wang et al. 2005 (H), Pertsev et al. 1998(I).

Table 4.4 *List of parameters used for PZT at 298 K in the literature.*

	A	B	C	D	E	F	Unit
x	0.6	0.6	0.6	0.6	0.3	0.6	
α_1	(−8.343)	(−14.875)	(−10.325)	–	–	–	10^7 J m C^{-2}
α_{11}	0.361	0.241	(0.053)	–	–	–	10^8 J m^5 C^{-4}
α_{12}	3.233	3.464	1.270	–	–	–	10^8 J m^5 C^{-4}
α_{111}	0.186	0.336	0.257	–	–	–	10^9 J m^9 C^{-6}
α_{112}	0.850	1.180	0.695	–	–	–	10^9 J m^9 C^{-6}
α_{123}	−4.063	−7.022	1.313	–	–	–	10^9 J m^9 C^{-6}
G_{11}	–	–	–	–	–	–	10^{-10} J m^3 C^{-2}
G_{12}	–	–	–	–	–	–	10^{-10} J m^3 C^{-2}
G_{44}	–	–	–	–	–	–	10^{-10} J m^3 C^{-2}
Q_{11}	–	–	8.4	8.116	–	–	10^{-2} m^4 C^{-2}
Q_{12}	–	–	−2.5	−2.950	–	–	10^{-2} m^4 C^{-2}
Q_{44}	–	–	3.5	6.710	–	–	10^{-2} m^4 C^{-2}
C_{11}	–	–	–	–	1.792	‡1.696	10^{11} J m^{-3}
C_{12}	–	–	–	–	0.793	‡0.819	10^{11} J m^{-3}
C_{44}	–	–	–	–	0.482	‡0.472	10^{11} J m^{-3}
ε_B	–	–	–	–	–	–	

Temperature-dependent parameters are in round brackets. Components of the stiffness elastic tensor $C_{\alpha\beta}$ marked with ‡ were calculated from the compliance elastic tensor in the literature using Equation (4.11). Data in columns after: Haun et al. 1989d (A), Iwata and Ishibashi 2013 (B), Heitmann and Rossetti 2014 (C), Haun et al. 1989c (D), Pertsev et al. 2003 (F). Data in column E were obtained by interpolating elastic constants from Pertsev et al. 2003 and Tagantsev et al. 2013.

Let us stress that although the GLD model is conceived as a systematic expansion scheme, all practical applications assume finite values of the order parameters, comparable to spontaneous ones. In fact, numerical coefficients are adjusted to match quantities like spontaneous polarization. This implies that whenever the model is extended by additional terms, the numerical coefficients of lower order might need to be revised. For example, usually, the most attention and development is focused on the Landau energy itself. Some authors have repeatedly argued that in order to extend the range of its applicability, the expansion has to be extended up to the 8th or even 12th order (Vanderbilt and Cohen 2001; Sergienko et al. 2002; Li et al. 2013). The truncation of

Table 4.5 List of parameters used for PZN-PT (A,B) and PMN-PT (C–H) at 298 K in the literature.

	A	B	C	D	E	F	G	H	Unit
x	0.15	0.15	0.30	0.42	0.70	0.30	0.42	0.70	
α_1	(−4.871)	(−4.046)	(−2.520)	(−4.343)	(−9.902)	—	—	—	10^7 J m C^{-2}
α_{11}	(−0.413)	0.556	(0.345)	(0.192)	(−0.166)	—	—	—	10^8 J m^5 C^{-4}
α_{12}	−0.384	3.370	0.608	1.085	2.198	—	—	—	10^8 J m^5 C^{-4}
α_{111}	0.257	0.323	0.257	0.257	0.257	—	—	—	10^9 J m^9 C^{-6}
α_{112}	0.695	1.020	0.695	0.695	0.695	—	—	—	10^9 J m^9 C^{-6}
α_{123}	2.703	−1.210	1.313	1.313	1.313	—	—	—	10^9 J m^9 C^{-6}
G_{11}	—	1.056	—	—	—	—	—	—	10^{-10} J m^3 C^{-2}
G_{12}	—	−1.056	—	—	—	—	—	—	10^{-10} J m^3 C^{-2}
G_{44}	—	1.056	—	—	—	—	—	—	10^{-10} J m^3 C^{-2}
Q_{11}	8.4	8.9	8.4	8.4	8.4	8.4	8.4	8.4	10^{-2} m^4 C^{-2}
Q_{12}	−2.5	−3.0	−2.5	−2.5	−2.5	−2.5	−2.5	−2.5	10^{-2} m^4 C^{-2}
Q_{44}	3.5	3.4	3.5	3.5	3.5	3.5	3.5	3.5	10^{-2} m^4 C^{-2}
C_{11}	—	‡0.909	—	—	—	‡0.329	‡1.149	‡1.632	10^{11} J m^{-3}
C_{12}	—	‡0.545	—	—	—	‡0.188	‡0.249	‡0.970	10^{11} J m^{-3}
C_{44}	—	‡0.500	—	—	—	‡0.714	‡0.285	‡0.364	10^{11} J m^{-3}
ε_B	—	100	—	—	—	—	—	—	

Temperature-dependent parameters are in round brackets. Components of the stiffness elastic tensor $C_{\alpha\beta}$ marked with ‡ were calculated from the compliance elastic tensor in the literature using Equation (4.11). Data in columns after: Heitmann and Rossetti 2014 (A,C–E), Li et al. 2017 (B), Khakpash et al. 2015 (F–H).

the order parameter expansion as well as the choice of temperature-independent coefficients strongly influences the coefficients of the Landau potential expansion, even if the overall landscape of the potential around the physically relevant area of the order parameter space might look rather similar. This can be nicely illustrated by comparison of the numerical coefficients of various Landau potentials for $BaTiO_3$ with the one expanded up to the eighth order (see Table 4.2). Some of the most recently proposed potentials (Kim et al. 2018a, 2018b; Pitike et al. 2019) have not been systematically tested by us yet, except for the ab initio predicted potential for $PbTiO_3$ (Pitike et al. 2019), which provided quite similar spontaneous bulk and domain wall properties as the Landau potential models given here in Table 4.3.

4.4 Full Sets of Ginzburg–Landau–Devonshire Model Parameters

In order to appreciate energy costs of an arbitrary order parameter spatial profile, or to perform unconstrained domain-structure optimization using phase-field simulation strategies, a complete set of parameters is needed. As can be seen from the data summarized in Tables 4.2, 4.3, 4.4, and 4.5, the original literature resources often yield data for individual terms of the potential, but almost never derive the complete set of the full KGLD model parameters. Therefore, a complete set of model parameters is usually obtained by combination of data from different authors; here, such complete sets are proposed in Tables 4.6 and 4.7. In Table 4.6 we propose three plausible parameter sets for $BaTiO_3$, one parameter set for $KNbO_3$, two models for $PbTiO_3$, and three different models for tetragonal PZT60. Table 4.7 summarizes two models for PZN-15PT and PMN-PT with various concentrations. Let us stress that $BaTiO_3$ and $KNbO_3$ have the rhombohedral ground state and their ferroelectrically active ions are in the B-site of the perovskite ABO_3 formula, while $PbTiO_3$ and PZT60 are examples of A-site-driven tetragonal perovskite ferroelectrics and PMN-PT and PZN-PT are chemically disordered relaxor ferroelectrics. Thus, a reader interested in quantitative estimations of polarization and strain profiles within oxide ferroelectric films or heterostructures of any desired geometry can find here complete material-specific descriptions of several of the most typical and still very distinct representatives of ferroelectric oxides.

Let us note that for most materials some coefficients are not known at all. In these cases, we have selected an arbitrary plausible value or the value of another similar material in the perovskite family. Certainly, there are other ways to combine the available data into complete parameter sets, and definitely there is a lot of scope for model improvements, but, at least, the choices made here allow us to make some rough estimates and appreciate some general trends.

The spread of collected parameter values, in particular for Landau potential expansion coefficients, suggests that these parameters have to be considered with caution, as the

Table 4.6 Selected models for $BaTiO_3$, $KNbO_3$, $PbTiO_3$, and $PZT60$ from Tables 4.2, 4.3, and 4.4.

Model	BaTiO₃ I	BaTiO₃ II	BaTiO₃ III	KNbO₃ I	PbTiO₃ I	PbTiO₃ II	PZT60 I	PZT60 II	PZT60 III	Unit
α_1	-2.772	-2.772	-3.712	-15.047	-17.090	-17.887	-8.343	-14.875	-10.325	$10^7\,\mathrm{J\,m\,C^{-2}}$
α_{11}	-6.476	-6.476	-2.097	-6.360	-0.725	-0.550	0.361	0.241	0.053	$10^8\,\mathrm{J\,m^5\,C^{-4}}$
α_{12}	3.230	3.230	7.974	9.660	7.500	3.390	3.233	3.464	1.270	$10^8\,\mathrm{J\,m^5\,C^{-4}}$
α_{111}	8.004	8.004	1.294	2.810	0.261	0.257	0.186	0.336	0.257	$10^9\,\mathrm{J\,m^9\,C^{-6}}$
α_{112}	4.470	4.470	-1.950	-1.990	0.610	0.695	0.850	1.810	0.695	$10^9\,\mathrm{J\,m^9\,C^{-6}}$
α_{123}	4.910	4.910	-2.500	6.030	-3.660	1.313	-4.063	-7.022	1.313	$10^9\,\mathrm{J\,m^9\,C^{-6}}$
α_{1111}	–	–	3.863	1.740	–	–	–	–	–	$10^{10}\,\mathrm{J\,m^{13}\,C^{-8}}$
α_{1112}	–	–	2.529	0.599	–	–	–	–	–	$10^{10}\,\mathrm{J\,m^{13}\,C^{-8}}$
α_{1122}	–	–	1.637	2.500	–	–	–	–	–	$10^{10}\,\mathrm{J\,m^{13}\,C^{-8}}$
α_{1123}	–	–	1.367	-1.170	–	–	–	–	–	$10^{10}\,\mathrm{J\,m^{13}\,C^{-8}}$
G_{11}	5.100	0.138	0.093	5.100	1.000	1.000	1.000	1.000	1.000	$10^{-10}\,\mathrm{J\,m^3\,C^{-2}}$
G_{12}	-0.200	0	-0.093	-0.200	-1.000	-1.000	-1.000	-1.000	-1.000	$10^{-10}\,\mathrm{J\,m^3\,C^{-2}}$
G_{44}	0.200	0.138	0.093	0.200	1.000	1.000	1.000	1.000	1.000	$10^{-10}\,\mathrm{J\,m^3\,C^{-2}}$
q_{11}	14.200	†14.140	†11.245	†22.178	†11.413	†10.698	†8.934	†8.934	†10.152	$10^9\,\mathrm{J\,m\,C^{-2}}$
q_{12}	-0.740	†-0.740	†0.310	†-8.172	†0.460	†0.317	†-0.773	†-0.773	†0.591	$10^9\,\mathrm{J\,m\,C^{-2}}$
q_{44}	3.140	†1.575	†3.538	†4.684	†7.500	†3.889	†3.165	†3.165	†1.651	$10^9\,\mathrm{J\,m\,C^{-2}}$
Q_{11}	†11.045	11.000	10.000	12.000	8.900	8.400	8.116	8.116	8.400	$10^{-2}\,\mathrm{m^4\,C^{-2}}$
Q_{12}	†-4.518	-4.500	-3.400	-5.300	-2.600	-2.500	-2.950	-2.950	-2.500	$10^{-2}\,\mathrm{m^4\,C^{-2}}$
Q_{44}	†5.783	2.900	2.900	5.200	6.750	3.500	6.710	6.710	3.500	$10^{-2}\,\mathrm{m^4\,C^{-2}}$
C_{11}	2.750	2.750	1.780	‡2.559	‡1.746	‡1.746	‡1.696	‡1.696	‡1.696	$10^{11}\,\mathrm{J\,m^{-3}}$
C_{12}	1.790	1.790	0.964	‡0.804	‡0.794	‡0.794	‡0.819	‡0.819	‡0.819	$10^{11}\,\mathrm{J\,m^{-3}}$
C_{44}	0.543	0.543	1.220	‡0.901	‡1.111	‡1.111	‡0.472	‡0.472	‡0.472	$10^{11}\,\mathrm{J\,m^{-3}}$
ε_B	7.35	1	1	7.35	1	1	1	1	1	1

The symbol † indicates parameters which were evaluated from published data, e.g. $q(Q, C)$ or $Q(q, C)$. Components of the stiffness elastic tensor $C_{\alpha\beta}$ marked with ‡ were calculated from the compliance elastic tensor in literature using Equation (4.11).

Table 4.7 *Selected models for PZN-PT and PMN-PT from Table 4.5.*

Model	PZN–PT I	PZN–PT II	PMN–PT I	PMN–PT II	PMN–PT II	PMN–PT II	Unit
x	0.15	0.15	0.42	0.30	0.42	0.70	
α_1	-4.046	-4.871	-4.343	-2.520	-4.343	-9.902	$10^7 \mathrm{J\,m\,C}^{-2}$
α_{11}	0.556	-0.413	0.192	0.345	0.192	-0.166	$10^8 \mathrm{J\,m}^5\,\mathrm{C}^{-4}$
α_{12}	3.370	-0.384	1.084	0.608	1.085	2.198	$10^8 \mathrm{J\,m}^5\,\mathrm{C}^{-4}$
α_{111}	0.323	0.257	0.257	0.257	0.257	0.257	$10^9 \mathrm{J\,m}^9\,\mathrm{C}^{-6}$
α_{112}	1.020	0.695	0.695	0.695	0.695	0.695	$10^9 \mathrm{J\,m}^9\,\mathrm{C}^{-6}$
α_{123}	-1.210	1.313	2.703	1.313	1.313	1.313	$10^9 \mathrm{J\,m}^9\,\mathrm{C}^{-6}$
G_{11}	1.056	1.000	1.000	1.000	1.000	1.000	$10^{-10}\mathrm{J\,m}^3\,\mathrm{C}^{-2}$
G_{12}	-1.056	-1.000	-1.000	-1.000	-1.000	-1.000	$10^{-10}\mathrm{J\,m}^3\,\mathrm{C}^{-2}$
G_{44}	1.056	1.000	1.000	1.000	1.000	1.000	$10^{-10}\mathrm{J\,m}^3\,\mathrm{C}^{-2}$
g_{11}	†4.818	†10.698	†10.698	†1.823	†8.408	†8.860	$10^9 \mathrm{J\,m\,C}^{-2}$
g_{12}	†0.491	†0.317	†0.317	†0.286	†-1.403	†1.641	$10^9 \mathrm{J\,m\,C}^{-2}$
g_{44}	†1.700	†3.889	†3.889	†2.500	†0.997	†1.273	$10^9 \mathrm{J\,m\,C}^{-2}$
Q_{11}	8.900	8.400	8.400	8.400	8.400	8.400	$10^{-2}\mathrm{m}^4\,\mathrm{C}^{-2}$
Q_{12}	-3.000	-2.500	-2.500	-2.500	-2.500	-2.500	$10^{-2}\mathrm{m}^4\,\mathrm{C}^{-2}$
Q_{44}	3.400	3.500	3.500	3.500	3.500	3.500	$10^{-2}\mathrm{m}^4\,\mathrm{C}^{-2}$
C_{11}	‡0.909	‡1.746	1.746	‡0.329	‡1.149	‡1.632	$10^{11}\mathrm{J\,m}^{-3}$
C_{12}	‡0.545	‡0.794	0.794	‡0.188	‡0.249	‡0.970	$10^{11}\mathrm{J\,m}^{-3}$
C_{44}	‡0.500	‡1.111	1.111	‡0.714	‡0.285	‡0.364	$10^{11}\mathrm{J\,m}^{-3}$
ε_B	100	1	1	1	1	1	

The symbol † indicates parameters which were evaluated from published data, e.g. $q(Q, C)$ or $Q(q, C)$. Components of the stiffness elastic tensor $C_{\alpha\beta}$ marked with ‡ were calculated from the compliance elastic tensor in the literature using Equation (4.11).

potentials have not been constructed to provide a very accurate description of nonlinear physical properties. Nevertheless, the derived values of spontaneous polarization as well as the linear response quantities, like dielectric permittivity and piezoelectric tensor components, seem to give comparable predictions for the same compounds (see Tables 4.8 and 4.9).

Table 4.8 *Physical properties of selected ferroelectrics at 298 K in Table 4.6.*

Model	BaTiO$_3$ I	BaTiO$_3$ II	BaTiO$_3$ III	KNbO$_3$ I	PbTiO$_3$ I	PbTiO$_3$ II	PZT60 I	PZT60 II	PZT60 III
Domain state	[001]	[001]	[001]	[101]	[001]	[001]	[001]	[001]	[001]
P_s (C/m^2)	0.26523	0.26523	0.25999	0.45342	0.75459	0.74690	0.57215	0.60092	0.59912
e_{11}	−0.00318	−0.00317	−0.00230	0.00689	−0.01480	−0.01395	−0.00966	−0.01065	−0.00897
e_{22}	−0.00318	−0.00317	−0.00230	−0.01090	−0.01480	−0.01395	−0.00966	−0.01065	−0.00897
e_{33}	0.00777	0.00774	0.00676	0.00689	0.05068	0.04686	0.02657	0.02931	0.03015
e_{13}	0	0	0	0.00267	0	0	0	0	0
κ_{11}	3298.34	3298.34	3599.8	115.71	124.40	249.28	497.43	265.92	1771.65
κ_{22}	3298.34	3298.34	3599.8	984.07	124.40	249.28	497.43	265.92	1771.65
κ_{33}	192.66	192.67	187.84	115.71	66.54	67.37	197.17	100.817	139.283
κ_{13}	0	0	0	−57.89	0	0	0	0	0
d_{31} (pC/N)	−40.88	−40.72	−29.51	−74.26	−23.12	−22.28	−58.93	−31.65	−36.94
d_{32} (pC/N)	−40.88	−40.72	−29.51	−17.40	−23.12	−22.28	−58.93	−31.65	−36.94
d_{33} (pC/N)	99.95	99.54	86.36	96.26	79.13	74.84	162.14	87.07	124.13
d_{15} (pC/N)	447.92	224.63	240.32	8.54	56.10	57.70	169.10	94.94	328.93
d_{24} (pC/N)	447.92	224.63	240.32	145.26	56.10	57.70	169.10	94.94	328.93

All properties were calculated in the [100], [010], [001] coordinate system for the indicated domain states. We used analytical relations given in Haun et al. 1989b. Properties of BaTiO$_3$ III and KNbO$_3$ I models, with Landau potential of the eighth order, were calculated from phase–field simulations by the program Ferrodo (Marton and Hlinka 2006).

Table 4.9 *Physical properties of selected ferroelectric solid solutions at 298 K in Table 4.7.*

Model	PZN–PT I	PZN–PT II	PMN–PT II	PMN–PT II	PMN–PT II
x	0.15	0.15	0.30	0.42	0.70
Domain state	[001]	[001]	[111]	[001]	[001]
P_s (C/m^2)	0.39353	0.55707	0.38746	0.46221	0.61672
e_{11}	−0.00465	−0.00776	0.00170	−0.00534	−0.00951
e_{33}	0.01378	0.02606	0.00170	0.01795	0.00320
e_{13}	0	0	0.00088	0	0
κ_{11}	1560.3	8961.4	19172	4926.3	663.58
κ_{33}	443.26	229.44	19172	358.96	134.01
κ_{13}	0	0	−9240.5	0	0
d_{31} (pC/N)	−92.67	−56.58	−4058.4	−73.45	−36.59
d_{33} (pC/N)	274.92	190.12	8209.9	246.79	122.94
d_{14} (pC/N)	0	0	−1281.2	0	0
d_{15} (pC/N)	184.85	1547.06	688.50	705.63	126.82

All properties were calculated in the [100], [010], [001] coordinate system for the indicated domain states. All the analytical relations are summarized in Haun et al., 1989b.

4.5   Phase Diagrams and Bulk Property Diagrams

The most interesting aspect of the phenomenological approach is its capacity to describe systematic variations of the properties with thermodynamical variables such as temperature or chemical composition. In fact, all the listed GLD models have at least one of the Landau potential coefficients expressed as an analytic function of temperature. Consequently, these models allow one to trace the temperature dependence of the corresponding material properties. For example, one can calculate the temperature dependence of the spontaneous polarization by determining the minimum of the stress-free Landau potential. Examples of the resulting polarization curves for some of the models considered here are shown in Figure 4.2.

For most of the potentials designed for PZT, PMN-PT, and PZN-PT solid solutions, expressions for the chemical composition dependence of model parameters have also been proposed. In the spirit of Curie's and Vegard's laws, one could expect just a linear dependence on temperature and chemical composition, but in order to reproduce

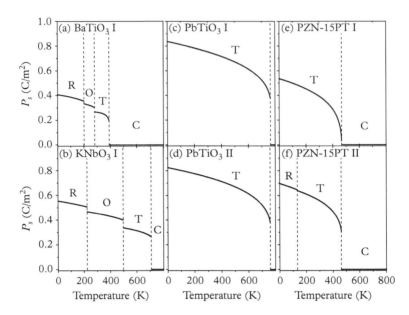

Figure 4.2 *Temperature dependence of spontaneous polarization in selected materials. Present phases: C-paraelectric cubic, T-ferroelectric tetragonal, O-ferroelectric orthorhombic, R-ferroelectric rhombohedral phase. The ferroelectric phases are single domain states. The vertical dash lines indicate T_C.*

the full range of the phase diagrams, one often needs to either go beyond the linear approximation, or to consider multiple variable coefficients. The available models from the literature allow calculation of temperature-concentration phase diagrams, for instance, for PZT, PMN-PT, and PZN-PT. For example, the position of the MPB can be estimated by comparing Gibbs free energies of their optimized rhombohedral and tetragonal ferroelectric states (there are no monoclinic phases present in these models; see the top panels of Figure. 4.3).

Similarly, one can explore the temperature-concentration dependence of various other properties, such as spontaneous polarization (see the bottom panels of Figure 4.3). As can be expected, the models show an increased transverse dielectric permittivity (the diagonal tensor element perpendicular to the spontaneous polarization) near the vicinity of the MPB (see the top of Figure 4.4). Interestingly, in PZT model I, the longitudinal component is also somewhat enhanced near MPB (see the bottom of Figure 4.4). The difference between the models is even more pronounced when comparing various components of the bulk piezoelectric tensor (Figure 4.5).

Readers interested in employing these models might appreciate the collection of explicit formulas for the temperature- and chemical-composition-dependent coefficients involved in the models listed in Tables 4.2, 4.3, 4.4, and 4.5. They read as follows (with units as in the corresponding tables and T expressed in the Kelvin scale):

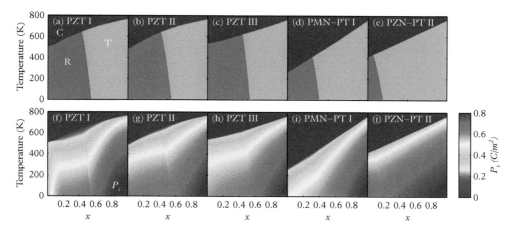

Figure 4.3 *(a–e) Phase diagrams and (f–j) spontaneous polarization of PZT, PMN-PT, and PZN-PT solid solutions. Color code of the phases: dark blue-paraelectric cubic (C), light blue-ferroelectric tetragonal (T), red-ferroelectric rhombohedral (R) phase. The ferroelectric phases are single domain states. Panels (a–c) and (f–h) refer to the same PZT material but the Zr/Ti concentration dependence is described with a different degree of detail. The phase diagram is traced for the PZT models I, II and III of Table 4.6 corresponding to potentials in Table 4.4 with composition-dependent coefficients resumed in the text of Section 4.5. One can clearly see that the paraelectric-ferroelectric phase transition boundary has a convex profile in PZT III, a concave profile in PZT II and a mixed one in PZT I (probably best matching the experimental observations). Interestingly, the models also differ in the experimentally most relevant concentration-dependent property, spontaneous polarization, even at room temperature. Another remarkable difference between PZT and relaxor ferroelectrics is that the morphotropic phase boundary between tetragonal and rhombohedral phases seems to be much more pronounced in models for ferroelectric PZT.*

- $BaTiO_3$ in column A of Table 4.2; after Bell and Cross 1984; Bell 2001:

$$\alpha_1(T) = 3.34 \, (T - 381) \times 10^5,$$
$$\alpha_{11}(T) = 4.69 \, (T - 393) \times 10^6 - 2.02 \times 10^8,$$
$$\alpha_{111}(T) = -5.52 \, (T - 393) \times 10^7 + 2.76 \times 10^9.$$

- $BaTiO_3$ in column B of Table 4.2; after Li et al. 2005:

$$\alpha_1(T) = 4.124 \, (T - 388) \times 10^5.$$

- $KNbO_3$ in column I of Table 4.2; after Liang et al. 2009:

$$\alpha_1(T) = 4.273 \, (T - 650.15) \times 10^5.$$

- $PbTiO_3$ in column A of Table 4.3; after Haun et al. 1987:

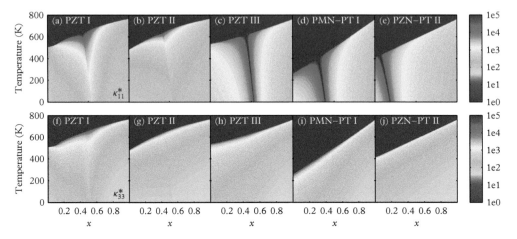

Figure 4.4 *Effective (a–e) transverse and (f–j) longitudinal dielectric susceptibility in PZT, PMN-PT, and PZN-PT solid solutions for the exactly same models as in the preceding Figure 4.3. The coefficients were calculated with respect to [111] and [001] single domain states in rhombohedral and tetragonal phases, respectively. It means that, for example, the longitudinal coefficient κ_{33}^* in the rhombohedral phase was calculated with probing field applied along the [111] direction. The data are plotted in the decimal logarithmic scale. These results demonstrate that the differences among various models for these materials are strongly reflected in the bulk material transverse dielectric susceptibility. Apart from the PZT I model, the longitudinal permittivity seems to be very little influenced by the morphotropic phase boundary.*

$$\alpha_1(T) = 3.7647\,(T - 751.95) \times 10^5.$$

- PbTiO$_3$ in column B of Table 4.3; after Heitmann and Rossetti 2014:

$$\alpha_1(T) = 3.975\,(T - 748) \times 10^5,$$
$$\alpha_{11}(T) = -3.775\,(T - 748) \times 10^4 - 7.2 \times 10^7.$$

- PZT in column A of Table 4.4; after Haun et al. 1989d:

$$\alpha_1(x, T) = \tfrac{1}{2\epsilon_0 C_0(x)}\,[T - T_0(x)],$$
$$\alpha_{11}(x) = \tfrac{1}{C_0(x)}\left(10.612 - 22.655x + 10.955x^2\right) \times 10^{13},$$
$$\alpha_{12}(x) = \tfrac{1}{3}\zeta(x) - \alpha_{11},$$
$$\alpha_{111}(x) = \tfrac{1}{C_0(x)}\left(12.026 - 17.296x + 9.179x^2\right) \times 10^{13},$$
$$\alpha_{112}(x) = \tfrac{1}{C_0(x)}\left(4.2904 - 3.3754x + 58.804e^{-29.397x}\right) \times 10^{14},$$
$$\alpha_{123}(x) = \xi(x) - 3\alpha_{111} - 6\alpha_{112},$$

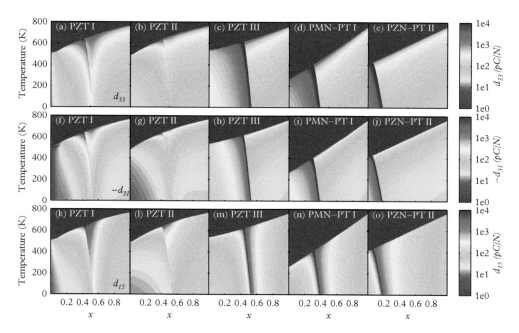

Figure 4.5 *Piezoelectric properties of single domain models considered in the previous two Figures 4.3 and 4.4. Panels (a-e) d_{33} (f-j), $-d_{31}$, and (k-o) d_{15} show predicted piezoelectric coefficients of monodomain PZT, PMN-PT, and PZN-PT solid solutions. The coefficients were calculated in the [100], [010], and [001] coordinate system and single domain states were parallel to [111] and [001] directions in rhombohedral and tetragonal phases, respectively. The data are plotted in the decimal logarithmic scale. One can see that the degree of enhancement of the bulk electromechanical response near the phase transition lines is not described in the same way even if all included models have seemingly similar phase diagram.*

where

$$T_0(x) = 462.63 + 843.4x - 2105.5x^2 + 4041.8x^3 - 3828.3x^4 + 1337.8x^5,$$

$$C_0(x) = \left[\frac{2.1716}{1+500.05(x-0.5)^2} + 0.131x + 2.01\right] \times 10^5 \quad when \quad x \in \langle 0, 0.5\rangle,$$

$$= \left[\frac{2.8339}{1+126.56(x-0.5)^2} + 1.4132\right] \times 10^5 \quad when \quad x \in (0.5, 1\rangle,$$

$$\zeta(x) = \frac{1}{C_0(x)}\left[2.6213 + 0.42743x - (9.6 + 0.012501x)e^{-12.6x}\right] \times 10^{14},$$

$$\xi(x) = \frac{1}{C_0(x)}\left[0.887 - 0.76973x + (16.225 - 0.088651x)e^{-21.255x}\right] \times 10^{15},$$

and ϵ_0 is the vacuum permittivity. C_0 and T_0 are corrected as in Rabe et al. (2007) to obtain published values by Haun et al. (1989d).

- PZT in column B of Table 4.4; after Iwata and Ishibashi (2013):

$$\alpha_1(x, T) = 3.7665 \, [T - T_C(x)] \times 10^5,$$
$$\alpha_{11}(x) = (1.6892 - 2.4131x) \times 10^8,$$
$$\alpha_{12}(x) = (-2.5798 + 10.073x) \times 10^8,$$
$$\alpha_{111}(x) = (4.4992 - 1.9037x) \times 10^8,$$
$$\alpha_{112}(x) = (3.6123 - 3.0038x) \times 10^9,$$
$$\alpha_{123}(x) = (-12.078 + 8.4262x) \times 10^9,$$

where $T_C(x) = 480.45 + 477.9x - 206.3x^2$.

- PZT in column C of Table 4.4; after Heitmann and Rossetti (2014):

$$\alpha_1(x, T) = (1.2 + 2.775x) \, [T - T_C(x)] \times 10^5,$$
$$\alpha_{11}(x, T) = (0.86175 - 1.575x) \times 10^8 - 3.775 \, [T - T_C(x)] \times 10^4,$$
$$\alpha_{12}(x) = (-1.91 + 5.3x) \times 10^8,$$

where $T_C(x) = 748x + 524 \, (1 - x)$.

- PZT in column D of Table 4.4; after Haun et al. (1989c):

$$Q_{11} = \frac{0.029578}{1 + 200(x - 0.5)^2} + 0.042796x + 0.045624,$$
$$Q_{12} = \frac{-0.026568}{1 + 200(x - 0.5)^2} - 0.012093x - 0.013386,$$
$$Q_{44} = \frac{0.025325}{1 + 200(x - 0.5)^2} + 0.020857x + 0.046147.$$

- PZN-100xPT in column A of Table 4.5; after Heitmann and Rossetti (2014):

$$\alpha_1(x, T) = (2.935 + 1.04x) \, [T - T_C(x)] \times 10^5,$$
$$\alpha_{11}(x, T) = (-4.23 - 3.3x) \times 10^7 - 3.775 \, [T - T_C(x)] \times 10^4,$$
$$\alpha_{12}(x) = (-1.05 + 4.44x) \times 10^8,$$

where $T_C(x) = 748x + 404 \, (1 - x)$.

- PZN-15PT in column B of Table 4.5; after Li et al. (2017):

$$\alpha_1(T) = 2.38 \, (T - 468) \times 10^5.$$

- PMN-100xPT in columns C-E of Table 4.5; after Heitmann and Rossetti (2014):

$$\alpha_1(x, T) = (1.575 + 2.4x)\,[T - T_C(x)] \times 10^5,$$
$$\alpha_{11}(x, T) = (0.7425 - 1.4625x) \times 10^8 - 3.775\,[T - T_C(x)] \times 10^4,$$
$$\alpha_{12}(x) = (-0.585 + 3.975x) \times 10^8.$$

where $T_C(x) = 748x + 262\,(1 - x)$.

4.6   Domain Wall Properties

In order to estimate the basic characteristics of domain walls, it is necessary to calculate the spatial variation of the order parameters on the real space trajectory across the domain wall (see also Chapter 1). This can be done by solving Euler–Lagrange equations for the GLD model under consideration with appropriate boundary conditions corresponding to the selected domain pair, and by imposing a suitable domain wall normal. The procedure has been well described in the literature (Zhirnov 1958; Cao and Cross 1991; Hlinka and Marton 2006; Marton et al. 2010). For the charge-neutral and mechanically compatible walls, the procedure can be greatly simplified by the requirement of a straight path interpolation in the primary order parameter space (straight path approximation, sometimes also called one-dimensional approximation) (Hlinka and Marton 2006; Marton et al. 2010). Another option is to use simulated annealing of a suitably prepared initial configuration using a phase-field simulation strategy. We have verified many times that the differences between the two approaches are usually rather small. Here we have used the former approach to determine the domain wall profiles and energies as a function of the temperature and composition. The result of these calculations is shown in Figures 4.6 and 4.7. The latter approach was applied only at selected temperatures to verify independently the stability and accuracy of the solutions obtained within the straight path approximation.

The basic characteristics of ferroelectric domain walls that enter various considerations are obviously the thickness and the energy density of the wall. As usual, we define the domain wall thickness 2ξ as the distance between the intersections of the horizontal asymptotes of the primary order-parameter polarization profile with the oblique asymptote of the polarization profile in the center of the wall, and the domain wall energy density Σ is the specific excess of the GLD energy above the single domain density (Hlinka and Marton 2006; Marton et al. 2010; Taherinejad et al. 2012). Results obtained for $BaTiO_3$ demonstrate how important the polarization gradient terms and their anisotropy are for both these quantities (see panels (a) and (c) in Figure 4.6, as well as panels (b) and (d) in the same figure). In the case of rhombohedral PZT, it is also interesting to note that different flavors of the potential can lead both to convex and concave temperature dependence of domain wall energies even though the energy hierarchy of different domain types is fairly similar in all three models (see Figure 4.7).

Obviously, most interesting is a comparison of domain wall properties in systems with MPBs. To what extent is the presence or absence of the anomalies of bulk properties

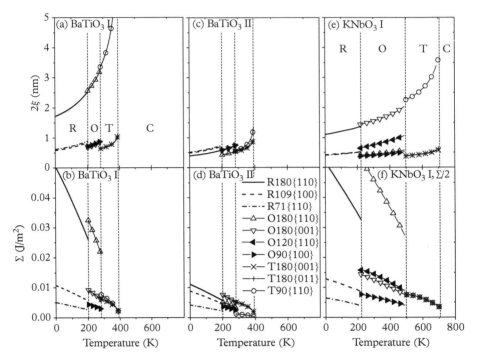

Figure 4.6 *Domain wall thickness and energy as a function of temperature in (a–d) BaTiO₃ and (e,f) KNbO₃. Data were calculated from the one-dimensional approximation used by Marton et al. (2010) apart from KNbO₃ data which were obtained from simulations with the program Ferrodo (Marton and Hlinka 2006). The O120 wall only calculated for KNbO₃. The vertical dash lines indicate T_C.*

at the MPB influencing domain wall properties? The numerical results displayed in Figures 4.8 and 4.9 demonstrate that anomalies at the MPB are actually rather moderate. Systematically, the 180° domain walls are considerably wider in the rhombohedral phase than in the tetragonal phase, so that they can be expected to be more mobile on the rhombohedral side of the MPB. Also, in general, rhombohedral domain walls have lower energy densities, so that finer domain structures are expected in the rhombohedral phases. Broadening near the MPB is also marked in the PZT I potential, but almost absent in the PZN-PT potential. Overall, all trends are rather smooth, and no dramatic enhancement in the near vicinity of the MPB was observed.

Together with this rather extensive survey of domain wall calculations performed in the straight path approximation, we have also explored all the models with phase-field calculations using the computational protocols described in our earlier works. Among others, this latter approach allows one to identify whether the lowest-energy domain wall of a given type passes through the state of zero polarization (what we denote as an Ising-like wall) or whether an additional transverse polarization appears there (we call these

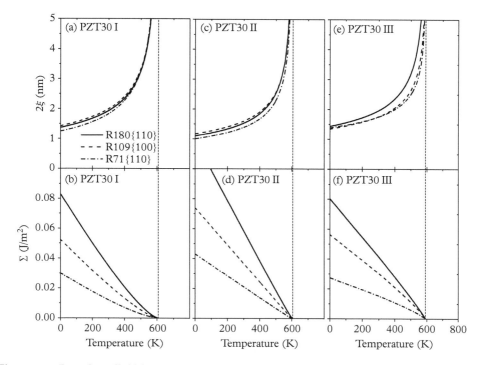

Figure 4.7 *Domain wall thickness and energy as a function of temperature in PZT30: (a,b) model I, (c,d) model II, and (e,f) model III. The values were calculated within the one-dimensional approximation used by Marton et al. (2010). The vertical dash lines indicate T_C.*

walls Bloch walls) as illustrated in Chapter 3, Figure 3.2 (Hlinka et al. 2016). Previously, we and others have reported the existence of these anomalous Bloch wall solutions in low-temperature phases of $BaTiO_3$ (Hlinka et al. 2011; Taherinejad et al. 2012; Stepkova et al. 2012, 2015; Stepkova and Hlinka 2017, 2018). In general, in all these favorable cases, the Bloch solutions have only a very slightly lower energy density than the corresponding Ising solutions, but they could be considerably wider and thus intrinsically more mobile. These Bloch walls are expected to be present also in $PbTiO_3$ and MPB compounds (Wojdeł and Íñiguez 2014; Yudin et al. 2013). Somewhat unexpectedly, within the GLD potentials studied here, we have found them quite rarely in $PbTiO_3$ and MPB compounds. Examples of such Bloch walls in tetragonal PZT are shown in Figure 4.10. Unfortunately, the Bloch components are rather minor in comparison with the spontaneous polarization. They are not able to induce a significant domain wall broadening and they are present only at the lowest temperatures. In particular, the lowest-energy ambient-temperature solutions show the usual Ising-like profiles. It remains to be understood whether the Bloch walls do not play any role in MPB materials, or whether the models are still too approximate to describe the underlying domain wall phenomena.

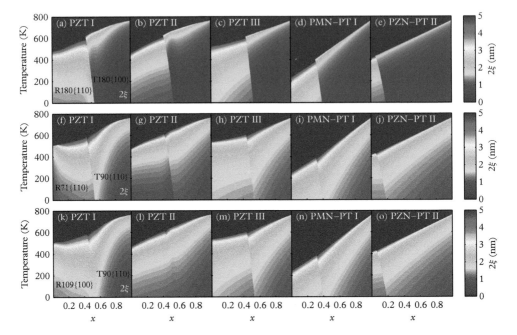

Figure 4.8 *Domain wall thickness (2ξ) of selected domain walls in PZT, PMN–PT, and PZN–PT solid solutions for the same models as used in Figures 4.3, 4.4, and 4.5 with model parameters listed in Tables 4.6 and 4.7. Panels in the first row refer to 180-degree domain walls, panels in the second row allow to compare R71 and T90 domain walls and panels in the bottom row allow to compare R109 with T90 domain walls. The values were calculated within the one-dimensional approximation used by Marton et al. (2010). Note that R180 domain walls are systematically broader than the T180 domain walls. Thicknesses of ferroelastic T90, R71, and R109 walls are also sensitive to the model parameters. Unfortunately, the differences are only enhanced in the vicinity of the paraelectric–ferroelectric phase transition boundary.*

4.7 Concluding Remarks

This chapter introduced the basic ideas, tools, and capabilities of the phenomenological KGLD model, which is a suitable instrument for material-specific quantitative theoretical investigations of domain structures and domain wall phenomena, at least in those perovskite ferroelectrics in which the influences of structural defects and mobile charge carriers can be neglected. The mathematical form of the model could be considered as minimal, since it consistently includes all the key ingredients that determine domain wall properties, but still describes them with the lowest-order terms in the order parameters only. Moreover, in the spirit of Landau theory, all symmetry requirements are automatically taken into account.

In the past decade, there have been multiple important results published in the field of perovskite ferroelectrics with a support of phase-field simulations. Certain predictions,

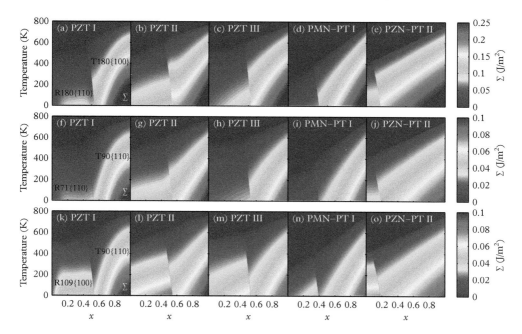

Figure 4.9 *Domain wall excess energy density (Σ) calculated for the same set of domain walls as in Figure 4.8. Results show that in all walls, all models and all materials, the concentration dependence of the domain wall excess energy density is much stronger in the tetragonal phase than in the rhombohedral phase. The overall resulting trend is such that a higher density of tetragonal domain walls should be formed in materials with a composition closer to the morphotropic phase boundary.*

such as the existence of Bloch walls in $BaTiO_3$ or vortex structures in $PbTiO_3$-$SrTiO_3$ superlattices, have been verified by atomistic or ab initio calculations (Taherinejad et al. 2012; Aguado-Puente and Junquera 2012). Also, the results reproduced here demonstrate the capabilities of GLD modeling. The mathematical framework is certainly computationally affordable and can be extended even to rather complicated tasks like simulation of nanoscale domain wall patterns in ferroelectric nanoparticles and nanocomposites. Examples of such challenging tasks are quantitative predictions of polarization vortices in ferroelectric nanoparticles (Pitike et al. 2018; Luk'yanchuk et al. 2019; Wang et al. 2019; Tikhonov et al. 2020), ferroelectric vortex and skyrmionic textures in superlattices (Hlinka and Ondrejkovic 2019; Wang et al. 2019), or hypothetic bulk ferroelectric Bloch skyrmion phases in chiral ferroelectrics (Erb and Hlinka 2019).

The scatter of the parameter values in the tables summarized in this chapter raises a concern about the methodology used to determine the model parameters. Because of the phenomenological nature of the theory, relatively little attention has been usually paid to the exact procedure of finding these numerical coefficients. On the one hand, it is generally considered that different parameter sets or levels of expansion could be suitable for different purposes. Authors of the individual models had often emphasized a different aspect of the model. On the other hand, it would perhaps be desirable to

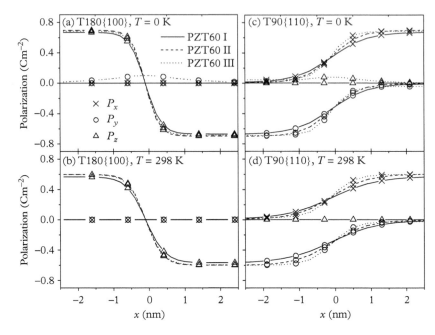

Figure 4.10 *(a,b) T180{100} and (c,d) T90{110} domain wall profiles of in PZT60 at 0 K and 298 K. The profiles were obtained from phase-field simulations with the program Ferrodo (Marton and Hlinka 2006).*

derive these coefficients systematically for a broad selection of perovskite oxides using some unified computational or experimental protocol. Soon it should be possible to derive the models completely from first principles (Marton et al. 2017; Pitike et al. 2019). It seems that systematic designing of GLD models is currently the main bottleneck for their future prospects, and it also represents the biggest challenge of phase-field-modeling-aided materials-design strategies in perovskite ferroelectrics.

Acknowledgments

This work was supported by the Czech Science Foundation (Project EXPRO 19-28594X).

··

REFERENCES

Aguado-Puente P., Junquera J., "Structural and energetic properties of domains in PbTiO₃/SrTiO₃ superlattices from first principles," *Phys. Rev. B* 85, 184105 (2012).
Ahluwalia R., Lookman T., Saxena A., Cao W., "Domain-size dependence of piezoelectric properties of ferroelectrics," *Phys. Rev. B* 72, 014112 (2005).

Bednyakov P. S., Sturman B. I., Sluka T., Tagantsev A. K., Yudin P. V., "Physics and applications of charged domain walls," *npj Comput. Mater.* 4(1), 65 (2018).

Behera R. K., Lee C.-W., Lee D., Morozovska A. N., Sinnott S. B., Asthagiri A., Gopalan V., Phillpot S. R., "Structure and energetics of $180°$ domain walls in $PbTiO_3$ by density functional theory," *J. Phys. Condens. Mat.* 23(17), 175902 (2011).

Bell A. J., "Phenomenologically derived electric field-temperature phase diagrams and piezoelectric coefficients for single crystal barium titanate under fields along different axes," *J. Appl. Phys.* 89(7), 3907–3914 (2001).

Bell A. J., Cross L. E., "Cross: a phenomenological Gibbs function for $BaTiO_3$ giving correct E-field dependence of all ferroelectric phase-changes," *Ferroelectrics* 59(1), 197–203 (1984).

Berlincourt D., Jaffe H., "Elastic and piezoelectric coefficients of single-crystal Barium Titanate," *Phys. Rev.* 111, 143–148 (1958).

Cao W., Cross L. E., "Theory of tetragonal twin structures in ferroelectric perovskites with a first-order phase transition," *Phys. Rev. B* 44, 5–12 (1991).

Catalan G., Seidel J., Ramesh R., Scott J. F., "Domain wall nanoelectronics," *Rev. Mod. Phys.* 84(1), 119 (2012).

Chen L.-Q., "Phase-field method of phase transitions/domain structures in ferroelectric thin films: a review," *J. Am. Ceram. Soc.* 91(6), 1835–1844 (2008).

Devonshire A. F., "XCVI. Theory of barium titanate: Part 1. *The London, Edinburgh, and Dublin Philosophical Magazine and Journal of Science* 40(309), 1040–1063 (1949).

Eliseev E. A., Morozovska A. N., Kalinin S. V., Li Y., Shen J., Glinchuk M. D., Chen L.-Q., Gopalan V., "Surface effect on domain wall width in ferroelectrics," *J. Appl. Phys.* 106(8), 084102 (2009).

Eliseev E. A., Yudin P. V., Kalinin S. V., Setter N., Tagantsev A. K., Morozovska A. N., "Structural phase transitions and electronic phenomena at 180-degree domain walls in rhombohedral $BaTiO_3$," *Phys. Rev. B* 87, 054111 (2013).

Erb K. C., Hlinka J., "Vector, bidirector and bloch skyrmion phases induced by structural crystallographic symmetry breaking," *arXiv preprint arXiv:1910.00075* (2019).

Fousek J., Janovec V., "The orientation of domain walls in twinned ferroelectric crystals," *J. Appl. Phys.* 40(1), 135–142 (1969).

Gu Y., Li M., Morozovska A. N., Wang Y., Eliseev E. A., Gopalan V., Chen L.-Q., "Flexoelectricity and ferroelectric domain wall structures: phase-field modeling and DFT calculations," *Phys. Rev. B* 89, 174111 (2014).

Haun M. J., Furman E., Halemane T. R., Cross L. E., "Thermodynamic theory of the lead zirconate-titanate solid solution system, part IV: tilting of the oxygen octahedra," *Ferroelectrics* 99(1), 55–62 (1989a).

Haun M. J., Furman E., Jang S. J., McKinstry H. A., Cross L. E., "Thermodynamic theory of $PbTiO_3$," *J. Appl. Phys.* 62(8), 3331–3338 (1987).

Haun M. J., Furman E., McKinstry H. A., Cross L. E., "Thermodynamic theory of the lead zirconate-titanate solid solution system, part II: tricritical behavior," *Ferroelectrics* 99(1), 27–44 (1989b).

Haun M. J., Zhuang Z. Q., Furman E., Jang S.-J., Cross L. E., "Electrostrictive properties of the lead zirconate titanate solid-solution system," *J. Am. Ceram. Soc.* 72(7), 1140–1144 (1989c).

Haun M. J., Zhuang Z. Q., Furman E., Jang S. J., Cross L. E., "Thermodynamic theory of the lead zirconate-titanate solid solution system, part III: Curie constant and sixth-order polarization interaction dielectric stiffness coefficients," *Ferroelectrics* 99(1), 45–54 (1989d).

Heitmann A. A., Rossetti G. A., "Thermodynamics of ferroelectric solid solutions with morphotropic phase boundaries," *J. Am. Ceram. Soc.* 97(6), 1661–1685 (2014).

Hlinka J., "Domain walls of BaTiO$_3$ and PbTiO$_3$ within Ginzburg-Landau-Devonshire model," *Ferroelectrics* 375(1), 132–137 (2008).

Hlinka J., Marton P., "Phenomenological model of a 90° domain wall in BaTiO$_3$-type ferroelectrics," *Phys. Rev. B* 74, 104104 (2006).

Hlinka J., Ondrejkovic P., "Skyrmions in ferroelectric materials," in *Solid State Physics*, Volume 70 (Academic Press, pp. 143–169, 2019).

Hlinka J., Ondrejkovic P., Marton P., "The piezoelectric response of nanotwinned BaTiO$_3$," *Nanotechnology* 20(10), 105709 (2009).

Hlinka J., Stepkova V., Marton P., Ondrejkovic P., In *Topological Structures in Ferroic Materials*, Volume 228 (Springer, Switzerland, pp. 161–180, 2016).

Hlinka J., Stepkova V., Marton P., Rychetsky I., Janovec V., Ondrejkovic P., "Phase-field modelling of 180° Bloch walls in rhombohedral BaTiO$_3$," *Phase Transit.* 84(9–10), 738–746 (2011).

Hong Z., Damodaran A. R., Xue F., Hsu S.-L., Britson J., Yadav A. K., Nelson C. T., Wang J., Scott J. F., Martin L. W., Ramesh R., Chen L.-Q., "Stability of polar vortex lattice in ferroelectric superlattices," *Nano Lett.* 17(4), 2246–2252 (2017).

Hu H.-L., Chen L.-Q., "Three-dimensional computer simulation of ferroelectric domain formation," *J. Am. Ceram. Soc.* 81(3), 492–500 (1998).

Iwata M., Ishibashi Y., "Coexistence States near the morphotropic phase boundary: II. The scaling of the free energy in PbZr$_{1-x}$Ti$_x$O$_3$," *Jpn. J. Appl. Phys.* 52(9S1), 09KF07 (2013).

Jin Y. M., Wang Y. U., Khachaturyan A. G., Li J. F., Viehland D., "Conformal miniaturization of domains with low domain-wall energy: monoclinic ferroelectric states near the morphotropic phase boundaries," *Phys. Rev. Lett.* 91, 197601 (2003).

Khakpash N., Khassaf H., Rossetti G. A., Alpay S. P., "Misfit strain phase diagrams of epitaxial PMN–PT films," *Appl. Phys. Lett.* 106(8), 082905 (2015).

Kim I., Jang K., Kim I., Li L., "Higher-order Landau phenomenological models for perovskite crystals based on the theory of singularities: a new phenomenology of BaTiO$_3$," *Phase Transit.* 91(3), 239–253 (2018a).

Kim I.-H., Kim I.-H., Jang K.-O., Sin K.-R., and Kim C.-J., "A new higher-order Landau–Ginzburg–Devonshire theory for KNbO$_3$ crystal," *Phase Transit.* 91(12), 1189–1205 (2018b).

Li F., Zhang S., Xu Z., Chen L.-Q., "The contributions of polar nanoregions to the dielectric and piezoelectric responses in domain-engineered relaxor-PbTiO$_3$ crystals," *Adv. Funct. Mater.* 27(18), 1700310 (2017).

Li L., Kim I., Jang K., Ri K., Cha J., "Higher-order approximation of the phase transition in the Pb-based perovskite oxide solid-solutions," *J. Appl. Phys.* 114(3), 034104 (2013).

Li Y. L., Chen L. Q., "Temperature-strain phase diagram for BaTiO$_3$ thin films," *Appl. Phys. Lett.* 88(7), 072905 (2006).

Li Y. L., Cross L. E., Chen L. Q., "A phenomenological thermodynamic potential for BaTiO$_3$ single crystals," *J. Appl. Phys.* 98(6), 064101 (2005).

Li Y. L., Hu S. Y., Liu Z. K., Chen L. Q., "Effect of substrate constraint on the stability and evolution of ferroelectric domain structures in thin films," *Acta Mater.* 50(2), 395–411 (2002).

Liang L., Li Y. L., Chen L.-Q., Hu S. Y., Lu G.-H., "A thermodynamic free energy function for potassium niobate," *Appl. Phys. Lett.* 94(7), 072904 (2009).

Lines M. E., Glass A. M., *Principles and Applications of Ferroelectrics and Related Materials* (Clarendon Press, Oxford, 1977).

Luk'yanchuk I., Tikhonov Y., Razumnaya A., Vinokur V. M., "Hopfions emerge in ferroelectrics," *arXiv preprint arXiv:1907.03866* (2019).

Marton P., Hlinka J., "Simulation of domain patterns in BaTiO$_3$," *Phase Transit.* 79(6–7), 467–483 (2006).

Marton P., Klíč A., Paściak M., Hlinka J., "First-principles-based Landau-Devonshire potential for BiFeO$_3$," *Phys. Rev. B* 96, 174110 (2017).

Marton P., Rychetsky I., Hlinka J., "Erratum: Domain walls of ferroelectric BaTiO$_3$ within the Ginzburg-Landau-Devonshire phenomenological model," *Phys. Rev. B* 81, 144125 (2010).

Meier D. "Functional domain walls in multiferroics," *J. Phys.: Condens. Mat.* 27(46), 463003 (2015).

Ondrejkovic P., Marton P., Guennou M., Setter N., Hlinka J., "Piezoelectric properties of twinned ferroelectric perovskites with head-to-head and tail-to-tail domain walls," *Phys. Rev. B* 88, 024114 (2013).

Pertsev N. A., Kukhar V. G., Kohlstedt H., Waser R., "Phase diagrams and physical properties of single-domain epitaxial Pb($Zr_{1-x}Ti_x$)O$_3$ thin films," *Phys. Rev. B* 67, 054107 (2003).

Pertsev N. A., Zembilgotov A. G., Tagantsev A. K., "Effect of mechanical boundary conditions on phase diagrams of epitaxial ferroelectric thin films," *Phys. Rev. Lett.* 80, 1988–1991 (1998).

Pitike K. C., Mangeri J., Whitelock H., Patel T., Dyer P., Alpay S. P., Nakhmanson S. "Metastable vortex-like polarization textures in ferroelectric nanoparticles of different shapes and sizes," *J. Appl. Phys.* 124, 064104 (2018).

Pitike K. C., Khakpash N., Mangeri J., Rossetti G. A., Nakhmanson S. M., "Landau–Devonshire thermodynamic potentials for displacive perovskite ferroelectrics from first principles." *J. Mater. Sci.* 54, 8381–8400 (2019).

Rabe K. M., Ahn C. H., Triscone J.-M., *Physics of Ferroelectrics: A Modern Perspective*, Volume 105 (Springer Science + Business Media, Berlin Heidelberg, 2007).

Semenovskaya S., Khachaturyan A. G., "Development of ferroelectric mixed states in a random field of static defects," *J. Appl. Phys.* 83(10), 5125–5136 (1998).

Sergienko I. A., Gufan Yu. M., Urazhdin S., "Phenomenological theory of phase transitions in highly piezoelectric perovskites," *Phys. Rev. B* 65, 144104 (2002).

Shin Y., Grinberg I., Chen I., Rappe A. M., "Nucleation and growth mechanism of ferroelectric domain-wall motion," *Nature* 449, 881–884 (2007).

Sluka T., Bednyakov P., Yudin P., Crassous A., Tagantsev A., In *Topological Structures in Ferroic Materials*, Volume 228 (Springer, Switzerland, pp. 103–138, 2016).

Sluka T., Tagantsev A. K., Damjanovic D., Gureev M., Setter N., "Enhanced electromechanical response of ferroelectrics due to charged domain walls," *Nat. Commun.* 3, 748 (2012).

Slutsker J., Artemev A., Roytburd A., "Phase-field modeling of domain structure of confined nanoferroelectrics," *Phys. Rev. Lett.* 100, 087602 (2008).

Stepkova V., Hlinka J., "On the possible internal structure of the ferroelectric Ising lines in BaTiO$_3$," *Phase Transit.* 90(1), 11–16 (2017).

Stepkova V., Hlinka J., "Pinning of a ferroelectric Bloch wall at a paraelectric layer," *Beilstein J. Nanotechnol.* 9, 2356–2360 (2018).

Stepkova V., Marton P., Hlinka J., "Stress-induced phase transition in ferroelectric domain walls of BaTiO$_3$," *J. Phys. Condens. Mat.* 24(21), 212201 (2012).

Stepkova V., Marton P., Hlinka J., "Ising lines: natural topological defects within ferroelectric Bloch walls," *Phys. Rev. B* 92, 094106 (2015).

Stepkova V., Marton P., Setter N., Hlinka J., "Closed-circuit domain quadruplets in BaTiO$_3$ nanorods embedded in a SrTiO$_3$ film," *Phys. Rev. B* 89, 060101(R) (2014).

Su Y., Landis C. M., "Interaction of domain walls with defects in ferroelectric materials," *J. Mech. Phys. Solids* 55(2), 280–305 (2007).

Tagantsev A. K., Cross L. E., Fousek J., *Domains in Ferroic Crystals and Thin Films* (Springer Science + Business Media, LLC, New York, 2010).

Tagantsev A. K., Vaideeswaran K., Vakhrushev S. B., Filimonov A. V., Burkovsky R. G., Shaganov A., Andronikova D., Rudskoy A. I., Baron A. Q. R., Uchiyama H. et al., "The origin of antiferroelectricity in PbZrO₃," *Nat. Commun.* 4, 2229 (2013).

Taherinejad M., Vanderbilt D., Marton P., Stepkova V., Hlinka J., "Bloch-type domain walls in rhombohedral BaTiO₃," *Phys. Rev. B* 86, 155138 (2012).

Tikhonov Y., Kondovych S., Mangeri J., Pavlenko M., Baudry L., Sené A., Galda A., Nakhmanson S., Heinonen O., Razumnaya A., Luk'yanchuk I., "Controllable skyrmion chirality in ferroelectrics," *arXiv preprint arXiv:2001.01790* (2020).

Vanderbilt D., Cohen M. H., "Monoclinic and triclinic phases in higher-order Devonshire theory," *Phys. Rev. B* 63, 094108 (2001).

Wang J., Li Y., Chen L.-Q., Zhang T.-Y., "The effect of mechanical strains on the ferroelectric and dielectric properties of a model single crystal – phase field simulation," *Acta Mater.* 53(8), 2495–2507 (2005).

Wang J. J., Wang B., Chen L. Q., "Understanding, predicting, and designing ferroelectric domain structures and switching guided by the phase-field method," *Annu. Rev. Mater. Res.* 49, 127–152 (2019).

Wojdeł J. C., Íñiguez J., "Ferroelectric transitions at ferroelectric domain walls found from first principles," *Phys. Rev. Lett.* 112, 247603 (2014).

Yamada T., "Electromechanical properties of oxygen-octahedra ferroelectric crystals," *J. Appl. Phys.* 43(2), 328–338 (1972).

Yudin P. V., Tagantsev A. K., Eliseev E. A., Morozovska A. N., Setter N., "Bichiral structure of ferroelectric domain walls driven by flexoelectricity," *Phys. Rev. B* 86, 134102 (2012).

Yudin P. V., Tagantsev A. K., Setter N., "Bistability of ferroelectric domain walls: morphotropic boundary and strain effects," *Phys. Rev. B* 88, 024102 (2013).

Zhirnov V. A., "On the theory of the domain walls in ferroelectrics," *Zh. Eksp. Teor. Fiz.* 35, 1175–1180 (1958).

5

Introduction to Domain Boundary Engineering

E. K. H. Salje[1,2] and G. Lu[2]

[1]*Department of Earth Sciences, University of Cambridge, United Kingdom*
[2]*Laboratory for Mechanical Behavior of Materials, Xi'an Jiaotong University, Xi'an 710049, China*

Research on functional domain boundaries is first introduced. Ever since the discovery of superconducting twin boundaries in the 1990s, highly conducting, polar, photovoltaic, magnetic, etc. domain boundaries have been discovered while the same bulk material displays none of these properties. Domain boundaries constitute planar templates for device applications with thicknesses of approximately 1 nm. Domains within domains are then the next step in miniaturization with Bloch lines within domain walls and Bloch points between Bloch lines.

In the overwhelming majority of cases, the geometrical template for the functional domain boundaries stems from the ferroelastic domain structure, while anti-phase boundaries are equally potential template providers. Complex structures are a particular case because they add vortices and skyrmions to the template topology. Correlations between such sub-structures maintain features like polarity and piezoelectricity in randomized samples where structural averages would not allow macroscopic polar effects.

The dynamics of the change of functionality is often much faster than the speed with which twin boundaries move. The novel information carrier is the kink inside twin walls, which moves with supersonic speed.

5.1 Introduction

It is generally understood since the 1990s that interfaces like twin boundaries or anti-phase boundaries can be structurally, physically, and chemically very different from the adjacent bulk material. They do not necessarily represent a simple "interpolation" between two domains with slightly different orientations but are more akin to a thin sheet of another phase, which is sandwiched between the adjacent domains. Such functional domain boundaries have emerging properties that are of great interest for many applications, and a new field of "domain boundary engineering" was formulated in 2010 to discover, improve, and apply such boundaries (Salje and Zhang 2009; Salje

E. K. H. Salje and G. Lu, *Introduction to Domain Boundary Engineering* In: *Domain Walls: From Fundamental Properties to Nanotechnology Concepts.* Edited by: Dennis Meier, Jan Seidel, Marty Gregg, and Ramamoorthy Ramesh, Oxford University Press (2020).
© E. K. H. Salje and G. Lu.
DOI: 10.1093/oso/9780198862499.003.0005

2010). The main functional wall properties are their conductivity, polarity, magnetism, and those emerging properties that are related to complex wall configurations (like tweed patterns). Historically, the first breakthrough was achieved for metal/insulator transitions in WO_3 when twin boundaries were shown to be superconducting (Aird and Salje 1998; Aird et al. 1998; Aird and Salje 2000; Kim et al. 2010; Hamdi et al. 2016) while the matrix is an insulator. In addition, domain wall intersections usually form needle domains next to a perpendicular wall (Salje and Ishibashi 1996) so that the resulting complex wall patterns automatically contain large arrays of needles in comb configurations (Salje et al. 2014a; Salje and Carpenter 2015; Wang et al. 2018). They naturally constitute an array of high-density Josephson junctions. Such Josephson junctions are at the heart of magnetic brain scanners. They are also used in astrophysics and paleomagnetism. The superconducting transition temperature of WO_3-based interfaces is so low (~3K) that devices need to operate at rather low temperatures until better compounds are found. Nevertheless, the potentially great economic gain of such devices stimulated much research on producing workable Josephson devices in thin films (Salje and Parlinski 1991; Yun et al. 2015). Today, device materials based on thin films exist, but Josephson devices have not been systematically exploited.

Enhanced electronic conductivity and photovoltaic currents in domain walls are commonly reported and constitute a wide field for future research (Salje 1992; Salje et al. 1993, 2009, 2010, 2011; Catalan et al. 2012) mirrored by research on magnetism inside domain walls (Daraktchiev et al. 2010). Interestingly, strong ionic conductivity along domain boundaries is commonly reported in natural minerals. The majority of minerals on our planet are ferroelastic, so that ferroelastic microstructures are very common and persist over geological times (Salje 1987, 2012). Such microstructures do not represent thermodynamic equilibrium states, but their lifetime can exceed 10^8 years. This observation proves that twins, tweed, arrays of anti-phase boundaries, and other microstructures are indeed metastable with extremely long lifetimes (Salje 2015b). Phenomena such as ionic transport—which usually occur over a much shorter time scale—are hence ideally studied on such "old" materials, and many of the initial discoveries of defect migration onto twin boundaries were first made in minerals (Salje 1987; Hayward et al. 1996; Hayward and Salje 2000). Equally, martensites generate characteristic domain wall patterns, which are candidates for domain boundary applications (Ball et al. 1992; Gallardo et al. 2010; Kustov et al. 2017a), including materials with re-entrant spin glass transitions (Kustov et al. 2017b).

Besides conductivity, most other research focuses on the ferroic properties of domain walls. Ferroic materials typically result from modifications of a high-symmetry structure (prototype). They are classified according to their primary order parameters, which include strain ε (ferroelastic), polarization P (ferroelectric), and magnetization M (ferromagnetic), where the term "ferro" designates the uniform alignment of the spontaneous moments in neighboring unit cells. Ferri- and anti-ferro phases can also exist when the order parameter is locally rotated against a crystallographic axis, or when the relevant wavevector (or wavevectors) associated with the structural instability occur(s) at special points at the surface of the Brillouin zone. Incommensurable phase transitions require a complex order parameter with a repetition unit that is not commensurate with the underlying crystal structure. In both of these cases, the translational invariance of the order parameter (incommensurate or commensurate) is preserved throughout the crystal (Salje 2015a; Pöttker and Salje 2016). Additionally, a novel class of ferroic materials has been garnering increasing attention; these so-called multiferroics feature more than one spontaneous primary order parameter, where the order parameters can be coupled or remain independent depending on the desired application (Marais et al. 1991; Bratkovsky et al. 1994; Salje et al. 2016a). This includes a pinning process of mixed origin such as ferroelectric walls pinning magnetic walls, and so forth (Fontcuberta et al. 2015). There are several classic textbooks on ferroics (Lines and Glass 1979; Khachaturyan 1983; Salje 1993; Jiles 1998; Scott 2000; Bhattacharya 2004; Sidorkin 2006; Rabe et al. 2007; Cullity and Graham 2011; Pavarini et al. 2013) and numerous reviews on multiferroics (Houchmandzadeh et al. 1991; Přívratská and Janovec 1999; Calleja et al. 2003; Fiebig 2005; Spaldin and Fiebig 2005; Eerenstein et al. 2006; Kadomtseva et al. 2006; Ramesh and Spaldin 2007; Goncalves-Ferreira et al. 2008, 2010; Conti et al. 2011; Guennou et al. 2011; Salje 2012; Van Aert et al. 2012; Bhatnagar et al. 2013; Ondrejkovic et al. 2013). In many cases, ferroic emerging properties can be found in domain boundaries, and in cases where they do not exist in the bulk, they constitute useful materials for domain boundary engineering.

5.2   The Concept

The translational invariance of a ferroic order parameter is often broken by cooling during phase transformation without the application of the corresponding conjugate ordering field (electric field E for P, magnetic field H for M, and stress σ for ε). Specifically, this can occur as a result of the nucleation and growth of low-symmetry distorted structures. The resulting regions, which are characterized by a uniform order parameter, are referred to as domains. In a single crystal, the domains within a poly-domain structure are oriented along its symmetry-equivalent directions.

For example, in a cube, the domains will orient along the cubic {111} and {001} directions in rhombohedral and tetragonal structures, respectively. Without the application of the conjugate field, the distribution of domains may be nearly equal between equivalent variants, which would result in a zero net value for the order parameter integrated over the crystal volume. Neighboring ferroelastic domains, which are separated by coherent twin boundaries, are related to each other via symmetry operations that are lost from the parent prototype phase during a low-symmetry phase transformation. While both theoretical reports and experimental observations have long described ferroic domain sizes in typical ranges over many decades, domains smaller than 500 nm are reported less frequently. Such is the case of classical domains that are commonly found in ferroics and has been discussed in numerous books and review papers cited above.

Domain boundaries in ferroic materials have two functional degrees of freedom. First, domain boundaries generate their intrinsic emerging property (conductivity, polarity, magnetism, etc.). Second, they are able to control the response through their motion. One such example is the polar domain boundary (toy model of an anti-phase boundary)

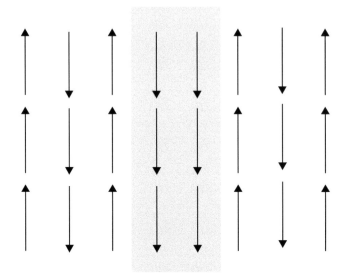

Figure 5.1 *Polar domain boundary in an anti-ferroelectric matrix (Viehland and Salje 2014).*

shown in Figure 5.1. An anti-ferroelectric array of dipoles in the bulk will usually contain polar domain boundaries that change the phase of the up–down pattern of the bulk. As indicated in Figure 5.1, which depicts the domain boundary region in gray, it is geometrically obvious that this domain boundary is polar with a dipole moment in the down direction. We expect, therefore, that anti-ferroelectric materials will contain polar domain boundaries that may be switchable under appropriate electric fields, thereby giving rise to a ferroelectric response. Furthermore, the location of the gray region can be changed by fields or stress, which represents the second, switchable property of the domain wall.

Another typical example involving polar domain boundaries involves ferroelastic twinning of the bulk, which is related to shear strain generated by a structural phase transition. This phenomenon is shown in a simple model in Figure 5.2, where the shaded areas represent the region of structural distortion within the twin boundary. Note that all interatomic distances between nearest neighbors are exactly the same in the twin boundary and the bulk. They are different only for the next nearest neighbor distances, as well as larger distances spanning the twin boundary. Consequently, in most instances, one might expect only very small structural changes as a weak perturbation of the strain profile of the two ferroelastic domains with spontaneous strains $+e$ and $-e$. In the case of a second-order phase transition, the trajectory is $e(x) \sim \tanh(x/w)$, where the locus of the twin boundary is $x = 0$ and the thickness of the twin boundary is $2w$ (Salje 1993, 2012; Bhatnagar et al. 2013). This scenario can change dramatically if the material is thermodynamically close to several structural phase transitions, thereby enabling it to flip easily from one structural configuration to another. The wall position can easily be changed by stress.

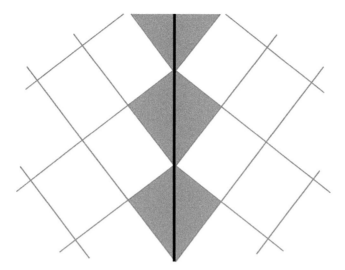

Figure 5.2 *A simple model of a twin boundary where the exchange between short and long repetition lengths leads to a lattice distortion inside the shaded area (Viehland and Salje 2014).*

In particular, when one of the incipient phases is polar, it is likely that the boundaries and interfaces will tend to be polar as well. The following four physical conditions typically generate polarity:

(1) Polarity and octahedral rotation repel each other in most perovskite structures. In such cases, the octahedral rotations often vanish inside twin walls, which generates polarity within the wall while remaining suppressed in the bulk (Houchmandzadeh et al. 1991; Conti et al. 2011). In addition, strong strain gradients promote polarity via the flexoelectric effect (Tagantsev 1986; Zubko et al. 2007; Majdoub et al. 2008; Eliseev et al. 2009; Zubko et al. 2013; Yudin and Tagantsev 2013). The fundamental difference between the biquadratic order parameter coupling and gradient coupling in flexoelectricity (Stengel 2016) is the symmetry of polar reversal. Biquadratic coupling has two energy minima for orientations of the polarity that are equal in opposite directions. Flexoelectricity does not allow any polarization reversal. This reversal is critical for many new ideas such as the use of Bloch lines as memory devices (Salje and Scott 2014). Simulations and observations seem to confirm the possibility of the desired reversibility and hence some aspect of biquadratic coupling (Salje et al. 2013a; Zykova-Timan and Salje 2014; Frenkel et al. 2017; Casals et al. 2018).

(2) The distances between lattice planes across a twin boundary are usually larger by approximately 1% compared with the lattice repetition unit in the bulk. This slight enlargement facilitates the formation of new structural elements in the wall—and with it polarity—because there is more space in this portion of the material compared with the bulk. This structural feature also tends to soften the elasticity of the wall, meaning that lattice relaxations are easier to achieve inside twin boundaries than in the bulk. A typical

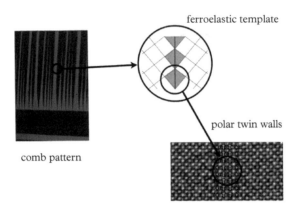

Figure 5.3 *Example of a complex domain structure with emerging properties. Polarity emerges from inside twin boundary in CaTiO$_3$ (Van Aert et al. 2012). They form displacements of the Ti atomic columns from the center of the four neighboring Ca atomic columns, indicated by arrows and displacements of the Ti atomic columns in the x- and y-directions averaged along and in mirror operation with respect to the twin wall. These displacements occur in twin walls, which, in turn, form comb patterns of needle domains and thereby form a hierarchical structure (Salje et al. 2014a; Salje and Carpenter 2015). If such comb configurations could be produced in WO$_3$, the emerging superconducting properties will lead to high-density arrays of Josephson junctions.*

example of this phenomenon can be seen in the structure of CaTiO$_3$ (Salje et al. 2015) (Figure 5.3).

(3) Ionic conductivity is significantly enhanced in the twin boundaries for the same reason. It is possible, therefore, to inject defects and dopants into the twin boundaries while the bulk remains largely undoped. This effect generates novel chemical compounds within twin boundaries, which can then be polar. Such targeted chemical doping is well known for oxygen vacancies and ions such as Ca, Li, and Na, but can also be expected for transition metals (Calleja et al. 2003; Goncalves-Ferreira et al. 2008, 2010). It should be noted, however, that such chemical enrichment of twin boundaries can occur inadvertently and is also often unavoidable in complex oxide materials. "Clean" domain boundaries are an exception, not the rule.

(4) Twin walls can also be electrically charged, as in the case of BaTiO$_3$. If the polar vectors in ferroelastic twin boundaries are arranged in a head-to-head or tail-to-tail configuration—rather than in a head-to-tail arrangement, which would correspond to a lower energy state—the outcome is a charged twin boundary, leading to high conductivity, which can modify (short-circuit) the polar properties of the boundaries. Moreover, charged twins can interact strongly among each other, which leads to complex patterns, which are also charged (Přívratská and Janovec 1999; Ondrejkovic et al. 2013). Charged walls also exist in 180° walls because such walls may contain kinks and meanders even in macroscopically straight walls where meandering leads to significant local charging in the case of LiNbO$_3$ (Schröder et al. 2012; Gonnissen et al. 2016). The functionality observed for twin boundaries mirrors the effects seen in the

interfaces between domains of different materials with different cation valence states (e.g. $SrTiO_3$ and $LaAO_3$). In such cases, the interface is invariably charged and mobile carriers with differing effective masses can be injected throughout the interface (Ohtomo and Hwang 2004; Siemons et al. 2007). Similarly, charged interfaces have also been described in instances when the interfaces between two phases of the same material (~phase boundaries) generate polarity (Wang et al. 2013). These examples confirm that numerous types of functionality can exist in domain boundaries, despite the fact that the same functionalities do not occur in the bulk. Other functionalities may include ferromagnetism, electronic transport, chemical transport, and chemical transformations to novel compounds, which, again, can exist within the domain boundary but are absent in the bulk (Bibes et al. 2002; Brinkman et al. 2007; Nabi et al. 2010; Carpenter et al. 2012; Oravova et al. 2013).

5.3   The Ferroelastic Template

Functionalities are in most cases anchored onto the ferroelastic template. When the template changes, the functionalities change with them. The two major elements are hence the wall property and the position of the wall. External fields can influence both entities. For example, polar domain walls can be moved under some circumstances by electric fields while the electric field also changes the local polarization in length and orientation. In other cases, elastic stress changes the ferroelastic template by moving twin boundaries and hence carries the polarization inside the twin boundaries with it. Changing the ferroelastic template falls into the regime of pattern formation (see Chapter 8), and many particular templates have been realized. The most common are single domain walls (Aird and Salje 1998; Kim et al. 2010; Salje 2012; Schröder et al. 2012; Van Aert et al. 2012; Gonnissen et al. 2016), corners and intersections (Kim et al. 2010; Salje et al. 2016b; Frenkel et al. 2017), needles (Salje and Ishibashi 1996), tweed (Wang et al. 2018) and secondary structures like Bloch lines (Salje and Scott 2014; Salje et al. 2014a; Zykova-Timan and Salje 2014; Salje and Carpenter 2015) (see Chapter 4). Highly complex templates will be discussed in the next chapter. Should a ferroelastic template approach be realized in industrial applications, the potential benefits are tremendous. Indeed, the list of potential candidate materials is very long since virtually all twin boundaries or anti-phase boundaries can be "functionalized" (see Chapter 15). This approach paves the way for nearly limitless opportunities to create novel materials that do not even need to be stable as bulk materials, but need only be stable within the confines of the twin boundary. Despite their application potential, collective research on functional-interface materials is still relatively young, since scholars have yet to fully understand several of their key properties (Catalan et al. 2012), in particular with respect to the mobility of the domain boundaries as twins are often highly mobile in materials like $SrTiO_3$ (Kityk et al. 2000; Harrison et al. 2004; Salje et al. 2013a).

An intrinsic advantage for applications (but still a bane for experimentalists) is that twin boundaries are very thin and the total number of atoms within them is usually less than 1 ppm (part per million) of the total number of atoms in a sample. Therefore, assessing boundary properties can be extremely difficult because any measurable macroscopic effects are generated by a tiny minority of atoms and, hence, are very small. Nevertheless, recent studies using computer simulation have shown that this percentage can be improved by more advanced production methods such as cold shearing (Salje et al. 2011; Ding et al. 2012; Salje et al. 2012; Salje et al. 2013b; Zhao et al. 2013).

5.4 Complexity

The ferroelastic template may strongly influence the performance of the emerging properties beyond the issue of pattern formation. Imagine two parallel flexoelectric domain walls. They will each possess equal polarizations but opposite orientation under flexoelectric coupling. This means that a stripe array of domain walls will have no macroscopic polarization despite the fact that each individual wall is polar. For most applications, this overall cancellation is irrelevant if the information is contained in individual walls, but it raises a new question: does functionality emerge from complex domain boundary patterns? This functionality should be related to the local functionality, but these two quantities are not expected to equal each other. With this question in mind, we immediately understand that simple domain structures, such as stripe arrays of domain boundaries, may annihilate local functionalities. This is not true for complex structures as first argued in Salje et al. (2016b). These authors developed a simple atomic model that contains as exclusive ingredient the flexoelectric effect to generate polarization. This is the extreme case, where polarization cannot emerge simply by inverting the polarization in a domain boundary by the application of an electric field.

Complex patterns are generated by "cold shearing" techniques (Ding et al. 2012). In single stress geometry, only horizontal twins are generated, but no vertical twins. The high twin density leads to kinks inside the twin walls, as shown in Figure 5.4(a), where only a small region (4 × 3 nm) out of a larger sample (10 × 10 nm) is displayed. The kinks generate additional features in the polarization patterns (Figure 5.4(b)) that are not present in stripe patterns. For example, the polarization locally acquires non-

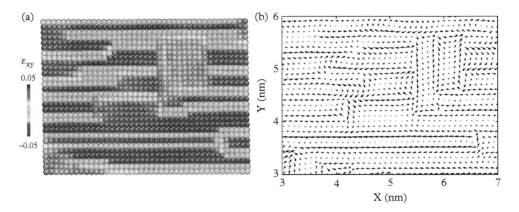

Figure 5.4 *Domain pattern with two orthogonal shear deformations. (a) Atomic image in view of ε_{xy} (b) Dipole configuration. Dipole displacements are amplified by a factor of 25.*

vanishing vertical components, which are forbidden by symmetry in the previously considered case of a straight domain boundary. Moreover, the interaction between neighboring walls gives rise to characteristic winding patterns reminiscent of a vortex structure. These new features could, in principle, be determined by long-range electrostatic interactions. Previously, polarization vortices near domain walls have been shown to form in ferroelectric closure domains under electric fields (Salje et al. 2015). The new features are also induced via the flexoelectric effect, the only difference being that the increased geometrical complexity in the deformation field now activates components of the strain-gradient tensor that were formerly inactive. Flexo-couplings associated with gradients of the diagonal components of the strain tensor can no longer be neglected, unlike the flat twin wall case where only the shear strain (ε_{xy}) is present. The shear strain now also displays strong gradients along the horizontal direction, which may contribute to the normal (vertical) components of P. The predominant new features in complex ferroelastic twin patterns are junctions and right angle corners between twin boundaries in different directions (Salje and Ishibashi 1996; Salje et al. 2012, 2016b). The junction density has previously been used as a measure for the complexity of a domain pattern (Ding et al. 2012). The resulting junction densities in these patterns are almost as high as in conventional tweed structures (Salje 2015b; Wang et al. 2018). Thus, flexoelectricity alone, together with the correlation term (which is responsible for the spatially nonlocal relationship between elastic and polar fields), is responsible for the complex patterns, which points to a remarkable conclusion: *polarization vortices can emerge from a complex ferroelastic structure via linear flexoelectricity alone.*

The macroscopic complex pattern now contains a macroscopic polarization because the local compensation mechanism in simple patterns no longer applies (Přívratská and Janovec 1999). This idea bears some similarities to the Garten–Trolier-McKinstry model (Garten and Trolier-McKinstry 2015), where some inherent polar instability is activated by the flexoelectric effect. The strain field plays the role of the primary order parameter, and flexoelectricity translates the macroscopic asymmetry in the ferroelastic structure into a net polarization. Levin et al. (2014) showed the off-centering of Ti in paraelectric $BaTiO_3$ as a typical case of spontaneous local polarization. If these off-centerings are correlated, they would exemplify flexoelectrically induced polarity. Equivalently, dipolar distortions are often induced by the Jahn–Teller effect in noncentrosymmetric systems, but they may also occur in both centrosymmetric and noncentrosymmetric systems as a result of the pseudo-Jahn–Teller effect (Carpenter 2015), which has similarities to our simulated case. The most direct comparison is with piezoresponse force microscopy (PFM) measurements of heterogeneously strained $LaAlO_3$, which is a centrosymmetric material with point symmetry -3m. PFM showed piezoelectric patches in the resulting tweed structure together with weak phonon shifts, which proves that the resulting strain pattern is indeed polar (Salje et al. 2016a).

We can now ask which other field-dependent parameters lead to emerging properties that are simply based on the complexity of the sample. In the following (Damjanovic and Demartin 1997; Taylor and Damjanovic 2000; Damjanovic 2005; Bassiri-Gharb et al. 2007), we focus on the piezoelectric response of a material where the ferroelastic domain boundaries are polar but where the random arrangement of the domain walls could suggest that the macroscopic piezoelectric effect self-compensates to zero. Previous work on fine quartz grains with random orientation already showed the opposite effect, namely, that piezoelectricity is reduced but remains strong for almost all samples with grain sizes larger than 10 nm (Aufort et al. 2015). This indicates that the randomization of the piezoelectric grains and domains is hardly ever achieved, and macroscopic strains can be observed under applied electric fields. Recent computer simulation experiments of complex systems in Salje et al. (2016b) confirm this idea. In Figure 5.5, field-induced shape changes are shown for a two-dimensional array of 3200 atoms. For a simple twin structure, the macroscopic inversion centrosymmetry is preserved, and no piezoelectric effect is observed. The field-induced strains are reflections of the electrostrictive effects

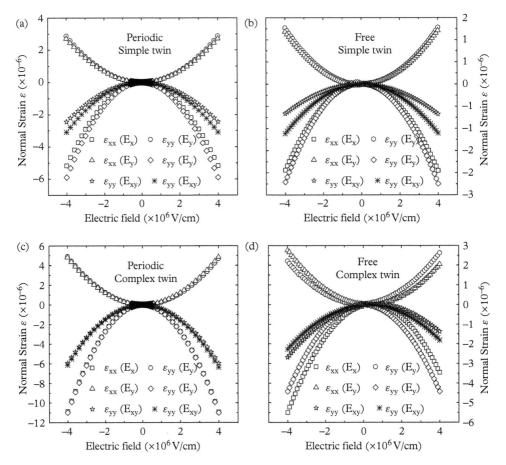

Figure 5.5 *Variation of strains ε_{xx} and ε_{bb} of a simple sandwich twin structure under electric fields in directions [10] (E_x), [11] (E_{xy}), and [01] (E_y) with (a) periodic boundary and (b) free boundary conditions. The simple twin shows only an electrostrictive effect. The field-induced strain changes in complex twin structure with periodic and free boundary conditions are shown in (c) and (d), respectively, and show piezoelectricity. The piezoelectric coefficients are 10^{-4} pm/V for the periodic boundary condition and 10^{-3} pm/V for the free boundary condition (Lu et al. 2019).*

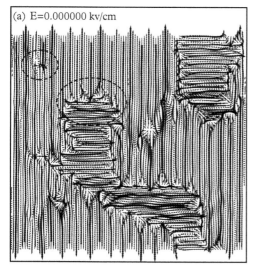
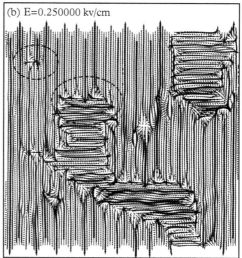

(a) E=0.000000 kv/cm

(b) E=0.250000 kv/cm

Figure 5.6 *Dipole configurations corresponding to (a) zero field and (b) small electric field (0.25 kV/cm).*

under both (a) periodic and (b) free boundary conditions. The field response changes dramatically for complex twin structures. Starting from a single domain state, a single domain crumbled into a complex structure with many fine domains (twin structures) owing to external stress. The intrinsic defects (junctions and intersections, etc.) generate polarity via the flexoelectric effect and can be system size dependent. The macroscopic strain generated by the piezoelectric effect is reduced by randomization in larger systems. Junctions, kinks, and intersections residing inside the complex system are the key factors for the presence of the piezoelectric effect while the simple system with stripe twin structures shows only an electrostriction effect, as indicated in Figures 5.5(c)–(d). Due to the presence of surface relaxation, the piezoelectric coefficient in complex twin structure under the free boundary condition (10^{-3} pm/V) are ten times larger than that under periodic boundary condition (10^{-4} pm/V).

Another potential application of such complex structures is the local polarization switching under an applied conjugate electric field. Some local switching in these complex twins (20,000 atoms), as indicated in Figure 5.6, was confirmed by simulation at a high temperature (10 K) with large polarization instabilities. Local switching was observed near the junctions and kinks where the strain states are more complex.

The coercive fields for these local switchings are generally below 0.4 kV/cm, which is several times smaller than for the normal ferroelectric materials (BaTiO$_3$, ~5 kV/cm). The emerging application for this local switching resides in the instabilities of non-centrosymmetric patches, which were generated by intrinsic defects (kinks and junctions, etc.) (Salje et al. 2016b). The polarization change in local patches (indicated by the black circles in Figure 5.6) is large, while the average value for the entire complex structure partially cancels out due to the complexity of the topology.

5.5   Functional Domain Boundaries and Crackling Noise

The need to work at much higher frequencies for telecommunications, cell phone applications, etc. requires very fast switching times. Ferroelectrics notoriously underperform in this field because the domain movements under electric fields are usually very slow. In domain boundary engineering, this problem can potentially be overcome. The fastest domain movement was reported for kinks in ferroelastic domain walls (Salje et al. 2017). These authors argued that devices operating at gigahertz frequencies can be based on ferroelectric kink domains moving at supersonic speed. They showed by computer simulation that kinks located inside ferroelastic twin boundaries are extremely mobile. Strong forcing generates velocities well above the speed of sound. Kinks are accelerated from $v = 0$ continuously with Döring masses in the order of skyrmion masses under constant strain rates. Moving kinks emit phonons at all velocities, and the emission cones coincide with the Mach cones at supersonic speed.

Kinks form avalanches with the emission of secondary kinks via a mother–daughter nucleation mechanism and may be observable in acoustic emission experiments. Supersonic kinks define a new type of material; while mobile domains are the key for ferroelastic and ferroelectric device applications at low frequencies, it is expected that fast kink movements replace such domain movements for materials applications at high frequencies. Experimental work (Li et al. 2004) has not yet progressed to such high speeds but has certainly bordered on the speed of sound.

Figure 5.7 shows a typical scenario where the initial kink (a) emits the strain field in (b). The moving kink is shown in the following six snapshots. The emitted strain waves are seen as clouds, which are left behind by the ultrasonic movements.

Most domain movements occur over a large scale of time 98 s where such kinks seem to constitute the upper cutoff for useful frequencies. Very slow movements are well studied and may proceed with movements of 1 mm over 24 h or even slower. However, experimental studies of the movement of needle domains revealed extensive scale invariance (Salje et al. 2014b; Puchberger et al. 2017). In Figure 5.8, the progression of a needle domain is shown together with the power law statistics of the emitted jerks related to the emission of kinetic energy.

Devices operating at gigahertz frequencies can be based on ferroelectric kink domains moving at supersonic speed. The kinks are located inside ferroelastic twin boundaries and are extremely mobile. Computer simulation shows that strong forcing generates velocities well above the speed of sound. Kinks are accelerated from v = 0 continuously with Döring masses in the order of skyrmion masses under constant strain rates. Moving kinks emit phonons at all velocities, and the emission cones coincide with the Mach cones at supersonic speed. Kinks form avalanches with the emission of secondary kinks via a mother–daughter nucleation mechanism and may be observable in acoustic emission experiments. Supersonic kinks define a new type of material, while mobile domains are the key for ferroelastic and ferroelectric device applications at low frequencies; it is expected that fast kink movements replace such domain movements for materials applications at high frequencies.

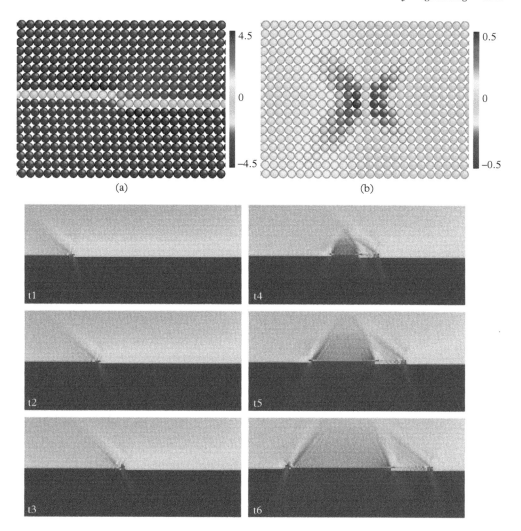

Figure 5.7 *A static kink shown by (a) the vertical shear angle and (b) the horizontal shear angle. The strain field in (b) is similar to those of shear dislocations. The six snapshots show mother–daughter mechanism for kinks with atoms colored by $|\theta v| -4.0 + \theta h$ and log10 (kinetic energy), respectively. The mother kink travels with supersonic speed to the right. At t3 two daughter kinks nucleate and travel subsonically in opposite directions (t4–t6). The mother kink shows a supersonic cone while the daughters travel at a lower speed (Salje et al. 2017).*

These results show that domain switching and needle movement occurs with a power law distribution of energy transfer of the jump probability $P(E)dE \sim EdE$. This power law means that the energy scale is scale invariant and that all domain boundary movements in random fields and in the jamming scenario are in fact independent of energy: the same physical principles are dependences that occur on all energy scales between a very low

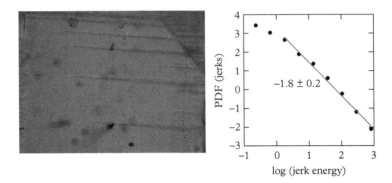

Figure 5.8 *(Left) The needle configuration and (right) the emitted energy per needle movement of one needle.*

cutoff (determined by the activation energy of the de-pinning process) and a high energy cutoff for supersonic domain movements. Such "crackling noise behaviour" (Sethna et al. 2001) is equivalent to Barkhausen noise in ferromagnets. Its detailed scaling and the consequences of the scale invariance were discussed in Salje and Dahmen (2014). A full analysis of the crackling noise was performed in $BaTiO_3$ (Salje et al. 2019) and PZT (Tan et al. 2019).

5.6   Conclusion

Domain boundary engineering depends on the emerging properties of domain boundaries. These properties are enhanced by the enormous flexibility of domain walls under the influence of stress, electric fields, or magnetic fields. This leads to a mixing of effects traditionally pursued in different fields of science: while the generation of emerging properties is typically a field of physics, the pattern formation of domain structures was for a long time the realm of material sciences and metallurgy, where the performance of materials depends largely on their microstructure. Typical examples are shape memory alloys (Romero et al. 2011; Faran et al. 2015; Kustov et al. 2017a), where the movement of domain boundaries has been explored in much detail.

In comparison, the length scale in domain boundary engineering is greatly reduced: the characteristic length is the thickness of the domain walls, which in non-magnetic materials is typically on the order of 1 nm. Bloch lines and other secondary domain boundaries are meandering lines with similar thicknesses. These lines have not yet been observed in computer simulations, and progress is needed in this field. Furthermore, the high density of the domain walls will lead to jamming and pinning by domain wall intersections. These effects lead to viscous movements of domain walls and the formation of domain glasses (Salje et al. 2014a; Salje and Carpenter 2015; Pesquera et al. 2018). Domain glasses are an almost virgin field of research and may well become prominent when the dynamics of domain walls become more relevant, such as in the production process of device materials.

....

REFERENCES

Aird A., Domeneghetti M. C., Mazzi F., Tazzoli V., Salje E. K. H., "Sheet superconductivity in WO_{3-x}: crystal structure of the tetragonal matrix," *J. Phys.: Condens. Mat.* 10, L569–L574 (1998).

Aird A., Salje E. K. H., "Sheet superconductivity in twin walls: experimental evidence of," *J. Phys.: Condens. Mat.* 10, L377–L380 (1998).

Aird A., Salje E. K. H., "Enhanced reactivity of domain walls in with sodium," *Eur. Phys. J. B: Condens. Mat. Complex Syst.* 15, 205–210 (2000).

Aufort J., Aktas O., Carpenter M. A., Salje E. K. H., "Effect of pores and grain size on the elastic and piezoelectric properties of quartz-based materials," *Am. Mineral.* 100, 1165–1171 (2015).

Ball J. M., James R. D., Smith F. T., "Proposed experimental tests of a theory of fine microstructure and the two-well problem," *Philos. Trans. Roy. Soc. London. Ser. A: Phys. Eng. Sci.* 338, 389–450 (1992).

Bassiri-Gharb N., Fujii I., Hong E., Trolier-Mckinstry S., Taylor D. V., DAmjanovic D., "Domain wall contributions to the properties of piezoelectric thin films," *J. Electroceram.* 19, 49–67 (2007).

Bhatnagar A., Roy Chaudhuri A., Heon Kim Y., Hesse D., Alexe M., "Role of domain walls in the abnormal photovoltaic effect in $BiFeO_3$," *Nat. Commun.* 4, 2835 (2013).

Bhattacharya K., *Microstructure of Martensite: Why It Forms and How It Gives Rise to Shape Memory Effects* (Oxford University Press, Oxford, 2004).

Bibes M., Valencia S., Balcells L., Martínez B., Fontcuberta J., Wojcik M., Nadolski S., Jedryka E., "Charge trapping in optimally doped epitaxial manganite thin films," *Phys. Rev. B* 66, 134416 (2002).

Bratkovsky A. M., Marais S. C., Heine V., Salje E. K. H., "The theory of fluctuations and texture embryos in structural phase transitions mediated by strain," *J. Phys.: Condens. Mat.* 6, 3679–3696 (1994).

Brinkman A., Huijben M., Van Zalk M., Huijben J., Zeitler U., Maan J. C., Van Der Wiel W. G., Rijnders G., Blank D. H. A., Hilgenkamp H., "Magnetic effects at the interface between non-magnetic oxides," *Nat. Mater.* 6, 493 (2007).

Calleja M., Dove M. T., Salje E. K. H., "Trapping of oxygen vacancies on twin walls of $CaTiO_3$: a computer simulation study," *J. Phys.: Condens. Mat.* 15, 2301–2307 (2003).

Carpenter M. A., "Static and dynamic strain coupling behaviour of ferroic and multiferroic perovskites from resonant ultrasound spectroscopy," *J. Phys.: Condens. Mat.* 27, 263201 (2015).

Carpenter M. A., Salje E. K. H., Howard C. J., "Magnetoelastic coupling and multiferroic ferroelastic/magnetic phase transitions in the perovskite $KMnF_3$," *Phys. Rev. B* 85, 224430 (2012).

Casals B., Schiaffino A., Casiraghi A., Hämäläinen S. J., López González D., Van Dijken S., Stengel M., Herranz G., "Low-temperature dielectric anisotropy driven by an antiferroelectric mode in $SrTiO_3$," *Phys. Rev. Lett.* 120, 217601 (2018).

Catalan G., Seidel J., Ramesh R., Scott J. F., "Domain wall nanoelectronics," *Rev. Mod. Phys.* 84, 119–156 (2012).

Conti S., Müller S., Poliakovsky A., Salje E. K. H., "Coupling of order parameters, chirality, and interfacial structures in multiferroic materials," *J. Phys.: Condens. Mat.* 23, 142203 (2011).

Cullity B. D., Graham C. D., *Introduction to Magnetic Materials* (John Wiley & Sons, United States, 2011).

Damjanovic D., "Contributions to the piezoelectric effect in ferroelectric single crystals and ceramics," *J. Am. Ceram. Soc.* 88, 2663–2676 (2005).

Damjanovic D., Demartin M., "Contribution of the irreversible displacement of domain walls to the piezoelectric effect in barium titanate and lead zirconate titanate ceramics," *J. Phys.: Condens. Mat.* 9, 4943–4953 (1997).

Daraktchiev M., Catalan G., Scott J. F., "Landau theory of domain wall magnetoelectricity," *Phys. Rev. B* 81, 224118 (2010).

Ding X., Zhao Z., Lookman T., Saxena A., Salje E. K. H., "High junction and twin boundary densities in driven dynamical systems," *Adv. Mater.* 24, 5385–5389 (2012).

Eerenstein W., Mathur N. D., Scott J. F., "Multiferroic and magnetoelectric materials," *Nature* 442, 759–765 (2006).

Eliseev E. A., Morozovska A. N., Glinchuk M. D., Blinc R., "Spontaneous flexoelectric/ flexomagnetic effect in nanoferroics," *Phys. Rev. B* 79, 165433 (2009).

Faran E., Salje E. K. H., Shilo D., "The exploration of the effect of microstructure on crackling noise systems," *Appl. Phys. Lett.* 107, 071902 (2015).

Fiebig M., "Revival of the magnetoelectric effect," *J. Phys. D: Appl. Phys.* 38, R123–R152 (2005).

Fontcuberta J., Skumryev V., Laukhin V., Granados X., Salje E. K. H., "Polar domain walls trigger magnetoelectric coupling," *Sci. Rep.* 5, 13784 (2015).

Frenkel Y., Haham N., Shperber Y., Bell C., Xie Y., Chen Z., Hikita Y., Hwang H. Y., Salje E. K. H., Kalisky B., "Imaging and tuning polarity at $SrTiO_3$ domain walls," *Nat. Mater.* 16, 1203 (2017).

Gallardo M. C., Manchado J., Romero F. J., Del Cerro J., Salje E. K. H., Planes A., Vives E., Romero R., Stipcich M., "Avalanche criticality in the martensitic transition of $Cu_{67.64}Zn_{16.71}Al_{15.65}$ shape-memory alloy: A calorimetric and acoustic emission study," *Phys. Rev. B* 81, 174102 (2010).

Garten L. M., Trolier-Mckinstry S., "Enhanced flexoelectricity through residual ferroelectricity in barium strontium titanate," *J. Appl. Phys.* 117, 094102 (2015).

Goncalves-Ferreira L., Redfern S. A. T., Artacho E., Salje E., Lee W. T., "Trapping of oxygen vacancies in the twin walls of perovskite," *Phys. Rev. B* 81, 024109 (2010).

Goncalves-Ferreira L., Redfern S. A. T., Artacho E., Salje E. K. H., "Ferrielectric twin walls in $CaTiO_3$," *Phys. Rev. Lett.* 101, 097602 (2008).

Gonnissen J., Batuk D., Nataf G. F., Jones L., Abakumov A. M., Van Aert S., Schryvers D., Salje E. K. H., "Direct observation of ferroelectric domain walls in $LiNbO_3$: wall-meanders, kinks, and local electric charges," *Adv. Funct. Mater.* 26, 7599–7604 (2016).

Guennou M., Bouvier P., Garbarino G., Kreisel J., Salje E. K. H., "Pressure-induced phase transitions in $KMnF_3$ and the importance of the excess volume for phase transitions in perovskite structures," *J. Phys.: Condens. Mat.* 23, 485901 (2011).

Hamdi H., Salje E. K. H., Ghosez P., Bousquet E., "First-principles reinvestigation of bulk WO_3," *Phys. Rev. B* 94, 245124 (2016).

Harrison R. J., Redfern S. A. T., Buckley A., Salje E. K. H., "Application of real-time, stroboscopic x-ray diffraction with dynamical mechanical analysis to characterize the motion of ferroelastic domain walls," *J. Appl. Phys.* 95, 1706–1717 (2004).

Hayward S. A., Chrosch J., Salje E. K. H., Carpenter M. A., "Thickness of pericline twin walls in anorthoclase: an X-ray diffraction study," *Eur. J. Mineral.* 8, 1301 (1996).

Hayward S. A., Salje E. K. H., "Twin memory and twin amnesia in anorthoclase," *Mineral. Mag.* 64, 195–200 (2000).

Houchmandzadeh B., Lajzerowicz J., Salje E., "Order parameter coupling and chirality of domain walls," *J. Phys.: Condens. Mat.* 3, 5163–5169 (1991).

Jiles D. *Introduction to Magnetism and Magnetic Materials* (Chapman & Hall, London, 1998).

Kadomtseva A. M., Popov Y. F., Pyatakov A. P., Vorob'ev G. P., Zvezdin A. K., Viehland D. (2006) "Phase transitions in multiferroic BiFeO$_3$ crystals, thin-layers, and ceramics: enduring potential for a single phase, room-temperature magnetoelectric 'holy grail,'" *Phase Transit.* 79, 1019–1042.

Khachaturyan A. G. *The Theory of Structural Transformations in Solids* (Wiley, New York, 1983).

Kim Y., Alexe M., Salje E. K. H., "Nanoscale properties of thin twin walls and surface layers in piezoelectric WO$_{3-x}$," *Appl. Phys. Lett.* 96, 032904 (2010).

Kityk A. V., Schranz W., Sondergeld P., Havlik D., Salje E. K. H., Scott J. F. (2000) "Low-frequency superelasticity and nonlinear elastic behavior of SrTiO$_3$ crystals," *Phys. Rev. B* 61, 946–956.

Kustov S., Cesari E., Liubimova I., Nikolaev V., Salje E. K. H., "Twinning in Ni–Fe–Ga–Co shape memory alloy: temperature scaling beyond the Seeger model," *Scripta Mater.* 134, 24–27 (2017a).

Kustov S., Torrens-Serra J., Salje E. K. H., Beshers D. N., "Re-entrant spin glass transitions: new insights from acoustic absorption by domain walls," *Sci. Rep.* 7, 16846 (2017b).

Levin I., Krayzman V., Woicik J. C., "Local structure in perovskite (Ba, Sr)TiO$_3$: Reverse Monte Carlo refinements from multiple measurement techniques," *Phys. Rev. B* 89, 024106 (2014).

Li J., Nagaraj B., Liang H., Cao W., Lee C. H., Ramesh R., "Ultrafast polarization switching in thin-film ferroelectrics," *Appl. Phys. Lett.* 84, 1174–1176 (2004).

Lines M. E., Glass A. M., *Principles and Applications of Ferroelectrics and Related Materials* (Clarendon Press, Oxford, 1979).

Lu G., Li S., Ding X., Salje E. K. H., "Piezoelectricity and electrostriction in ferroelastic materials with polar twin boundaries and domain junctions," *Appl. Phys. Lett.* 114, 202901 (2019).

Majdoub M. S., Sharma P., Cagin T., "Enhanced size-dependent piezoelectricity and elasticity in nanostructures due to the flexoelectric effect," *Phys. Rev. B* 77, 125424 (2008).

Marais S., Heine V., Nex C., Salje E., Phenomena due to strain coupling in phase transitions. *Phys. Rev. Lett.* 66, 2480–2483 (1991).

Nabi H. S., Harrison R. J., Pentcheva R., "Magnetic coupling parameters at an oxide-oxide interface from first principles: Fe$_2$O$_3$-FeTiO$_3$," *Phys. Rev. B* 81, 214432 (2010).

Ohtomo A., Hwang H. Y., "A high-mobility electron gas at the LaAlO$_3$/SrTiO$_3$ heterointerface," *Nature* 427, 423–426 (2004).

Ondrejkovic P., Marton P., Guennou M., Setter N., Hlinka J., "Piezoelectric properties of twinned ferroelectric perovskites with head-to-head and tail-to-tail domain walls," *Phys. Rev. B* 88, 024114 (2013).

Oravova L., Zhang Z., Church N., Harrison R. J., Howard C. J., Carpenter M. A., "Elastic and anelastic relaxations accompanying magnetic ordering and spin-flop transitions in hematite, Fe$_2$O$_3$," *J. Phys.: Condens. Mat.* 25, 116006 (2013).

Pavarini E., Koch E., Schollwöck U., *Emergent Phenomena in Correlated Matter: Autumn School Organized by the Forschungszentrum Jülich and the German Research School for Simulation Sciences at Forschungszentrum Jülich 23–27 September 2013; Lecture Notes of the Autumn School Correlated Electrons 2013* (Forschungszentrum Jülich, 2013).

Pesquera D., Carpenter M. A., Salje E. K. H., "Glasslike dynamics of polar domain walls in cryogenic SrTiO$_3$," *Phys. Rev. Lett.* 121, 235701 (2018).

Pöttker H., Salje E. K. H., "Flexoelectricity, incommensurate phases and the Lifshitz point," *J. Phys.: Condens. Mat.* 28, 075902 (2016).

Přívratská J., Janovec V., "Spontaneous polarization and/or magnetization in non-ferroelastic domain walls: symmetry predictions," *Ferroelectrics* 222, 23–32 (1999).

Puchberger S., Soprunyuk V., Schranz W., Tröster A., Roleder K., Majchrowski A., Carpenter M. A., Salje E. K. H., "The noise of many needles: jerky domain wall propagation in $PbZrO_3$ and $LaAlO_3$," *APL Mater.* 5, 046102 (2017).

Rabe K. M., Triscone J. M., Ahn C. H., *Physics of Ferroelectrics: A Modern Perspective* (Springer, Berlin, 2007).

Ramesh R., Spaldin N. A., "Multiferroics: progress and prospects in thin films," *Nat. Mater.* 6, 21 (2007).

Romero F. J., Manchado J., Martín-Olalla J. M., Gallardo M. C., Salje E. K. H., "Dynamic heat flux experiments in $Cu_{67.64}Zn_{16.71}Al_{15.65}$: Separating the time scales of fast and ultra-slow kinetic processes in martensitic transformations," *Appl. Phys. Lett.* 99, 011906 (2011).

Salje E., "Thermodynamics of plagioclases I: Theory of the $P1c - I1c$ phase transition in anorthite and Ca-rich plagioclases," *Phys. Chem. Mineral.* 14, 181–188 (1987).

Salje E., Parlinski K., "Microstructures in high Tc superconductors," *Supercond. Sci. Technol.* 4, 93–97 (1991).

Salje E., Zhang H., "Domain boundary engineering," *Phase Transit.* 82, 452–469 (2009).

Salje E. K., "Multiferroic domain boundaries as active memory devices: trajectories towards domain boundary engineering," *Chemphyschem* 11, 940–950 (2010).

Salje E. K. H., "Hard mode spectroscopy: experimental studies of structural phase transitions," *Phase Transit.* 37, 83–110 (1992).

Salje E. K. H., *Phase Transitions in Ferroelastic and Co-elastic Crystals* (Cambridge University Press, Cambridge, 1993).

Salje E. K. H., "Ferroelastic materials," *Annu. Rev. Mater. Res.* 42, 265–283 (2012).

Salje E. K. H., "Modulated minerals as potential ferroic materials," *J. Phys.: Condens. Mat.* 27, 305901 (2015a).

Salje E. K. H., "Tweed, twins, and holes," *Am. Mineral.* 100, 343–351 (2015b).

Salje E. K. H., Aktas O., Carpenter M. A., Laguta V. V., Scott J. F., "Domains within domains and walls within walls: evidence for polar domains in cryogenic $SrTiO_3$," *Phys. Rev. Lett.* 111, 247603 (2013a).

Salje E. K. H., Alexe M., Kustov S., Weber M. C., Schiemer J., Nataf G. F., Kreisel J., "Direct observation of polar tweed in $LaAlO_3$," *Sci. Rep.* 6, 27193 (2016a).

Salje E. K. H., Carpenter M. A., "Domain glasses: twin planes, Bloch lines, and Bloch points," *Phys. Status Solidi (b)* 252, 2639–2648 (2015).

Salje E. K. H., Dahmen K. A., "Crackling noise in disordered materials," *Annu. Rev. Condens. Mat. Phys.* 5, 233–254 (2014).

Salje E. K. H., Ding X., Aktas O., "Domain glass," *Phys. Status Solidi (b)* 251, 2061–2066 (2014a).

Salje E. K. H., Ding X., Zhao Z., "Noise and finite size effects in multiferroics with strong elastic interactions," *Appl. Phys. Lett.* 102, 152909 (2013b).

Salje E. K. H., Ding X., Zhao Z., Lookman T., "How to generate high twin densities in nano-ferroics: Thermal quench and low temperature shear," *Appl. Phys. Lett.* 100, 222905 (2012).

Salje E. K. H., Ding X., Zhao Z., Lookman T., Saxena A., "Thermally activated avalanches: jamming and the progression of needle domains," *Phys. Rev. B* 83, 104109 (2011).

Salje E. K. H., Graeme-Barber A., Carpenter M. A., Bismayer U., Lattice parameters, spontaneous strain and phase transitions in $Pb_3(PO_4)_2$. *Acta Crystall. B* 49, 387–392 (1993).

Salje E. K. H., Ishibashi Y., "Mesoscopic structures in ferroelastic crystals: needle twins and right-angled domains," *J. Phys.: Condens. Mat.* 8, 8477–8495 (1996).

Salje E. K. H., Li S., Stengel M., Gumbsch P., Ding X., "Flexoelectricity and the polarity of complex ferroelastic twin patterns," *Phys. Rev. B* 94, 024114 (2016b).

Salje E. K. H., Li S., Zhao Z., Gumbsch P., Ding X., "Polar twin boundaries and nonconventional ferroelectric switching," *Appl. Phys. Lett.* 106, 212907 (2015).

Salje E. K. H., Scott J. F., "Ferroelectric Bloch-line switching: a paradigm for memory devices?" *Appl. Phys. Lett.* 105, 252904 (2014).

Salje E. K. H., Wang X., Ding X., Scott J. F., "Ultrafast switching in avalanche-driven ferroelectrics by supersonic kink movements," *Adv. Funct. Mater.* 27, 1700367 (2017).

Salje E. K. H., Wang X., Ding X., Sun J., "Simulating acoustic emission: the noise of collapsing domains," *Phys. Rev. B* 90, 064103 (2014b).

Salje E. K. H., Xue D., Ding X., Dahmen K. A., Scott J. F., "Ferroelectric switching and scale invariant avalanches in $BaTiO_3$," *Phys. Rev. Mater.* 3, 014415 (2019).

Schröder M., Haußmann A., Thiessen A., Soergel E., Woike T., Eng L. M., "Conducting domain walls in lithium niobate single crystals," *Adv. Funct. Mater.* 22, 3936–3944 (2012).

Scott J. F., *Ferroelectric Memories* (Springer, Berlin, 2000).

Seidel J., Fu D., Yang S.-Y., Alarcón-Lladó E., Wu J., Ramesh R., Ager J. W., "Efficient photovoltaic current generation at ferroelectric domain walls," *Phys. Rev. Lett.* 107, 126805 (2011).

Seidel J., Maksymovych P., Batra Y., Katan A., Yang S. Y., He Q., Baddorf A. P., Kalinin S. V., Yang C. H., Yang J. C., Chu Y. H., Salje E. K., Wormeester H., Salmeron M., Ramesh R., "Domain wall conductivity in La-doped $BiFeO_3$," *Phys. Rev. Lett.* 105, 197603 (2010).

Seidel J., Martin L. W., He Q., Zhan Q., Chu Y. H., Rother A., Hawkridge M. E., Maksymovych P., Yu P., Gajek M., Balke N., Kalinin S. V., Gemming S., Wang F., Catalan G., Scott J. F., Spaldin N. A., Orenstein J., Ramesh R., "Conduction at domain walls in oxide multiferroics," *Nat. Mater.* 8, 229–234 (2009).

Sethna J. P., Dahmen K. A., Myers C. R., "Crackling noise," *Nature* 410, 242–250 (2001).

Sidorkin A. S., *Domain Structure in Ferroelectrics and Related Materials* (Cambridge International Science Publishing, Cambridge, 2006).

Siemons W., Koster G., Yamamoto H., Harrison W. A., Lucovsky G., Geballe T. H., Blank D. H. A., Beasley M. R., "Origin of charge density at $LaAlO_3/BaTiO_3$ heterointerfaces: Possibility of intrinsic doping," *Phys. Rev. Lett.* 98, 196802 (2007).

Spaldin N. A., Fiebig M., "The renaissance of magnetoelectric multiferroics," *Science*, 309, 391 (2005).

Stengel M., "Unified ab initio formulation of flexoelectricity and strain-gradient elasticity," *Phys. Rev. B* 93, 245107 (2016).

Tagantsev A. K., "Piezoelectricity and flexoelectricity in crystalline dielectrics," *Phys. Rev. B* 34, 5883–5889 (1986).

Tan C. D., Flannigan C., Gardner J., Morrison F. D., Salje E. K. H., Scott J. F., "Electrical studies of Barkhausen switching noise in ferroelectric PZT: critical exponents and temperature dependence," *Phys. Rev. Mater.* 3, 034402 (2019).

Taylor D. V., Damjanovic D., "Piezoelectric properties of rhombohedral $Pb(Zr, Ti)O_3$ thin films with (100), (111), and 'random' crystallographic orientation," *Appl. Phys. Lett.* 76, 1615–1617 (2000).

Van Aert S., Turner S., Delville R., Schryvers D., Van Tendeloo G., Salje E. K. H., "Direct observation of ferrielectricity at ferroelastic domain boundaries in $CaTiO_3$ by electron microscopy," *Adv. Mater.* 24, 523–527 (2012).

Viehland D. D., Salje E. K. H., "Domain boundary-dominated systems: adaptive structures and functional twin boundaries," *Adv. Phys.* 63, 267–326 (2014).

Wang D., Salje E. K. H., Mi S.-B., Jia C.-L., Bellaiche L., "Multidomains made of different structural phases in multiferroic $BiFeO_3$: a first-principles-based study," *Phys. Rev. B* 88, 134107 (2013).

Wang X., Salje E. K. H., Sun J., Ding X., "Glassy behavior and dynamic tweed in defect-free multiferroics," *Appl. Phys. Lett.* 112, 012901 (2018).

Yudin P. V., Tagantsev A. K., "Fundamentals of flexoelectricity in solids," *Nanotechnology* 24, 432001 (2013).

Yun S., Woo C.-S., Kim G.-Y., Sharma P., Lee J. H., Chu K., Song J. H., Chung S.-Y., Seidel J., Choi S.-Y., Yang C.-H., "Ferroelastic twin structures in epitaxial WO_3 thin films," *Appl. Phys. Lett.* 107 (2015).

Zhao Z., Ding X., Lookman T., Sun J., Salje E. K. H., "Mechanical loss in multiferroic materials at high frequencies: Friction and the evolution of ferroelastic microstructures," *Adv. Mater.* 25, 3244–3248 (2013).

Zubko P., Catalan G., Buckley A., Welche P. R. L., Scott J. F., "Strain-gradient-induced polarization in $SrTiO_3$ single crystals," *Phys. Rev. Lett.* 99, 167601 (2007).

Zubko P., Catalan G., Tagantsev A. K., "Flexoelectric effect in solids," *Annu. Rev. Mater. Res.* 43, 387–421 (2013).

Zykova-Timan T., Salje E. K. H., "Highly mobile vortex structures inside polar twin boundaries in $SrTiO_3$," *Appl. Phys. Lett.* 104, 082907 (2014).

6

Improper Ferroelectric Domain Walls

D. M. Evans[1], Ch. Cochard[2], R. G. P. McQuaid[2], A. Cano[3], J. M. Gregg[2], and D. Meier[1]

[1]*Department of Materials Science and Engineering, Norwegian University of Science and Technology (NTNU), Trondheim, Norway*
[2]*Centre for Nanostructured Media, School of Mathematics and Physics, Queen's University Belfast, Belfast, United Kingdom*
[3]*Institut Néel, CNRS & Univ. Grenoble Alpes, Grenoble, France*

Interfaces in oxide materials offer amazing opportunities for fundamental and applied research, giving a new dimension to functional properties, such as magnetism, multiferroicity, and superconductivity. Ferroelectric domain walls have recently emerged as an intriguing type of interface, where the distinct features of these walls introduce the important element of spatial mobility, allowing for the real-time adjustment of position, density, and orientation (see Chapters 11 to 13). This mobility adds an additional degree of flexibility that enables domain walls to take an active role in future devices and hold great potential as functional 2D systems for electronics (Catalan et al. 2012; Meier 2015; Bednyakov et al. 2018).

In this chapter, we will focus on the specific physical properties at domain walls in ferroelectric materials where the spontaneous electric polarization appears as a by-product of a structural or magnetic phase transition, and not as its primary order parameter. This is fundamentally different compared to the *proper* ferroelectrics discussed in previous chapters—e.g. $BaTiO_3$, $LiNbO_3$, and $Pb(Zr_xTi_{1-x})O_3$. The systems of interest to us are referred to as *improper* ferroelectrics (Levanyuk and Sannikov 1974) and display special properties that are remarkably different from *proper* ferroelectrics. In Section 6.1, we will begin with a short introduction to the fundamentals of improper ferroelectricity, followed by a discussion of emergent functional domain wall properties in different improper ferroelectric model systems. Section 6.2 covers the broad variety of electronic states and application opportunities associated with improper ferroelectric domain walls in hexagonal manganites. Section 6.3 addresses electronic transport and manipulation of domain walls in boracites, and in Section 6.4, we will present additional magnetoelectric coupling phenomena that arise when the interaction of magnetic spins and electric charges gives rise to improper ferroelectricity. A perspective regarding future research and application opportunities of improper ferroelectric domain walls is given in Section 6.5.

D. M. Evans, Ch. Cochard, R. G. P. McQuaid, A. Cano, J. M. Gregg, and D. Meier, *Improper Ferroelectric Domain Walls* In: *Domain Walls: From Fundamental Properties to Nanotechnology Concepts.* Edited by: Dennis Meier, Jan Seidel, Marty Gregg, and Ramamoorthy Ramesh, Oxford University Press (2020). © D. M. Evans, Ch. Cochard, R. G. P. McQuaid, A. Cano, J. M. Gregg, and D. Meier.
DOI: 10.1093/oso/9780198862499.003.0006

6.1 Basic Background

As in standard ferroelectric domain walls (see Chapter 1), the domain walls in improper ferroelectrics separate regions with different orientations of the corresponding order parameter, and they are spatially mobile and can be injected, positioned, and erased on demand. The important difference is that the primary order parameter of improper ferroelectrics is not the electric polarization, which simply appears as a symmetry-enforced "collateral" effect (Levanyuk and Sannikov 1974; Tolédano and Tolédano 1987; Strukov and Levanyuk 1998). A classic example of an improper ferroelectric material is gadolinium molybdate, $Gd_2(MoO_4)_3$ (Keve et al. 1971), where the structural instability of the paraelectric parent phase results in a doubling of the unit cell volume, which leads to the formation of a polar axis—which is then accompanied by a spontaneous polarization. An outstanding magnetic counterpart is $TbMnO_3$ (Kimura et al. 2003), where the instability is with respect to a spiral magnetic order that also exhibits a polar axis.

From the symmetry point of view, the primary order parameter of improper ferroelectrics always breaks more symmetries than the electric polarization P (otherwise the system is referred to as *pseudoproper* (Tolédano and Tolédano 1987; Strukov and Levanyuk 1998; Wadhawan 2000)). This means that P alone does not suffice to describe the maximal polar subgroup of the corresponding paraelectric parent phase, so that, in this sense, the corresponding transition can be regarded as a meta-ferroelectric transition. In fact, even if the system is uniaxial from the ferroelectric point of view and P points along one crystallographic direction only, improper ferroelectricity requires a primary order parameter q with two or more components. Thus, the reduced transformation properties of some nth product of q with $n \geq 2$, rather than q itself, matches those of P: $P \propto q^n$, from which the nontrivial coupling between these two quantities can be read. For example, in the important case of the hexagonal manganites discussed below, $q = (q_1, q_2)$ is associated with a particular translation symmetry breaking of the lattice for which $P_z \propto (q_1^3 - 3q_1 q_2^2)$ (Artyukhin et al. 2013; Cano 2014). In the case of spiral magnets such as $TbMnO_3$ or $MnWO_4$, in contrast, $q = (q_1, q_2)$ is associated with the cycloidal arrangement of the Mn spins, and the above product involves the gradient also: $P_z \propto (q_1 \nabla_x q_2 - q_2 \nabla_x q_1)$ for an xz-cycloid, implying a coupling term similar to a Lifshitz invariant that arises from spin–orbit interactions (Tolédano 2009; Cano and Levanyuk 2010; Tolédano et al. 2010; Sando et al. 2013).

Since the primary order parameter of improper ferroelectrics is not the electric polarization, different—generally less stringent—boundary conditions apply regarding the elastic and mechanical compatibility of associated domain walls (Tagantsev et al. 2010). Thus, a wide variety of unconventional—and otherwise energetically very unfavorable—electronic domain wall configurations arise in the as-grown state. In particular, charged domain walls with positive or negative domain wall bound charges can readily form (see Chapter 3, Figure 3.2(c) for an illustration) (Meier et al. 2012). This is a crucial and important difference compared to proper ferroelectrics, where rather special preparation procedures, such as frustrated electrical poling (Sluka et al. 2013), electrical trailing fields (Crassous et al. 2015), and photo-excitation of mobile carriers (Bednyakov et al. 2015), are typically required to introduce such charged walls. In improper ferroelectrics, on the contrary, charged domain walls are naturally observed as reported for, e.g. $RMnO_3$

(R = Sc, Y, In, Dy to Lu) (Meier et al. 2012; Wu et al. 2012; Meier et al. 2017), $Cu_3B_7O_{13}Cl$ (McQuaid et al. 2017), $(Ca,Sr)_3Ti_2O_7$ (Oh et al. 2015), $Mn_{1-x}Co_xWO_4$ (Leo et al. 2015), and $TbMnO_3$ (Matsubara et al. 2015a).

Although improper ferroelectrics usually exhibit a smaller spontaneous polarization than proper ferroelectrics (e.g. $\approx$ 5.5 $\mu C/cm^2$ in $YMnO_3$ (Fujimura et al. 1996; Van Aken et al. 2004) and $\lesssim 0.1$ $\mu C/cm^2$ in $TbMnO_3$ (Kimura et al. 2003) compared to $\approx$ 20 $\mu C/cm^2$ in $BaTiO_3$ (Lines and Glass 2001; Merz 1949)), the locally diverging electrostatic potentials at the charged walls still require screening and hence a spatial redistribution of mobile carriers. For example, for $YMnO_3$, considerable carrier densities ($= 2P/q$) on the order of up to ~6 $\times$ 10^{13} per cm^2 are expected to arise to screen the bound charges at 180° tail-to-tail walls (Meier et al. 2012). Because of this accumulation of mobile carriers, the domain wall conductivity at improper ferroelectric domain walls can be significantly higher than in the surrounding bulk as we explain in more detail in Sections 6.2 and 6.3. Thus, analogous to proper ferroelectric domain walls, improper ferroelectric domain walls develop unusual electronic properties, which makes them interesting as functional nanoscale entities for future electronics. Key advantages compared to conventional ferroelectrics are the abundance and stability of the charged domain walls, and the generally higher flexibility to host domain walls with different charge states, enabled by the subordinate nature of the induced ferroelectric order.

Several extended articles and review papers are now published that discuss the screening at charged domain walls (e.g. Eliseev et al. 2008; Catalan et al. 2012; Meier 2015; Bednyakov et al. 2018; Schoenherr et al. 2019), where the reader can find details about stability limits, screening mechanisms, and theoretical models. In the following, we will focus on instructive examples from the recent literature, highlighting the interesting physics and unique functionalities of improper ferroelectric domain walls based on selected materials.

6.2 Functional Domain Walls in Hexagonal Manganites

6.2.1 Domain Wall Structure

Probably the most intensively studied system with improper ferroelectric domain walls are the hexagonal manganites, $RMnO_3$ (R = Sc, Y, In, Dy to Lu). Hexagonal manganites exhibit ferroelectric order below a critical temperature $T_C \approx 1400$ K, with the polarization, P, pointing along the crystallographic c-axis (Van Aken et al. 2004; Lilienblum et al. 2015). Here, the primary symmetry-lowering order parameter is a unit-cell-tripling distortive mode, which breaks the P6$_3$/mmc symmetry of the paraelectric phase and induces a structural non-centrosymmetric distortion (Fennie and Rabe 2005). This distortion leads to periodic tilts of the corner-shared MnO_5 bipyramids in $RMnO_3$ together with displacements of the R ions as illustrated in Figure 6.1(a), resulting in the characteristic displacement pattern seen in the high-angle annular dark field scanning transmission electron microscopy (HAADF-STEM); see Figure 6.1(b). As explained in Holtz et al. (2017), these displacements can be described by

$$u(r_n) = u_0 + Q_1 \cos q \cdot r_n + Q_2 \cos q \cdot r_n, \tag{6.1}$$

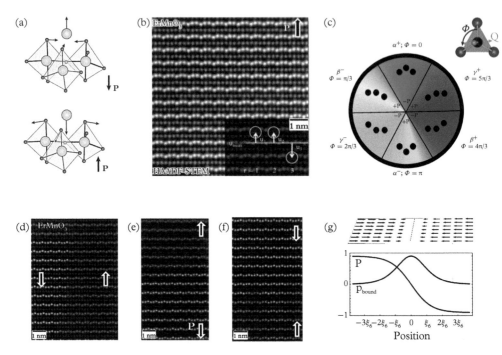

Figure 6.1 *Ferroelectric polarization and the domain walls in the hexagonal manganites.
(a) Schematic showing the origin of the geometrically driven ferroelectric polarization: the tilt and
deformation of MnO₅ bipyramids displaces the rare-earth ions leading to a spontaneous polarization
along the [001] axis. (b) High-angle annular dark field scanning transmission electron microscopy
(HAADF-STEM) of the ErMnO₃ viewed down the $\left[1\bar{1}0\right]$ zone axis; the bright atoms are Er, and the
grey atoms are Mn: the former shows the characteristic corrugation that leads to ferroelectricity in these
systems. (b) Schematic of the six possible domain configuration and colored to represent their differences
with respect to the order parameter φ— see Equation (6.1). (d)–(f) HAADF-STEM of the three different
domain wall orientations (neutral, tail-to-tail, and head-to-head), with the color overlay shown in (c) to
illustrate the change in φ. (g) Trimerization at a head-to-head domain wall separating two regions of
ferroic order. Note: Artyukhin et al. (2013) and Holtz et al. (2017) use φ or Φ, respectively, to describe
the order parameter: there is no physical distinction, and it is only a question of terminology.*

*Source: (a) Based on an original image from Fiebig et al. 2016. (b–g) are slightly adapted from Holtz et al. 2017, this
is an unofficial adaptation of an article that appeared in an ACS publication. ACS has not endorsed the content of
this adaptation or the context of its use.*

where u_0 denotes a polar distortion ($\propto P$), while Q_i ($i = 1, 2$) represent the components
of the primary structural order parameter $[Q = (Q_1, Q_2) = Q\cos\varphi, Q\sin\varphi]$ (Fennie and
Rabe 2005; Artyukhin et al. 2013). Thus, $u(r_n)$ can be directly linked to the canonical
Landau free energy of the system:

$$F(Q, T) = \frac{a(T)}{2} Q^2 + \frac{b}{4} Q^4 + \frac{1}{6}\left(c + c'\cos^2(3\varphi)\right)Q^6 + \frac{g}{2}\left[(\nabla Q)^2 + Q^2(\nabla\varphi)^2\right], \quad (6.2)$$

where the temperature dependence is given by the coefficient $a(T) = a'(T - T_C)$, with a', b, c, and g being positive constants (see Holtz et al. 2017 for details). Most importantly for the formation of domain walls in $RMnO_3$, the anisotropy term in Equation (6.2) leads to six symmetry-equivalent domain states, i.e. six distinct displacement patterns as sketched in Figure 6.1(c), corresponding to values $\varphi_n = \frac{n\pi}{3}$ ($n = 0, \ldots, 5$). These six displacement patterns (up-up-down, up-down-up, etc.) represent the structural trimerization domains/anti-phase domains that form below T_C. The improper ferroelectric polarization then arises due to a coupling to Q, linked via the relation $P \sim Q^3 \cos 3\varphi$ as shown by Fennie and Rabe (2005), Artyukhin et al. (2013), and Cano (2014).

The structural trimerization domains are separated by comparatively narrow domain walls across which the trimerization phase changes from φ_n to φ_{n+1} over a distance of about 5 to 10 Å as shown in Figure 6.1(d–f) (Holtz et al. 2017). Along with Q, the direction of the electric polarization changes across the wall by 180°, going through zero at the center of the wall. The latter is shown for the case of a head-to-head domain wall in Figure 6.1(g). This figure further shows the density of bound charges that arises at such 180° domain walls, leading to a redistribution of mobile carriers that will be addressed in Section 6.2.2. Whenever trimerization domain walls intersect, they do so in characteristic sixfold meeting points. This results in structural vortices where the phase φ of the trimerization order parameter changes around the meeting points from 0 to 2π in a discrete fashion, which have attracted significant attention as a test system for cosmology-related questions (for details, we refer the interested reader to Griffin et al. 2012, Huang and Cheong 2017, and Meier et al. 2017). This special distribution of the trimerization order parameter is enabled by an intrinsically weak sixfold anisotropy (Artyukhin et al. 2013), which is further weakened by the coupling to the electric polarization (Cano 2014) and by driving the system near the transition owing to its universality class (Lilienblum et al. 2015). Since these vortices are topologically protected, the meeting points (i.e. the vortex cores) serve as anchor points for the domain walls in $RMnO_3$, preventing the system from minimizing its energy by forming, e.g. simple stripe-like domains with neutral domain walls. This topology-driven effect explains why charged domain walls naturally occur in $RMnO_3$ in the as-grown state. Figure 6.2(a) presents the resulting distribution of improper ferroelectric domains, with a multitude of sixfold meeting points and freely meandering domain walls in between them (Schaab et al. 2014) (note that with respect to the ferroelectric order, these meeting points represent so-called vertex structures). The image is obtained by piezoresponse force microscopy (PFM, in-plane contrast) where the bright and dark colors show ferroelectric domains with opposite polarization direction. Most importantly for the study of electronic domain wall states, due to the meandering wall structure seen in Figure 6.2(a), the local electrostatic configuration continuously varies between positively charged (head-to-head), neutral (side-by-side), and negatively charged (tail-to-tail) domain wall states, thus including all fundamental orientations of ferroelectric 180° walls (see Figure 6.1(d–f) high-resolution images). The manifestation of this broad variety of neutral and charged domain wall states in one material is a key property of the hexagonal manganites, turning the system

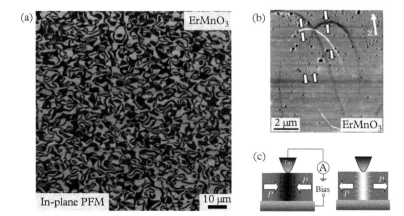

Figure 6.2 *Atomic force microscopy (AFM) images of domain and domain walls in the hexagonal manganites. (a) PFM image showing the ferroelectric domain structure with the iconic sixfold vertex meeting points. (b) Conductive atomic force microscopy (cAFM) image showing how conductivity depends on the domain walls' orientation relative to the unique polar axis: bright/dark areas show higher/lower conductivity compared to the bulk. (c) Schematic illustration of how the screening charges form at head-to-head and tail-to-tail domain walls. Simplistically, for p-type conductors, tail-to-tail walls have a build-up of majority carriers to screen the polar discontinuity, leading to an enhanced conductivity.*

Source: (a) Image slightly modified from original work by Schaab et al. (2014), with the permission of AIP Publishing. (b) and (c) are slightly modified from the original publication by Meier et al. (2012).

into an ideal playground for studying the physics at charged domain walls and the exploration of their functional properties.

6.2.2 Electronic Domain Wall Properties

The electronic transport at domain walls in hexagonal manganites is predominantly determined by the domain wall bound charges and the related electrostatic potential (Meier et al. 2012; Smaabraaten et al. 2018). This relation is also evident from the representative conductance map in Figure 6.2(b), which shows the local currents measured at the domain walls in hexagonal $ErMnO_3$. The data is recorded on the surface of a sample with in-plane-polarization ([110]-orientation) by conductive atomic force microscopy (cAFM), revealing tail-to-tail domain walls with enhanced conductance (bright) and head-to-head domain walls with suppressed conductance (dark) relative to the bulk. Between these two extreme cases, the domain wall conductance varies gradually, exhibiting bulk-like behavior at positions where the walls are oriented parallel to P and, hence, in a neutral configuration.

This emergence of anisotropic conductance at the domain walls and the general trend seen in the cAFM scan in Figure 6.2(b) can be understood based on the electronic bulk properties of $ErMnO_3$. The material is a p-type semiconductor with a small band gap of about 1.6 eV (Skjærvø et al. 2016; Schoenherr et al. 2019). The mobile majority carriers (holes) within the material experience the pronounced electrostatic fields associated with the domain wall bound charges and redistribute accordingly, leading to hole accumulation at tail-to-tail walls and hole depletion at head-to-head walls as sketched in Figure 6.2(c). Typical current values that are measured at the conducting tail-to-tail walls in as-grown samples are on the order of 10 to 100 pA for bias voltages of about 3–5 V and sample thicknesses of ≈ 0.5 to 1 mm. First studies have demonstrated, however, that the domain wall currents can be manipulated, for example, by aliovalent doping (Hassanpour et al. 2016; Schaab et al. 2016; Holstad et al. 2018)—following established strategies known from semiconductor research—or by oxidizing/reducing the material, i.e. by changing the concentration of interstitial oxygen, which is responsible for the p-type characteristics in RMnO$_3$ (Du et al. 2013; Skjærvø et al. 2016; Schaab et al. 2018). The latter reflects the general opportunity to control and optimize the electronic domain wall behavior, with the goal of achieving technologically relevant current densities in the nanoampere regime.

It is important to note, however, that a comprehensive characterization of the electronic transport at ferroelectric domain walls remains a challenging experimental task. While DC current measurements by cAFM provide valuable insight into relative changes in conductance with nanoscale spatial resolution, it is difficult to gain quantitative information from such scans. As cAFM is a two-probe measurement, it is crucial to carefully consider contributions from contact resistance and possible displacement currents. In addition, the typical width of ferroelectric domain walls (≈ 10 Å) is well below the resolution of cAFM ($\gtrsim 20$ nm) and the spread of the injected currents, so that the detected signals necessarily represent a convolution of both sample- and probe-tip-related parameters that need to be carefully disentangled. More recently, microscopy approaches such as electrostatic force microscopy (EFM) (Schoenherr et al. 2019) and photoemission electron microscopy (PEEM) have been applied to study improper ferroelectric domain walls (Schaab et al. 2014; Pawlik et al. 2017). These microscopy measurements were successful in demonstrating that the anomalous electronic transport at the walls—previously measured by cAFM—is indeed an intrinsic phenomenon and not just an effect related to induced domain wall movements or extrinsic tip–sample interactions.

A first breakthrough toward a quantitative analysis of electronic domain wall properties has been the recent realization of local Hall measurements. Campbell et al. (2016) used patterned thin film contact pads, on the top surface of YbMnO$_3$ single crystals, as source and drain electrodes to drive currents along conducting tail-tail domain walls. In the presence of an approximately perpendicular in-plane magnetic field, the scanned conducting tip in an AFM was used to monitor local surface potential changes that resulted from the Hall effect (Figure 6.3(a)). While the experiments were initially done using the AFM pick-up in contact mode (which required a separate calibration to obtain numerical values for Hall voltages), more refined subsequent experiments on ErMnO$_3$ by Turner et al. (2018) used non-contact Kelvin probe force microscopy (KPFM) to directly and quantitatively determine the developed Hall voltages (Figure 6.3(b)).

These two Hall effect studies revealed several important aspects of the carrier behavior in tail-to-tail manganite domain walls: first, that the active carriers are p-type, as expected;

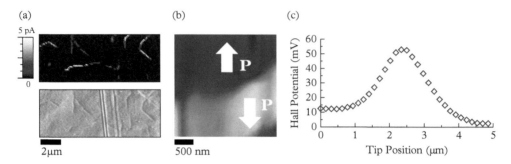

Figure 6.3 *(a) Current-carrying tail-to-tail domain walls in YbMnO$_3$ (top panel) show a distinct Hall voltage signal (bottom panel). Quantitative KPFM was used across the ErMnO$_3$ tail-to-tail domain wall shown in (b) to generate the Hall voltage peak seen in (c).*

Source: (a) is slightly modified from Campbell et al. 2016. (b,c) are slightly adapted from Turner et al. 2018.

second, that the number density of carriers active in transport (and contributing to the Hall effect) is much smaller than that needed to screen the bound charge associated with the polar discontinuity (between three and seven orders of magnitude smaller). This clearly suggests that either the walls are not fully screened, or that an extremely small subset of screening charges is active in conduction; third, that the carrier mobilities are surprisingly high (on the order of 100 cm^{2}V^{-1}s^{-1} at room temperature). Such mobilities are noteworthy, as they are among the highest reported in oxide systems to date. Dimensional confinement in 2D systems does often seem to be associated with elevated carrier mobility (in graphene and in lanthanum aluminate–strontium titanate interfaces, for example), but these observations in YbMnO$_3$ and ErMnO$_3$ demand confirmation and an extension of research to specifically examine carrier mobilities at low temperature.

In 2017, Mundy et al. (2017) demonstrated that charged head-to-head domain walls in ErMnO$_3$ could also deliver enhanced conduction over bulk, but only under significant electric fields. Turner and co-workers have very recently measured the Hall voltages associated with these conducting head-to-head walls. It is clear that, in these walls, the active carriers are n-type. This presents extremely interesting possibilities for domain wall electronics in the hexagonal manganites, as both p and n-type walls can be simultaneously active in carrying current. Moreover, the microstructure characteristic of the manganites is such that domain walls meet regularly at sixfold vertices, potentially creating arrays of 1D p-n junctions. There is a genuine opportunity to exploit such p-n junctions to allow domain walls to not only act as current carriers for mobile interconnects, but to also become active electronic devices in their own right.

6.2.3   Emulation of Electronic Components

In proper ferroelectrics, such as BiFeO$_3$, LiNbO$_3$, and BaTiO$_3$, conducting domain walls are very mobile and can readily be created, moved, and erased at moderate bias voltages (Seidel et al. 2009; Whyte et al. 2014). This degree of flexibility has triggered the idea of using them as active elements and control resistivity by writing and erasing conducting domain walls, realizing multi-configurational devices and non-volatile domain wall memory (Jiang et al. 2017; Sharma et al. 2017). In contrast to the easy-to-write/easy-to-erase type of domain walls in $BiFeO_3$, $LiNbO_3$, and $BaTiO_3$, the walls in hexagonal manganites stand out because of their extraordinary stability (Hassanpour et al. 2016; Schoenherr et al. 2019). Analogous to proper ferroelectric domain walls, it is possible to control the density of domain walls (Griffin et al. 2012; Lin et al. 2014; Meier et al. 2017) and set their position by electric-field poling (Han et al. 2013; Chen et al. 2016; Ruff et al. 2018). However, the domain wall density is adjusted at a high temperature ($\gtrsim$1000 K) and to induce domain wall movements, high electric fields, ca. 100 kV/cm at 120 K, are required (Ruff et al. 2018). Thus, within the temperate and voltage regimes relevant for applications in the field of nanoelectronics, the improper ferroelectric domain walls in $RMnO_3$ represent stable, stationary 2D objects. This stability opens up conceptually new opportunities for domain-wall-based applications—instead of using the walls as rewritable wires in classical two- or three-terminal device architectures, the wall itself becomes the device.

A first proof-of-concept study by Mundy et al. (2017) showed that domain wall currents can be gated by application of an electrical voltage, emulating the behavior of a digital switch as presented in Figure 6.4(a,b). While head-to-head domain walls in $RMnO_3$ are insulating with respect to the surrounding bulk at low voltage (Figure 6.2(b)), enhanced conductance is observed at higher voltages ($\gtrsim$5 V, inset to Figure 6.4(a)). This behavior can be explained by the formation of an electric inversion layer at the head-to-head walls as sketched in Figure 6.4(a). Because of the positive domain wall bound charges, the valence-band maximum (VBM) shifts downwards at the head-to-head walls, generating the aforementioned hole depletion region—see inset to Figure 6.4(a). In addition, strong band bending can cause the conduction-band

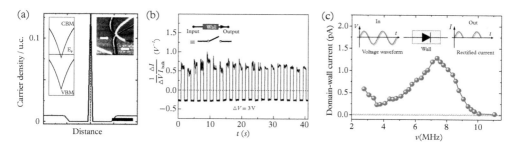

Figure 6.4 *Domain walls as functional circuit elements. (a) Calculated band bending at a head-to-head domain wall, inset is cAFM image showing local inversion layer at head-to-head domain wall—light colors represent higher conductivity. (b) Normalized current at a domain wall as a function of time, demonstrating how domain walls can be used as functional elements: in this case, as switches. (c) The frequency-dependent response of a single neutral domain wall in hexagonal $ErMnO_3$, with insets illustrating the diode-like properties of the wall.*

Source: (a) and (b) are slightly modified from original work by Mundy et al. (2017). (c) is slightly modified from original work by Schaab et al. (2018).

minimum to dip below the Fermi energy, resulting in the representative carrier density distribution shown in Figure 6.4(a). The realization of this carrier distribution in $RMnO_3$ single crystals is corroborated by the cAFM scan taken at 9 V in the inset to Figure 6.4(a), showing enhanced conductance right at the wall (bright), framed by a region with conductance lower than the surrounding bulk (dark). The observation of insulating head-to-head walls at low voltages can thus be understood in terms of hole depletion, which overrides the emergent electron accumulation. In contrast, at higher voltages, electrons dominate the domain wall transport, giving rise to enhanced conductance relative to the bulk. This effect allows reversible switching between resistive and conductive behavior by applying a variable gate voltage, emulating the functionality of a binary switch as shown in Figure 6.4(b).

A second example of the functionality of improper ferroelectric domain walls in $RMnO_3$ is the diode-like behavior that arises when a conducting wall is brought in contact with a metallic probe tip. By applying AC voltages with frequencies in the kilo-to megahertz range, i.e. frequencies for which the domain walls are effectively pinned, Schaab and co-workers observed electrical half-wave rectification at the tip–wall contact when measuring the DC output signal (Schaab et al. 2018). The diode-like response is presented in Figure 6.4(c), showing the measured domain wall current (DC) as function of the frequency of the applied AC voltage. The data indicates rectifying domain wall properties for frequencies $\lesssim 10$ MHz for a voltage amplitude of 3 V. The effect was explained based on the distinct Schottky barrier forming at the tip–wall interface, which is higher at the domain wall than in the surrounding bulk whenever $\sigma_{wall} > \sigma_{bulk}$ is fulfilled. Thus, although half-wave rectification has been reported for neutral domain walls, it is independent of the underlying microscopic mechanism and, hence can be expected to generally occur at all domain walls that exhibit enhanced conductivity with respect to the bulk. With this, it becomes possible to use domain walls for alternating-to-direct (AC-to-DC) current conversion, facilitating the design of nanodiodes.

Additional functionality arises from the structural dynamics related to the domain walls in $RMnO_3$, which begin to play a role at even higher AC frequencies—toward the gigahertz regime (Wu et al. 2017). By applying local impedance spectroscopy measurements, Wu and co-workers discovered anomalous microwave AC conductivity (Wu et al. 2017). This anomalous AC conductivity was explained based on synchronized domain wall oscillations in the terahertz frequency regime—an effect that can potentially be exploited for harnessing AC conduction in systems with otherwise poor DC conduction at the nanoscale as proposed by Tselev et al. (2016).

The different examples of proof-of-concept experiments discussed in this section highlight the emergence of functionalities beyond just conductance at improper ferroelectric domain walls. Particularly interesting is the general possibility of emulating the behavior of electronic components by exploiting intrinsic domain wall properties. This possibility, together with the extraordinary stability and the small width of the walls (5–10 Å; see Figure 6.1), adds a new dimension to the field of domain wall nanoelectronics, foreshadow a promising pathway toward atomic-scale devices and circuitry.

6.3 Boracites

The history of scientific endeavor and discovery within the boracite mineral group has already been eloquently presented by Hans Schmid in a fascinating and engaging personal perspective article (Schmid 2012). In it, he describes how, on attending an international conference in Grenoble in 1958, he and his colleagues were inspired to think how a single-phase material might simultaneously be ferroelectric and ferromagnetic. Driving back to Geneva after the conference, they imagined a perovskite oxide, in which two opposing apical oxygens, in the characteristic MO_6 perovskite octahedra, might be replaced by halogens (or divalent paramagnetic ions): conceptually, this should generate a double-well potential, which could facilitate geometric cation off-centering and hence the possibility of ferroelectricity. Moreover, if the cations contained within such octahedra had a net magnetic moment, then magnetic ordering might also be expected. Their attempts to synthesize targeted oxyhalide perovskites were, however, unsuccessful and so, to keep their hopes alive, they looked at literature to see what oxyhalide compounds containing mixed-ligand octahedra were already known to exist. One such compound was the naturally occurring mineral boracite ($Mg_3B_7O_{13}Cl$).

It was certainly immediately apparent that boracite was crystallographically polar at room temperature: it was one of the systems in which the Curie brothers had established the phenomenon of piezoelectricity in the nineteenth century (Curie and Curie 1880). Moreover, structure determination had shown that the high-temperature (>538 K) high-symmetry cubic and the room-temperature orthorhombic space groups were both non-centrosymmetric piezoelectric ($\overline{F43}c$ and $Pca2_1$, respectively). Ferroelectricity seemed plausible: other nineteenth-century works (Mack 1884) described how sulfur and lead oxide powders were attracted to the surfaces of heat-cycled boracite crystals in unusual patterns, which hinted at both the existence of domains and, importantly from the perspective of this book, charged domain walls. Direct evidence through hysteretic P-E loop measurement was, however, not easily obtained. Nevertheless, Schmid and co-workers went on to synthesize and characterize a number of different oxyhalide boracites, eventually discovering the first known ferromagnetic-ferroelectric single crystal material ($Ni_3B_7O_{13}I$) (Ascher et al. 1966).

6.3.1 Improper Ferroelectricity in Boracites

Subsequent research (often in which Schmid maintained a leading role) has shown that the boracites are, in fact, improper ferroelectrics. The primary order parameter associated with the paraelectric cubic ($\overline{F43}c$) to ferroelectric orthorhombic ($Pca2_1$) transition is still seemingly not completely understood (Meyer et al. 1982; Pascual et al. 2009; Feng et al. 2018). This phase transition is accompanied by a modest uniaxial shearing of the high-temperature cubic unit cell and a slight internal redistribution of atoms, such that a net electrical dipole of order ~1 μCcm^{-2} is developed, which lies parallel to the original cubic <100> shear axis. In most cases, the magnitude of the spontaneous polarization scales with the shear angle of the unit cell, but it must be said that these angles are rather small. Indeed, some early reports on the structural

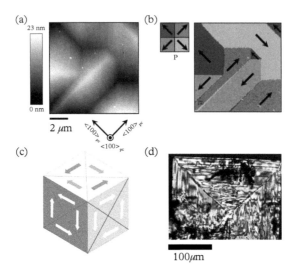

Figure 6.5 *The spontaneous shear associated with the onset of ferroelectricity in boracites is modest (order 0.1°); nevertheless, topographic surface corrugations, induced by different domain variants, are clearly seen using atomic force microscopy (AFM) (a). Such corrugations clearly map onto the different polar orientations revealed by piezoresponse force microscopy (PFM) (b). 90° domain walls show polar discontinuities, suspected as commonplace in naturally occurring Mg-Cl boracite crystals by Schmid (c). Transmission optical microscopy of naturally formed Mg-Cl boracite (d) explicitly shows the quadrant structures illustrated schematically in (c), but also shows that the internal domain microstructure within each quadrant is more complex than given in the schematic.*

Source: (a) and (b) are taken from McQuaid et al. 2017. (c) is loosely adapted from Schmid 2012. (d) is previously unpublished data.

characterization of boracites (Dowty and Clark 1972, 1973; Ito et al. 1951) gave unit cell parameters from diffraction experiments that do not resolve the shear at all. Atomic force microscopy can, however, readily map the surface corrugations that form when a pristine polished crystal is heated and cooled through the ferroelectric phase transition: shear angles at the 0.1° level are easily seen (see Figure 6.5(a)).

There are six equivalent ways in which symmetry can be broken, each generating a different shear strain variant with only one specific direction in polarization. Hence, only six possible polar domain states can form as a result of the symmetry breaking (Dvořák and Petzelt 1971; Erhart 2004; Hlinka et al. 2016), and these can, in principle, be directly mapped using PFM, as can be seen in Figure 6.5(b). However, few aspects of boracite behavior are straightforward to understand, and PFM domain mapping is no exception: in all other known ferroelectrics examined to date, the electric-field-induced movement of the tip–sample contact point lies parallel to the polarization axis. In boracites (or at least in the $Cu_3B_7O_{13}Cl$ crystals studied recently (McQuaid et al. 2017)), conventional vector PFM generates polar directions at 90° to those allowed by symmetry, and hence data must be interpreted differently to give physically meaningful insight. Dvořák (1972) has suggested that the electromechanical coupling equations in improper ferroelectrics may differ from their usual form in proper ferroelectrics, and this may provide the platform upon which unusual PFM behavior might be fully rationalized.

As well as accounting for the domain states, symmetry considerations also dictate the nature and crystallographic orientation of the domain *walls* that can form in the boracites: {100}pseudocubic (pc) oriented walls delineate ferroelastic shear variants that have 180° anti-aligned polarization's on either side. These 180° walls are fully elastically compatible; moreover, the polarization axes, in the domains themselves, lie parallel to the walls, and hence the walls are uncharged. By contrast, group theory analysis and 0-lattice theory (Zimmermann et al. 1970; Erhart 2004) show that although $\{110\}_{pc}$ 90° walls are elastically and structurally viable, they are necessarily charged. In other words, all 90° domain walls show either head-to-head or tail-to-tail polar discontinuities. Although there is undoubtedly an electrostatic energy cost associated with the existence of such walls, they are surprisingly common in naturally formed crystals and have been deemed to be responsible for the microstructural quadrant patterns seen rather ubiquitously in naturally occurring boracite (Figure 6.5(c,d)). In synthetic crystals, according to Schmid [private communication], these charged walls are not seen as often. However, they can be introduced (injected) using point pressure as will be discussed below.

6.3.2 Injection and Motion of Conducting Domain Walls

As in the manganites, polar discontinuities across domain walls in boracites lead to the formation of sheets of bound charges. Near fields produced by these charges may distort the fundamental band structure of the material locally, such that conductivity either increases or decreases due to changes in the energies of intrinsic electronic states; equally, they may cause the aggregation of mobile charged defects that had otherwise been homogeneously distributed throughout the material. In either case, mobile carrier densities are expected to be affected at the charged domain walls, resulting in conductivities that are different from those of the domains themselves. This is verified in $Cu_3B_7O_{13}Cl$ by AFM observations (Figure 6.6a). Notwithstanding issues surrounding the use of PFM to identify polar directions in the Cu-Cl boracites, it seems that tail-to-tail 90° domain walls possess enhanced conductivity, while head-to-head walls show diminished conductivity, compared to bulk.

On cooling through the Curie temperature (~363 K in the case of $Cu_3B_7O_{13}Cl$), charged 90° domain walls do form spontaneously, but not often as extended microstructural features. However, applying point pressure when the crystal is held just below the Curie temperature (pressing with a needle or a pair of tweezers to generate up to 1 GPa locally) causes local deformation and the creation of quadrant features (Figure 6.6(b)). These are extremely similar to those seen in naturally formed Mg-Cl boracite single crystals (Figure 6.5(c,d)), with boundaries parallel to the crystallographic $\{110\}_{pc}$ planes. Although the details of the domain structures are more complex than in the schematic given by Schmid (Figure 6.5(c)), scanning probe microscopy on the Cu-Cl boracites shows that these quadrant boundaries are indeed charged 90° domain walls with enhanced or diminished conductivities. Often, however, these quadrant boundaries

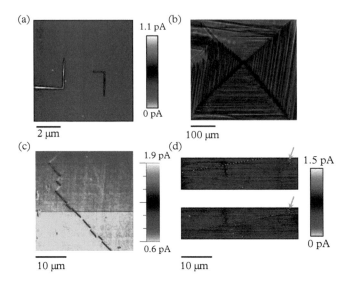

Figure 6.6 *Charged 90° domain walls in Cu₃B₇O₁₃Cl demonstrate different conduction behavior from the domains themselves (a) as revealed through conductive atomic force microscopy (cAFM), revealing tail-to-tail domain walls with enhanced conductance (bright) and head-to cAFM. Point pressure applied just below the Curie temperature can induce domain restructuring into quadrants reminiscent of those formed in naturally occurring Mg₃B₇O₁₃Cl crystals, revealed here (b) by optical microscopy. Quadrant boundaries are composed of charged 90° domain walls, but they can be mixed in nature and hence create discontinuous conduction pathways (c). Importantly, motion does not change the conducting nature of the walls (d), as can be seen in the cAFM maps before and after a wall has been displaced using an electric field. The orange arrows point to the signal from a topographic feature that allows the displacement of the wall between top and bottom panels to be appreciated.*
Source: *Panels (a), (b), and (d) reproduced from McQuaid et al. 2017; panel (c) is previously unpublished.*

are not entirely head-to-head or tail-to-tail, but instead contain sections of walls with opposite senses of polar discontinuity, and this is reflected in the conducting AFM information (Figure 6.6(c)).

With practice, pressing and dragging point probes can create rather long unbroken sections of 90° domain walls, up to hundreds of microns in length. Controlled motion of these unbroken walls is exciting, from the point of view of domain wall electronics, and it is noteworthy that they can be moved using applied electric fields and that the conductivity of the wall is preserved (Fig. 6.6(d)) (McQuaid et al. 2017). It seems that the planes of enhanced conductivity cannot be easily dissociated from the domain walls themselves, as has been the case in $BiFeO_3$, for example (Stolichnov et al. 2014). This possibly points toward a more intrinsic origin for the domain wall conductivity seen in the boracites. Very recent explorations suggest that the sense in which 90° domain walls move under applied electric fields can be unique, and work is ongoing to rationalize this unexpected behavior.

6.4   Spin-Driven Improper Ferroelectrics

6.4.1   Fundamentals

Magnetic order can break spatial inversion symmetry and thereby drive the emergence of a spontaneous electric polarization. This effect was first reported by Newnham et al. (1978) for Cr_2BeO_4 in 1978 and rationalized by Bar'yachtar et al. (1983) and Bar'yakhtar et al. (1985) in analogy with the flexoelectric effect in liquid crystals. Later on, this phenomenon attracted worldwide attention after Kimura et al. observed pronounced magnetoelectric couplings between spin-spiral order and magnetically induced improper ferroelectricity in $TbMnO_3$ in 2003 (Kimura et al. 2003). In the following years, a wide range of materials was demonstrated to exhibit spin-driven ferroelectric order, and different mechanisms were established as reviewed, for example, in Kimura (2007), Picozzi and Stroppa (2012), Tokura et al. (2014), and Bousquet and Cano (2016). Here, we will focus on the so-called *inverse* Dzyaloshinskii–Moriya (DM) interaction (Katsura et al. 2005; Mostovoy 2006; Sergienko and Dagotto 2006), which is relevant for the domain wall phenomena discussed in Section 6.4.2.

In non-centrosymmetric materials, the magnetic exchange interaction can contain an antisymmetric contribution due to spin–orbit coupling; the DM interaction. The respective energy contribution between neighboring spins S at positions i,j can be written as $E_{DM} \sim \sum_{ij} D_{ij} (S_i \times S_j)$, where D_{ij} is the DM vector $(D_{ij} = -D_{ji})$. Unlike simple symmetric exchange interactions $(\sim S_i \cdot S_j)$, the DM interaction favors a noncollinear arrangement of the spins, and hence can lead to helicoidal, conical, and cycloidal spin structures as illustrated in Figure 6.7(a). Magnetically induced ferroelectricity can be obtained due to the inverse effect. That is, if the spontaneous order of spins lacks inversion symmetry by itself, it can further induce a polar displacement as a secondary order parameter. A simple formula describing the relation between the ordered spins and the induced ferroelectric polarization via the inverse DM effect was derived by Mostovoy (2006), $P \sim e_{ij} \times (S_i \times S_j)$, indicating a one-to-one correlation between the vector chirality $(C = S_i \times S_j)$ of the spin structure and the orientation of P (e_{ij} is the unit vector between neighboring spins). This is illustrated in the lower part of Figure 6.7(a), showing that spin cycloids of opposite chirality induce opposite polar displacements $(+C \leftrightarrow +P$ and $-C \leftrightarrow -P)$.

Because of the rigid correlation between magnetism and ferroelectricity, magnetically induced ferroelectrics develop intriguing hybrid domains with inseparably entangled magnetic and electric order parameters, which was first demonstrated in nonlinear optical microscopy measurements $MnWO_4$ (Meier et al. 2009a, 2009b). On the bulk level, such correlations are now well understood, and readers looking for a comprehensive coverage of spin-driven ferroelectricity are referred to more extensive reviews and books on microscopic mechanisms in multiferroics (Kimura 2007; Tokura et al. 2014; Fiebig et al. 2016; Stojanovic 2018). Most importantly for this chapter, recent investigations at the local scale highlight the promising potential of the strong magnetoelectric correlations in spin-driven improper ferroelectrics for the field of domain wall nanoelectronics, as we will discuss in the next section.

6.4.2   Configurational Domain Wall Control with Magnetic Fields and Light

Among the spin-driven improper ferroelectrics, $Mn_{0.95}Co_{0.05}WO_4$ is outstanding because its polarization P gradually rotates by about 90° (i.e. from the crystallographic b- to the a-axis) under application of a magnetic field (Leo et al. 2015). This effect is remarkably different from other spin-spiral systems, such as $TbMnO_3$, $DyMnO_3$, and $MnWO_4$, where magnetic fields induce a first-order flop of P. The continuous polarization rotation in $Mn_{0.95}Co_{0.05}WO_4$ gives the opportunity to gradually modify the configuration of domain wall bound charges as summarized in Figure 6.7(b,c). Figure 6.7(b) shows the magnetic/electric hybrid domains in $Mn_{0.95}Co_{0.05}WO_4$ at 5 K after zero-field cooling ($T_C = 12$ K). The optical second harmonic (SHG) image reflects the position of 180° domain walls by dark lines, across which both C and P reverse sign as explained in Manz et al. (2016). The image reveals that—analogous to the cases of $RMnO_3$ (Section 6.2) and $Cu_3B_7O_{13}Cl$ (Section 6.3)—domain wall sections with head-to-head and tail-to-tail configuration naturally arise in $Mn_{0.95}Co_{0.05}WO_4$. The presence of these walls reflects that the electric dipole–dipole interactions, which usually make such walls energetically unfavorable, are overruled by the energy gain associated with the underlying primary magnetic order parameter.

In spite of the secondary nature of the ferroelectric order, however, Leo et al. (2015) demonstrated that the domain walls in $Mn_{0.95}Co_{0.05}WO_4$ can be controlled by an electrical field. After preparing a sample into a single domain state, they managed to inject and position a side-by-side domain wall by application of an electric field, which remained stationary after removing the field. By application of a magnetic field, the team then gradually rotated the polarization, turning the side-by-side domain wall into a nominally charged wall as presented in Figure 6.7(c). This magnetoelectric manipulation shows the general possibility of controlling the electronic domain wall configuration in a deterministic and reversible process. The emergence of domain wall bound charges at domain walls $Mn_{0.95}Co_{0.05}WO_4$ suggested by the nonlinear optical data was corroborated by atomistic Landau–Lifshitz–Gilbert (LLG) simulations. The LLG simulations revealed that side-by-side domain walls as in Figure 6.7(c) (upper panels) are not simple Ising walls but have a finite Néel component, which is preserved for the induced head-to-head wall (Figure 6.7(c), lower panels). Because of the genuine Néel component that gives $\partial_a P_a = -\rho_{bound} < 0$ across the wall, these domain walls can be expected to get charged with a density of bound charges up to 10^5 C m^{-3}.

Performing similar measurements on the $TbMnO_3$, Matsubara et al. (2015b) showed that conversion of domain wall charge states is a rather general effect in spin-driven improper ferroelectrics and is not restricted to the case of continuous polarization rotations. Although the magnetically driven reorientation of the electric polarization in $TbMnO_3$ occurs through a first-order phase transition, the domain walls can be reversibly tuned between side-by-side and head-to-head/tail-to-tail configurations, pegged to their location by the magnetic order. Going beyond domain wall manipulation via electric and magnetic fields, it was demonstrated that the hybrid domain walls in spin-driven improper ferroelectrics can be controlled optically. By scanning a laser

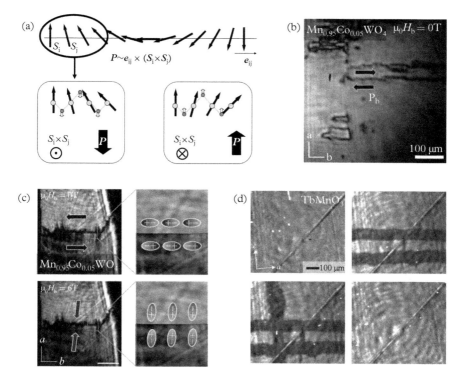

Figure 6.7 *SHG of spin driven multiferroics. (a) Schematic illustration of a spin cycloid. The spin structure violates inversion symmetry and, as a consequence, induces a spontaneous polarization P as explained in the main text. The chirality of the cycloid $\left(C = S_i \times S_j\right)$ determines the direction of P as illustrated in the lower part. Larger arrows represent localized spins S (e.g., at Mn sites), and smaller arrows show the spin-driven displacement of intermediary ions (e.g., O). (b) SHG image of ferroelectric domain structure at 5.5 K after cooling in zero field. The dark lines are the ferroelectric domain walls. (c) SHG image of the ferroelectric domain structure at a single wall under 0 T (top) and 6 T (bottom). The magnetic field is used to rotate the polarization direction so that neutral walls (top) have a charge discontinuity in the ferroelectric polarization. (d) SHG image of the ferroelectric domain structure in $TbMnO_3$ showing how ferroelectric domains can be written and erased with optical pulses.*

Source: (b) and (c) are slightly modified from original work by Leo et al. (2015). (d) is taken from Manz et al. 2016.

(continuous wave or pulsed) and adjusting the wave length, Manz et al. (2016) wrote and erased side-by-side and head-to-head/tail-to-tail walls in $TbMnO_3$, highlighting the possibility of optical domain wall engineering and patterning (Figure 6.7(d)).

The low value of the electric polarization and the cryogenic temperature at which current model systems, such as $Mn_{0.95}Co_{0.05}WO_4$ and $TbMnO_3$, exhibit hybrid magnetic/electric domain walls undermine the technological advantages. The expected domain wall conductance in $Mn_{0.95}Co_{0.05}WO_4$, for example, is on the order of a few femtoamperes, which is challenging to detect even under laboratory conditions. The successful development of spin-driven improper ferroelectrics with polarization values $> 1000 \; \mu C \; m^{-2}$ and higher ordering temperatures, however, is continuously improving these odds (see, e.g. Johnson et al. 2012; Rocquefelte et al. 2013).

6.5   Outlook

Conducting domain walls offer a complete paradigm shift in the way we think about the future of nanoelectronics: since walls can be injected, moved, and annihilated, functionality can, in principle, be dynamically deployed site-specifically. Thus "now-you-see-it-now-you-don't" ephemeral circuitry, which could exist in one form in one instant and then be completely reconfigured the next, could become a reality. Improper ferroelectrics are particularly exciting in this respect. In fact, one of their most outstanding features is that they naturally provide conducting, charged domain walls (those which support polar discontinuities) by the very nature of their (improper) spontaneous electric polarization. This is in sharp contrast to proper ferroelectrics, where charged walls are meta- or unstable.

The discovery of the simultaneous existence of both n- and p-type conduction in oppositely charged manganite domain walls is unique (to date) and profound in terms of its potential implications: active p-n junctions should be formed at domain wall vertices, and this opens the way for a plethora of semiconducting devices built *inside* the domain walls themselves. In one sense, this makes improper ferroelectrics the front-runners in terms of dynamic domain wall nanoelectronics.

However, the truth is that none of the improper systems discussed above is perfect: all are rather complex in their chemistry, and no single material offers the full gamut of complete domain reconfigurability combined with domain wall functionality that is really needed. Thus, there is a great deal of work yet to be done in modifying currently known materials to expand their properties and map new improper systems. This is not all: a great deal of potentially revolutionary domain wall physics has not yet been explored, which could turn out to be both useful and exotic. For example, genuine confinement effects in domain wall conduction properties (quantum Hall, cyclotron resonance, etc.) have not yet been established; large, potentially giant, carrier mobilities need to be verified and explored further as a function of temperature; opportunities and possibilities for spintronics along walls need to be investigated (Rashba effects, for example) and the basic band structures of domain walls as functional materials in their own right need to be mapped. The domain wall is, after all, a new kind of functional material which may be as exciting as any 2D material known to date.

In summary, improper ferroelectrics are a toolbox with extensive possibilities for the design of new nanoelectronics devices; however, challenges remain to be overcome in the choice and processing of these materials. Further studies of domain walls in improper ferroelectrics will lead to the discovery of new properties that may prove practically useful in these exotic two-dimensional systems.

....

REFERENCES

Artyukhin S., Delaney K. T., Spaldin N. A., Mostovoy M., "Landau theory of topological defects in multiferroic hexagonal manganites," *Nat. Mater.* 13, 42 (2013).

Ascher E., Rieder H., Schmid H., Stössel H., "Some properties of ferromagnetoelectric nickel-iodine boracite, $Ni_3B7O_{13}I$," *J. Appl. Phys.* 37(3), 1404–1405 (1966).

Bar'yachtar V. G., L'vov V. A., Yablonskii D. A., "Theory of inhomogeneous magnetoelectric effect," *Pisma Zh. Eksp. Teor. Fiz.* 37(12), 565–567 (1983).

Bar'yakhtar V. G., Stefanovskii E. P., Yablonskii D. A., "Theory of magnetic structure and electric polarization of the Cr_2BeO_4," *Pisma Zh. Eksp. Teor. Fiz.* 42(6), 258–260 (1985).

Bednyakov P. S., Sluka T., Tagantsev A. K., Damjanovic D., Setter N., "Formation of charged ferroelectric domain walls with controlled periodicity," *Sci. Rep.* 5, 15819 (2015).

Bednyakov P. S., Sturman B. I., Sluka T., Tagantsev A. K., Yudin P. V., "Physics and applications of charged domain walls," *Npj Comput. Mater.* 4(1), 65 (2018).

Bousquet E., Cano A., "Non-collinear magnetism in multiferroic perovskites," *J. Phys.: Condens. Mat.* 28(12), 123001 (2016).

Campbell M. P., McConville J. P. V. V., McQuaid R. G. P. P., Prabhakaran D., Kumar A., Gregg J. M., "Hall effect in charged conducting ferroelectric domain walls," *Nat. Commun.* 7, 13764 (2016).

Cano A., "Hidden order in hexagonal $RMnO_3$ multiferroics (R=Dy -Lu, In, Y, and Sc)," *Phys. Rev. B: Condens. Mat. Mater. Phys.* 89(21), 214107 (2014).

Cano A., Levanyuk, A. P., "Pseudoproper ferroelectricity in thin films," *Phys. Rev. B: Condens. Mat. Mater. Phys.* 81(17), 172105 (2010).

Catalan G., Seidel J., Ramesh R., Scott J., "Domain wall nanoelectronics," *Rev. Mod. Phys.* 84(1), 119–156 (2012).

Chen Z., Wang X., Ringer S. P., Liao X., "Manipulation of nanoscale domain switching using an electron beam with omnidirectional electric field distribution," *Phys. Rev. Lett.* 117(2), 27601 (2016).

Crassous A., Sluka T., Tagantsev A. K., Setter N., "Polarization charge as a reconfigurable quasi-dopant in ferroelectric thin films," *Nat. Nanotechnol.* 10(7), 614–618 (2015).

Curie J., Curie P., "Développement par compression de l'électricité polaire dans les cristaux hémièdres à faces inclinées," *Bull. Minér.* 3, 4, 90–93 (1880).

Dowty E., Clark J. E., "Crystal-structure refinements for orthorhombic boracite, $Mg_7ClB_7O_{13}$, and a trigonal, iron-rich analogue," *Zeitschrift für Kristallographie* 138(1–6), 64–99 (1973). ISSN (Online) 2196-7105, ISSN (Print) 2194-4946. DOI: https://doi.org/10.1524/zkri.1973.138.jg.64

Dowty E., Clark J. R., Atomic displacements in ferroelectric trigonal and orthorhombic boracite structures. *Solid State Commun.* 10(6), 543–548 (1972).

Du Y., Wang X., Chen D., Yu Y., Hao W., Cheng Z., Dou S. X., "Manipulation of domain wall mobility by oxygen vacancy ordering in multiferroic $YMnO_3$," *Phys. Chem. Chem. Phys.* 15(46), 20010–20015 (2013).

Dvořák V., "Boracites—an example of improper ferroelectrics," *Le J. Phys. Colloq.* 33(C2), C2-89–90 (1972).

Dvořák, V., Petzelt, J., "Symmetry aspect of the phase transitions in boracites," *Czech. J. Phys.* 21(11), 1141–1152 (1971).

Eliseev E. A., Morozovska A. N., Svechnikov G. S., Rumyantsev E. L., Shishkin E. I., Shur V. Y., Kalinin S. V., "Screening and retardation effects on 180°-domain wall motion in ferroelectrics: wall velocity and nonlinear dynamics due to polarization-screening charge interactions," *Phys. Rev. B* 78(24), 245409 (2008).

Erhart J., "Domain wall orientations in ferroelastics and ferroelectrics," *Phase Transit.* 77(12), 989–1074 (2004).

Feng J. S., Xu K., Bellaiche L., Xiang H. J., "Designing switchable near room-temperature multiferroics via the discovery of a novel magnetoelectric coupling," *New J. Phys.* 20(5), 53025 (2018).

Fennie C. J., Rabe K. M., "Ferroelectric transition in YMnO₃ from first principles," *Phys. Rev. B: Condens. Mat. Mater. Phys.* 72(10), 100103 (2005).

Fiebig M., Lottermoser T., Meier D., Trassin M., "The evolution of multiferroics," *Nat. Rev. Mater.* 1, 16046 (2016).

Fujimura N., Ishida T., Yoshimura T., Ito T., "Epitaxially grown YMnO₃ film: new candidate for nonvolatile memory devices," *Appl. Phys. Lett.* 69(7), 1011–1013 (1996).

Griffin S. M., Lilienblum M., Delaney K. T., Kumagai Y., Fiebig M., Spaldin N. A., "Scaling behavior and beyond equilibrium in the hexagonal manganites," *Phys. Rev. X* 2(4), 41022 (2012).

Han M.-G., Zhu Y., Wu L., Aoki T., Volkov V., Wang X., Chae S. C., Oh Y. S., Cheong S.-W., "Ferroelectric switching dynamics of topological vortex domains in a hexagonal manganite," *Adv. Mater.* 25(17), 2415–2421 (2013).

Hassanpour E., Wegmayr V., Schaab J., Yan Z., Bourret E., Lottermoser Th., Fiebig M., Meier D., "Robustness of magnetic and electric domains against charge carrier doping in multiferroic hexagonal ErMnO₃," *New J. Phys.* 18(4), 43015 (2016).

Hlinka J., Privratska J., Ondrejkovic P., Janovec V., "Symmetry guide to ferroaxial transitions," *Phys. Rev. Lett.* 116(17), 177602 (2016).

Holstad T. S., Evans D. M., Ruff A., Småbråten D. R., Schaab J., Tzschaschel Ch., Yan Z., Bourret E., Selbach S. M., Krohns S., Meier D., "Electronic bulk and domain wall properties in B-site doped hexagonal ErMnO₃," *Phys. Rev. B* 97(8), 85143 (2018).

Holtz M. E., Shapovalov K., Mundy J. A., Chang C. S., Yan Z., Bourret E., Muller D. A., Meier D., Cano A., "Topological defects in hexagonal manganites: inner structure and emergent electrostatics," *Nano Lett.* 17(10), 5883–5890 (2017).

Huang F.-T., Cheong S.-W., "Aperiodic topological order in the domain configurations of functional materials," *Nat. Rev. Mater.* 2(3), 17004 (2017).

Ito T., Morimoto N., Sadanaga R., "The crystal structure of boracite," *Acta Crystall.* 4(4), 310–316 (1951).

Jiang J., Bai Z. L., Chen Z. H., He L., Zhang D. W., Zhang Q. H., Shi J. A., Park M. H., Scott J. F., Hwang C. S., Jiang A. Q., "Temporary formation of highly conducting domain walls for nondestructive read-out of ferroelectric domain-wall resistance switching memories," *Nat. Mater.* 17, 49 (2017).

Johnson R. D., Chapon L. C., Khalyavin D. D., Manuel P., Radaelli P. G., Martin, C., "Giant improper ferroelectricity in the ferroaxial magnet CaMn₇O₁₂," *Phys. Rev. Lett.* 108(6), 67201 (2012).

Katsura H., Nagaosa N., Balatsky A. V., "Spin current and magnetoelectric effect in noncollinear magnets," *Phys. Rev. Lett.* 95(5), 57205 (2005).

Keve E. T., Abrahams S. C., Bernstein J. L., "Ferroelectric ferroelastic paramagnetic beta-Gd₂(MoO₄)₃ crystal structure of the transition-metal molybdates and tungstates. VI," *J. Chem. Phys.* 54(7), 3185–3194 (1971).

Kimura, T., "Spiral magnets as magnetoelectrics," *Annu. Rev. Mater. Res.* 37(1), 387–413 (2007).

Kimura T., Goto T., Shintani H., Ishizaka K., Arima T., Tokura Y., "Magnetic control of ferroelectric polarization," *Nature* 426(6962), 55–58 (2003).

Leo N., Bergman A., Cano A., Poudel N., Lorenz B., Fiebig M., Meier D., "Polarization control at spin-driven ferroelectric domain walls," *Nat. Commun.* 6, 4–9 (2015).

Levanyuk A. P., Sannikov D. G., "Improper ferroelectrics," *Soviet Phys. Uspekhi* 17(2), 199–214 (1974).

Lilienblum M., Lottermoser T., Manz S., Selbach S. M., Cano A., Fiebig M., "Ferroelectricity in the multiferroic hexagonal manganites," *Nat. Phys.* 11, 1070 (2015).

Lin S.-Z., Wang X., Kamiya Y., Chern G.-W., Fan F., Fan D., Casas B., Liu Y., Kiryukhin V., Zurek W. H., Batista C. D., Cheong S.-W., "Topological defects as relics of emergent continuous symmetry and Higgs condensation of disorder in ferroelectrics," *Nat. Phys.* 10, 970 (2014).

Lines M. E., Glass A. M., *Principles and Applications of Ferroelectrics and Related Materials. Oxford Classic Texts in the Physical Sciences* (Oxford University Press, Oxford, 2001). doi:10.1093/acprof:oso/9780198507789.001.0001

Mack K., "Ueber das pyroelektrische Verhalten des Boracits," *Z. Kristallogr.– Crystall. Mater.* 8, 503–522 (1884).

Manz, S., Matsubara M., Lottermoser T., Büchi J., Iyama A., Kimura T., Meier D., Fiebig M., "Reversible optical switching of antiferromagnetism in $TbMnO_3$," *Nat. Photon.* 10(10), 653–656 (2016).

Matsubara M., Manz S., Mochizuki M., Kubacka, T., Iyama A., Aliouane N., Kimura T., Johnson S., Meier D., Fiebig M., "Magnetoelectric domain control in multiferroic $TbMnO_3$," *Science* 348(6239), 1112–1115 (2015a).

Manz S., Matsubara M., Lottermoser T., Büchi J., Iyama A., Kimura T., Meier D., Fiebig M., "Reversible optical switching of antiferromagnetism in $TbMnO_3$," *Nat. Photon.* 10(10), 1112–1115 (2015b).

McQuaid R. G. P., Campbell M. P., Whatmore R. W., Kumar A., Gregg J. M., "Injection and controlled motion of conducting domain walls in improper ferroelectric Cu-Cl boracite," *Nat. Commun.* 8, 15105 (2017).

Meier D., "Functional domain walls in multiferroics," *J. Phys.: Condens. Mat.* 27(46) (2015). doi:10.1088/0953-8984/27/46/463003

Meier D., Leo N., Maringer M., Lottermoser Th., Fiebig M., Becker P., Bohatý L., "Topology and manipulation of multiferroic hybrid domains in $MnWO_4$," *Phys. Rev. B* 80(22), 224420 (2009a).

Meier D., Maringer M., Lottermoser T., Becker P., Bohatý L., Fiebig M., "Observation and coupling of domains in a spin-spiral multiferroic," *Phys. Rev. Lett.* 102(10), 107202 (2009b).

Meier D., Seidel J., Cano A., Seidel J., Cano A., Delaney K., Kumagai Y., Mostovoy M., Spaldin N. A., Ramesh R., Fiebig M., "Anisotropic conductance at improper ferroelectric domain walls," *Nat. Mater.* 11(4), 284–288 (2012).

Meier Q. N., Lilienblum M., Griffin S. M., Lilienblum M., Griffin S. M., Conder K., Pomjakushina E., Yan Z., Bourret E., Meier D., Lichtenberg F., Salje E. K. H., Spaldin N. A., Fiebig M., Cano A., "Global formation of topological defects in the multiferroic hexagonal manganites," *Phys. Rev. X* 7(4), 41014 (2017).

Merz W. J., "The electric and optical behavior of $BaTiO_3$ single-domain crystals," *Phys. Rev.* 76(8), 1221–1225 (1949).

Meyer G. M., Nelmes R. J., Thornley F. R., Stirling W. G., "An inelastic neutron-scattering study of the improper ferroelectric transition in copper chlorine boracite," *J. Phys. C: Solid State Phys.* 15(13), 2851–2866 (1982).

Mostovoy M., "Ferroelectricity in spiral magnets," *Phys. Rev. Lett.* 96(6), 67601 (2006).

Mundy J. A., Schaab J., Kumagai Y., Cano A., Stengel M., Krug I. P., Gottlob D. M., Dog Anay H., Holtz M. E., Held R., Yan Z., Bourret E., Schneider C. M., Schlom D. G., Muller D. A., Ramesh R., Spaldin N. A., Meier D., "Functional electronic inversion layers at ferroelectric domain walls," *Nat. Mater.* 16(6), 622 (2017).

Newnham R. E., Kramer J. J., Schulze W. A., Cross L. E., "Magnetoferroelectricity in Cr_2BeO_4," *J. Appl. Phys.* 49(12), 6088–6091 (1978).

Oh Y. S., Luo X., Huang F.-T., Wang Y., Cheong S.-W., "Experimental demonstration of hybrid improper ferroelectricity and the presence of abundant charged walls in $(Ca,Sr)_3Ti_2O_7$ crystals," *Nat. Mater.* 14, 407 (2015).

Pascual J., Íñiguez J., Iliev M. N., Hadjiev V. G., Meen J., "Phonons in the cubic phase of $Co_3B_7O_{13}$ X (X=Cl, Br, and I) boracites," *Phys. Rev. B: Condens. Mat. Mater. Phys.* 79(10), 104115 (2009).

Pawlik A.-S., Kämpfe T., Haußmann A., Woike T., Treske U., Knupfer M., Büchner B., Soergel E., Streubel R., Koitzsch A., Eng L. M., "Polarization driven conductance variations at charged ferroelectric domain walls," *Nanoscale*, 9(30), 10933–10939 (2017).

Picozzi S., Stroppa A., "Advances in ab-initio theory of multiferroics," *Eur. Phys. J. B* 85(7), 240 (2012).

Rocquefelte X., Schwarz K., Blaha P., Kumar S., Van Den Brink J., "Room-temperature spin-spiral multiferroicity in high-pressure cupric oxide," *Nat. Commun.* 4, 2511 (2013).

Ruff A., Li Z., Loidl A., Schaab J., Fiebig M., Cano A., Yan Z., Bourret E., Glaum J., Meier D., Krohns S., "Frequency dependent polarisation switching in h-$ErMnO_3$," *Appl. Phys. Lett.* 112(18), 182908 (2018).

Sando D., Agbelele A., Rahmedov D., Liu J., Rovillain P., Toulouse C., Infante I. C., Pyatakov A. P., Fusil S., Jacquet E., Carrétéro C., Deranlot C., Lisenkov S., Wang D., Le Breton J. M., Cazayous M., Sacuto A., Juraszek J., Zvezdin A. K., Bellaiche L., Dkhil B., Barthélémy A., Bibes M., "Crafting the magnonic and spintronic response of $BiFeO_3$ films by epitaxial strain," *Nat. Mater.* 12(7), 641–646 (2013).

Schaab J., Cano A., Lilienblum M., Yan Z., Bourret E., Ramesh R., Fiebig M., Meier D., "Optimization of electronic domain-wall properties by aliovalent cation substitution," *Adv. Electron. Mater.* 2(1), 1500195 (2016).

Schaab J., Krug I. P., Nickel F., Gottlob D. M., Doğanay H., Cano A., Hentschel M., Yan Z., Bourret E., Schneider C. M., Ramesh R., Meier D., "Imaging and characterization of conducting ferroelectric domain walls by photoemission electron microscopy," *Appl. Phys. Lett.* 104(23), 232904 (2014).

Schaab J., Skjærvø S. H., Krohns S., Dai X., Holtz M. E., Cano A., Lilienblum M., Yan Z., Bourret E., Muller D. A., Fiebig M., Selbach S. M., Meier D., "Electrical half-wave rectification at ferroelectric domain walls," *Nat. Nanotechnol.* 13(11), 1028–1034 (2018).

Schmid H., "The Dice—Stone der Würfelstein: some personal souvenirs around the discovery of the first ferromagnetic ferroelectric," *Ferroelectrics* 427(1), 1–33 (2012).

Schoenherr P., Shapovalov K., Schaab J., Yan Z., Bourret E. D., Hentschel M., Stengel M., Fiebig M., Cano A., Meier D., "Observation of uncompensated bound charges at improper ferroelectric domain walls," *Nano Lett.*, 19(3), 1659–1664 (2019).

Seidel J., Martin L. W., He Q., Zhan Q., Chu Y.-H., Rother A., Hawkridge M. E., Maksymovych P., Yu P., Gajek M., Balke N., Kalinin S. V., Gemming S., Wang F., Catalan G., Scott J. F., Spaldin N. A., Orenstein J., Ramesh R., "Conduction at domain walls in oxide multiferroics," *Nat. Mater.* 8(3), 229–234 (2009).

Sergienko I. A., Dagotto E., "Role of the Dzyaloshinskii-Moriya interaction in multiferroic perovskites," *Phys. Rev. B: Condens. Mat. Mater. Phys.* 73(9), 94434 (2006).

Sharma P., Zhang Q., Sando D., Lei C. H., Liu Y., Li J., Nagarajan V., Seidel J., "Nonvolatile ferroelectric domain wall memory," *Sci. Adv.* 3(6) (2017).

Skjærvø S. H., Wefring E. T., Nesdal S. K., Gaukås N. H., Olsen G. H., Glaum J., Tybell T., Selbach S. M., "Interstitial oxygen as a source of p-type conductivity in hexagonal manganites," *Nat. Commun.* 7, 13745 (2016).

Sluka T., Tagantsev A. K., Bednyakov P., Setter N., "Free-electron gas at charged domain walls in insulating $BaTiO_3$," *Nat. Commun.* 4(May), 1808 (2013).

Smaabraaten D. R., Meier Q. N., Skjærvø S. H., Inzani K., Meier D., Selbach S. M., "Charged domain walls in improper ferroelectric hexagonal manganites and gallates," *Phys. Rev. Mater.* 2(11), 114405 (2018).

Stojanovic B., *Magnetic, Ferroelectric, and Multiferroic Metal Oxides* (G. Korotcenkov, Ed.), 1st edn. (Elsevier, Amsterdam, Netherlands, 2018).

Stolichnov I., Iwanowska M., Colla E., Ziegler B., Gaponenko I., Paruch P., Huijben M., Rijnders G., Setter N., "Persistent conductive footprints of 109° domain walls in bismuth ferrite films," *Appl. Phys. Lett.* 104(13), 132902 (2014).

Strukov B. A., Levanyuk A. P., "General characteristics of structural phase transitions in crystals," *Ferroelectric Phenomena in Crystals* (B. A. Strukov and A. P. Levanyuk, Eds.), pp. 1–29. (Springer Berlin Heidelberg, Berlin, Heidelberg, 1998).

Tagantsev A. K., Cross L. E., Fousek J., *Domains in Ferroic Crystals and Thin Films. Domains in Ferroic Crystals and Thin Films* (Springer-Verlag, New York, 2010). doi:10.1007/978-1-4419-1417-0

Tokura Y., Seki S., Nagaosa N., "Multiferroics of spin origin," *Rep. Progr. Phys.* 77(7), 76501 (2014).

Tolédano J. C., Tolédano P., *The Landau Theory of Phase Transitions* (World Scientific, Singapore, 1987). doi:10.1142/0215

Tolédano P., "Pseudo-proper ferroelectricity and magnetoelectric effects in $TbMnO_3$," *Phys. Rev. B: Condens. Mat. Mater. Phys.* 79(9), 94416 (2009).

Tolédano P., Mettout B., Schranz W., Krexner G., "Directional magnetoelectric effects in $MnWO_4$: magnetic sources of the electric polarization," *J. Phys.: Condens. Mat.* 22(6), 65901 (2010).

Tselev A., Yu P., Cao Y., Dedon L. R., Martin L. W., Kalinin S. V., Maksymovych P., "Microwave A.C. conductivity of domain walls in ferroelectric thin films," *Nat. Commun.* 7(May), 11630 (2016).

Turner P. W., McConville J. P. V., McCartan S. J., Campbell M. H., Schaab J., McQuaid R. G. P., Amit Kumar J., Gregg J. M., "Large carrier mobilities in $ErMnO_3$ conducting domain walls revealed by quantitative hall-effect measurements," *Nano Lett.* 18(10), 6381–6386 (2018).

Van Aken B. B., Palstra T. T. M., Filippetti A., Spaldin N. A., "The origin of ferroelectricity in magnetoelectric $YMnO_3$," *Nat. Mater.* 3(3), 164–170 (2004).

Wadhawan V., *Introduction to Ferroic Materials* (CRC Press, London, 2000). doi:https://doi.org/10.1201/9781482283051

Whyte J. R., McQuaid R. G. P., Sharma P., Canalias C., Scott J. F., Gruverman A., Gregg J. M., "Ferroelectric domain wall injection," *Adv. Mater.* 26(2), 293–298 (2014).

Wu W., Horibe Y., Lee N., Cheong S.-W., Guest J. R., "Conduction of topologically protected charged ferroelectric domain walls," *Phys. Rev. Lett.* 108(7), 77203 (2012).

Wu X., Petralanda U., Zheng L., Hu R., Cheong S. W., Artyukhin S., Lai K., "Low-energy structural dynamics of ferroelectric domain walls in hexagonal rare-earth manganites," *Sci. Adv.* 3(5) (2017). doi:10.1126/sciadv.1602371

Zimmermann A., Bollmann W., Schmid H., "Observations of ferroelectric domains in boracites," *Phys. Status Solidi (A)* 3(3), 707–720 (1970).

7

Three-Dimensional Optical Analysis of Ferroelectric Domain Walls

A. Haußmann[1], L. M. Eng[1], and S. Cherifi-Hertel[2]

[1] *Institute of Applied Physics, Technische Universität Dresden, D-01062 Dresden, Germany*
[2] *Université de Strasbourg, CNRS, Institut de Physique et Chimie des Matériaux de Strasbourg, UMR 7504, 67000 Strasbourg, France*

7.1 Introduction

Domain walls (DWs) in ferroelectric materials are nanoscale interfaces separating regions with different orientations of the polarization. They have long been considered as imperfections affecting the overall macroscopic properties of ferroelectrics. However, the recently discovered rich and diverse local physical properties of ferroelectric DWs, such as the conductivity of the walls in an otherwise insulating oxide (Seidel et al. 2009; Kim et al. 2010; Guyonnet et al. 2011; Schröder et al. 2012) and conductivity-related extraordinary electronic properties of these DWs as addressed in Chapter 6 and 15 (Campbell et al. 2016; Mundy et al. 2017; Schaab et al. 2018; Xiao et al. 2018), or the enhancement of the photovoltaic effect (Chapter 9) at the domain walls (Yang et al. 2010, 2017), have transformed these domain boundary regions into individual nanostructures with significant fundamental interest and potentially viable application in nanoelectronic device components (Béa and Paruch 2009; Salje 2010; Catalan et al. 2012).

As discussed in Chapters 3 and 4, the exciting properties of ferroelectric domain walls can be related to a local change in the crystal symmetry of the parent phase due to the reorientation of the polarization across the domain walls (Janovec and Přívratská 2006; Tolédano et al. 2014), which can promote exotic polar topological structures and charge redistribution, as well as the dominance of an otherwise energetically less favorable state among different competing orders (Houchmandzadeh et al. 1991). From point group symmetry considerations, for example, a spontaneous polarization or magnetization is permitted specifically within DWs in materials where no such ordering is otherwise present (Přívratská 2007) (see Chapter 1). Elucidating the crucial role of the local

A. Haußmann, L. M. Eng, and S. Cherifi-Hertel, *Three-Dimensional Optical Analysis of Ferroelectric Domain Walls* In: *Domain Walls: From Fundamental Properties to Nanotechnology Concepts*. Edited by: Dennis Meier, Jan Seidel, Marty Gregg, and Ramamoorthy Ramesh, Oxford University Press (2020). © A. Haußmann, L. M. Eng, and S. Cherifi-Hertel.
DOI: 10.1093/oso/9780198862499.003.0007

symmetry and topology at domain wall regions and quantifying their interrelated physical and optical properties is thus of great importance. The extreme sensitivity of second-harmonic generation (SHG) to symmetry makes this method highly suitable for the analysis of changes in the ferroic order parameter at the sub-micron scale (Uesu et al. 1995; Fiebig et al. 2002a; Denev et al. 2011; Chauleau et al. 2017; Nordlander et al. 2018). Furthermore, the quadratic relation between the SHG intensity and the electric field amplitude of the fundamental wave provides high sensitivity to small local field discontinuities or enhancements, even if these effects are limited to a region with a size well below the lateral resolution limit (Valev et al. 2012).

Early studies based on observations by means of far field (Bozhevolnyi and Hvam 1998; Bozhevolnyi et al. 1998; Flörsheimer et al. 1998), near field (Bozhevolnyi et al. 1998), and Cherenkov-type SHG (Fragemann et al. 2004; Deng and Chen 2010; Sheng et al. 2010) have shown that ferroelectric DWs can indeed be imaged by means of nonlinear optical microscopy. The domain boundary regions can appear either as dark lines (Fiebig et al. 2002a), due to destructive interference of the SHG between opposite domains (180° phase shift), or as bright regions (Bozhevolnyi et al. 1998; Bozhevolnyi and Hvam 1998; Flörsheimer et al. 1998). In bulk ferroelectric materials such as lithium niobate crystals, DWs can show either dark (surface) or bright contrast (bulk) in the SHG confocal image, depending on the analysis depth (Berth et al. 2007).

Linear optical spectroscopy and microscopy methods have also proved to be suitable for ferroelectric domain studies. In spite of the limited lateral resolution of linear optical methods, they have facilitated the discovery important DW properties such as the coexistence of electric and magnetic DWs (Fiebig et al. 2002b), DW birefringence (Kim and Gopalan 2005a), defects at the domain boundary regions (Gopalan et al. 2007; Nataf et al. 2016), multiferroic DWs (Meier et al. 2009), as well as the observation of local strain, electric fields, and unusual internal structure at ferroelectric DWs (Rüsing et al. 2018).

In this chapter, we present the latest results demonstrating the flexibility and sensitivity of optical methods for the investigation of the physical properties of DWs in three dimensions (3D). We particularly emphasize the important contribution of both nonlinear and linear optical microscopy in different geometries (transmission, reflection, and noncollinear geometries) to access the detailed morphology of ferroelectric domain walls, obtain their 3D profile, access their internal structure, and establish correlations with their electronic properties.

7.2 Second-Harmonic Generation for Domain Wall Imaging

7.2.1 The Nonlinear Optical Process

Light travelling through a dielectric medium induces a charge separation, creating local electrical dipoles, i.e. a polarization P that is proportional to the electric field E of the incoming wave. While in a first approximation a linear dependence between the local field E and the induced polarization P can be assumed, nonlinear polarization become more pronounced if the incident electric field is large, i.e. when a high-intensity laser beam is used (Bloembergen 1996; Shen 1984). In this case, the material response can be accounted for by expanding the polarization in a power series of the incident electric field:

$$P_i = \sum_i \chi_{ij}^{(1)} E_j + \varepsilon_0 \sum_{j,k,\ldots} \left(D^{(2)} \chi_{ijk}^{(2)} E_j E_k + D^{(3)} \chi_{ijkl}^{(3)} E_j E_k E_l + \ldots \right) \tag{7.1}$$

The first term in Equation (7.1) represents the linear polarization, and the higher-order terms correspond to the nonlinear polarization response. ε_0 is the vacuum permittivity, D is a degeneracy factor, $\chi^{(m)}$ is the mth order optical susceptibility tensor, and the indices i, j, k refer to the Cartesian laboratory coordinates. The second term represents the second-harmonic generation (SHG) process occurring in non-centrosymmetric materials such as ferroelectrics. On a microscopic scale, it involves the coupling of two incident photons at frequency ω to produce a polarization oscillating with the double frequency $P^{2\omega}$. This nonlinear process arises in non-centrosymmetric crystals in which the second-order nonlinear optical susceptibility $\chi^{(2)}$ is nonzero. The susceptibility tensor is an optical property of the material that reflects the point group symmetry of the material. This comprises not only the symmetry of the crystal but also information about the ferroic order. Therefore, when the SHG process is spatially resolved, it allows for the observation of ferroic domains and domain walls, as will be discussed in this chapter.

The intrinsic permutation symmetry allows the susceptibility tensor to be replaced by a contracted d-tensor following the Voigt notation: $2d_{ij} = \chi_{ikl}^{(2)}$. The degeneracy factor is $D^{(2)} = 1/2$ for SHG and optical rectification. The number of the non-vanishing d_{ij} elements is further reduced when considering the point group symmetry of the material according to Neumann's principle.

The SHG intensity is given by $I^{\text{SHG}} = |P^{2\omega}|^2$ where $P^{2\omega}$ takes the following tonsorial form:

$$\begin{pmatrix} P_x(2\omega) \\ P_y(2\omega) \\ P_z(2\omega) \end{pmatrix} = \varepsilon_0 \begin{pmatrix} d_{11} & d_{12} & d_{13} & d_{14} & d_{15} & d_{16} \\ d_{21} & d_{22} & d_{23} & d_{24} & d_{25} & d_{26} \\ d_{31} & d_{32} & d_{33} & d_{34} & d_{35} & d_{36} \end{pmatrix} \begin{pmatrix} E_x^2(\omega) \\ E_y^2(\omega) \\ E_z^2(\omega) \\ 2E_y(\omega) E_z(\omega) \\ 2E_z(\omega) E_x(\omega) \\ 2E_x(\omega) E_y(\omega) \end{pmatrix} \tag{7.2}$$

The intensity of the second-harmonic emission increases quadratically with the power of the incident laser, in addition to the frequency doubling process. Therefore, nonlinear materials, which allow for both signal amplification and efficient energy transfer from one wave to another, are key elements in nanophotonic devices. Among these nonlinear materials, ferroelectric photonic crystals, in particular the lithium niobate material family, have attracted much attention. The possibility of engineering ferroelectric domains with

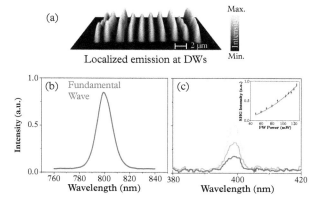

Figure 7.1 *Localized emission created at (a) ferroelectric domain walls in a Pb(Zr$_{0.20}$,Ti$_{0.80}$)O$_3$ ferroelectric films induced by an intense, linearly polarized laser beam. The spectral analysis of (b) the fundamental wave (FW), and (c) the local emission at the ferroelectric domain boundary regions show that the wavelength of the light emitted by the DWs corresponds to the half of the FW (or double frequency), and the local intensity shows a quadratic dependence on the power of the fundamental wave as shown in the inset displayed in panel (c).*

Source: Adapted from Cherifi-Hertel et al. 2017.

different shapes and orientations adds further functionalities to these optical systems (Ferraro et al. 2009). Ferroelectric domains are widely used for frequency conversion and power amplification Yu et al. 2005, as well as for light deflection and focusing Yamada et al. 1996, Gahagan et al. 1999.

Figure 7.1 demonstrates that the emission detected at the ferroelectric walls shows both a frequency doubling process and a quadratic variation of the emission intensity with the input power of the fundamental wave, which confirm the SHG process. Ferroelectric domain boundaries can thus be considered as nonlinear optical nanostructures. Remarkably, while the importance of domain engineering for photonic applications has been long established, the consideration of ferroelectric DWs as active elements in photonic devices is at its infancy. It has only been mentioned by Coppola et al. (2003). The possibility of writing and erasing, *à la carte*, light nano-emitters localized at domain boundaries is expected to pave the way for domain wall nanophotonics.

7.2.2 Noncollinear Cherenkov-Type SHG

The effect of SHG, or optical frequency doubling, does occur as a bulk property in ferroelectric materials due to their inherent lack of inversion symmetry in the crystal structure. However, in order to achieve a substantial yield of SHG intensity, i.e. conversion efficiency, phase matching conditions have to be fulfilled. The phase matching problem arises from the fact that, due to dispersion, the refractive indices at the fundamental and doubled frequency are in general different, preventing a monotonous increase of the SHG amplitude with the distance traversed through the crystal. An important and effective way of solving this issue at least approximately is the quasi-phase-matching method, which has led to the industrial-scale production of periodic domain structures in materials such as $KTiOPO_4$ and $LiNbO_3$ (Hu et al. 2013).

It is remarkable that the phase matching requirement can in fact be applied to restrict the SHG emission exclusively to the DWs (or their vicinity within a diffraction-limited focus size). A possible way to do so is the use of the so-called Cherenkov geometry, first applied to $LiNbO_3$ by Flörsheimer et al. (1998) and Sheng et al. (2010). Phase matching can be understood as photon momentum (or wavevector) conservation. The length of the k vectors for the fundamental (k_0) and SHG light (k_2) are given by

$$k_0 = n(\omega) \cdot \frac{\omega}{c} \qquad (7.3)$$

$$k_2 = n(2\omega) \cdot \frac{2\omega}{c} \qquad (7.4)$$

With $n(2\omega) > n(\omega)$ (normal dispersion), this results in $k_2 > 2k_0$. In the Cherenkov geometry, the momentum conservation is facilitated through a reciprocal lattice vector G aligned approximately perpendicular to k_0 (see Figure 7.2(a)):

$$k_2 = 2k_0 + G \qquad (7.5)$$

When the fundamental illumination is confined to a small focus volume,[1] and this focus encounters a single DW, reciprocal lattice vectors over a wide range of magnitudes, but always directed perpendicular to the DW, are provided. This allows for Cherenkov

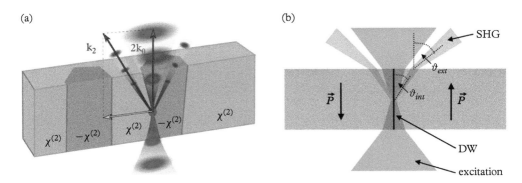

Figure 7.2 *(a) Fundamental principle of Cherenkov phase matching at weakly inclined DWs and for an incidence direction parallel to the polar axis (perpendicular to the sample surface), (b) relation between the internal (within crystal) and external (outside the crystal) Cherenkov angles when applying Cherenkov second-harmonic generation (CSHG) to a highly refractive medium.*

Source: Adapted from Kämpfe 2016.

---

[1]   In this case, a finite k_0 spread instead of a just a single k_0 will be incident on the DW. However, for simplicity, the discussion will only concentrate on plane waves. This approximation holds at least in qualitative terms if the NA of the fundamental illumination is not too high. It is anyway not advisable to use very high NAs in such experiments, as then the focus inside the crystal becomes increasingly distorted due to aberration, which leads to an undesirable decrease in both the signal intensity and spatial resolution with depth.

phase matching if the optical axis of the illumination coincides with the polar axis of the crystal and the DW inclinations are small, resulting in the emission of two SHG beams under the internal Cherenkov angle[2] $\cos(\vartheta_{int}) = n(\omega)/n(2\omega)$. This angular deviation from the incident direction is further increased by refraction at the crystal surface, finally resulting in the external Cherenkov angle[3] $\vartheta_{ext} = \arcsin(n \cdot \sin(\vartheta_{int}))$; see Figure 7.2(b). The name "Cherenkov" was chosen for this process because of the similarities with the well-known Cherenkov effect in nuclear physics, which describes the photon emission during the passage of a particle through a medium at greater velocity than the speed of light in this medium. In the SHG experiment, the excitation (the fundamental wave) likewise travels faster than the emitted (SHG) radiation, as seen from the relation $n(2\omega) > n(\omega)$.

DW imaging by Cherenkov SHG (CSHG) microscopy has been successfully applied to various transparent ferroelectrics such as lithium niobate, strontium barium niobate, triglycine sulfate, barium titanate, or lead hydrogen phosphate. The high data acquisition speed facilitated by the use of galvanometric laser focus scanners also allows investigation of DW dynamics via CSHG at phase transitions and under external electric fields (Ayoub et al. 2017; Wehmeier et al. 2017). Here, the application to electrically poled domains in z-cut LiNbO$_3$ crystals will be discussed in more detail, as these results are especially relevant for the analysis and modeling of DW conductivity in these crystals (Kämpfe et al. 2014). For the measurements, we have used a laser scanning microscopy setup originally designed for investigations of biological samples (see Figure 7.3). Suitable parameters for the excitation were pulse energies of 13 nJ during pulse lengths of 100 fs at a repetition rate of 80 MHz, with center wavelengths to be chosen from a range of 780 nm to 990 nm. We thereby ensured that both the fundamental as well as the SHG radiation lie well within the transparency range of our LiNbO$_3$ crystals. The straightforward detection geometry is to collect the CSHG emitted in the forward direction by a condenser and guide it to the detector. However, in our experiments it turned out that detection in backward direction through the illumination objective is also possible; i.e. some CSHG light is backscattered by an independent process, likely at the sample surface.

A typical result obtained from periodically poled LiNbO$_3$ (PPLN) doped with 5% Mg (supplier Crystal Technology, Inc.) is shown in Figure 7.4. The green color indicates the spatial regions in which the SHG emission intensity exceeded a certain threshold. Especially in these Mg-doped crystals, the industrial production of periodic structures suffers from a certain amount of imperfections, such as nonuniform stripe widths or the merging of domains. Moreover, the end regions of the stripes show a characteristic evolution from sharp to round caps from one z surface of the crystal to the other. All these effects can easily be studied in detail by CSHG.

From the recorded 3D stacks of the CSHG intensity maps, it is possible to reconstruct the DW position by fitting the data with suitable point-spread functions such as Gaussian distributions. Thereby, an accuracy beyond the diffraction limit can be achieved (sometimes called superresolution). In the next step, a mesh of triangular elements is fitted to the position data, and normal vectors are calculated for each triangle. From

---

[2] This formula holds exactly for non-inclined DWs in an optically isotropic medium. For real-world situations, such as electrically poled DWs in z-cut LiNbO$_3$, this is an acceptable approximation.

[3] Again assuming an isotropic medium and, furthermore, refraction into air or vacuum below the critical angle of total internal reflection.

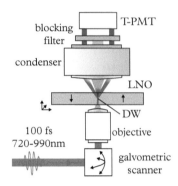

Figure 7.3 *Typical laser scanning microscopy setup for Cherenkov second-harmonic generation (CSHG) measurements. Infrared femtosecond pulses are focused into the ferroelectric sample; the lateral and z-focus positions are adjusted using galvanometric scanners and a z-translation stage for the objective (not shown), respectively. The Cherenkov SHG emission is then collected in forward scattering geometry using a condenser lens, and detected by a photomultiplier tube after blocking of the fundamental radiation.*

Source: *Adapted from Kämpfe 2016.*

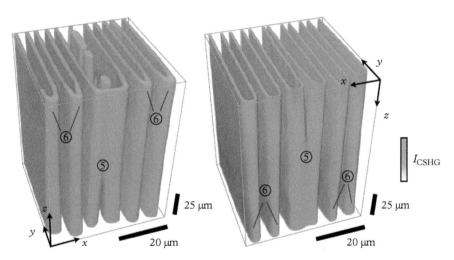

Figure 7.4 *Cherenkov second-harmonic generation (CSHG) inspection of 5% Mg:PPLN. Both illustrations show exactly the same sample region, but providing a clear view of either the sample bottom (left) or sample top surface (right) (the picture was simply displayed upside down). CSHG allows the monitoring of many imperfections of the poling process, such as the merging of domains [marked with the number 5] as well as non-parallel domain wall alignment across the PPLN single crystal [marked with the number 6]. The nonuniform poling field is represented by the rounded caps at the top (poling) sample surface and pointed domain wall tips at the bottom crystal surface.*

Source: *Adapted from Kämpfe 2016.*

the direction of the normal vectors, the inclination of each triangle with respect to the polar axis is then easily calculated, and can be overlaid on a projected view of the DW as a false-color scaling (see Figure 7.5). For as-poled domains (grown quasi-stochastically by the application of super-coercive homogeneous fields over liquid electrodes), only low inclinations below 0.5° naturally occur, with even lower values for a lower Mg doping content (Kämpfe et al. 2014).

Both the conductivity and mean inclination of a DW can be drastically enhanced by post-treatment using sub-coercive voltages after poling. This process can be precisely studied and even tracked in real time by CSHG. Figure 7.6 shows the final state of a previously hexagonal domain after such a treatment. As seen, the shape is distorted into a rounded triangular cross section, and high positive (head-to-head) inclinations can be found locally, especially in the region at about a third of the full extent from the bottom face.

Displaying DWs in their true geometrical shape (or with a compressed scaling in z direction) has, however, the disadvantage that multiple projections are needed to show the full DW interface. This limitation can be overcome, at least for simple domain shapes, by "unfolding" the DW onto the φ, z plane, with φ representing the azimuthal angle measured from the center axis of the domain in all x, y planes. Such "world map projections" of pre- and post-enhancement states are depicted in Figures 7.7(a,b), impressively showing the overall increase in DW inclination achieved through this process. Remarkably, the negative (tail-to-tail) inclinations (shown in blue) are also enhanced at certain positions. Considerable head-to-head inclinations reaching the upper z surface are most prominently found near the corners of the triangular domain. From the projections, it is straightforward to calculate histograms of the inclination values as depicted in Figures 7.7(c,d), indicating a nearly Gaussian distribution before enhancement and a strongly skewed one afterward.

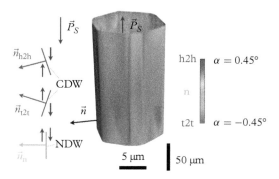

Figure 7.5 *Geometrical reconstruction of the DW around an isolated, as-poled hexagonal domain in 5% Mg-doped congruent lithium niobate. The color scale represents the inclination toward the polar axis, indicating positively charged ("head-to-head"), negatively charged ("tail-to-tail"), and neutral segments. CDW stands for charged domain wall and NDW for neutral domain wall.*

Source: Adapted with permission from Godau et al. 2017. Copyright (2017) American Chemical Society.

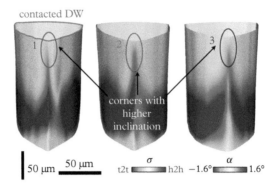

Figure 7.6 *Geometrical reconstruction of the DW around an isolated domain in 5% Mg-doped congruent lithium niobate, after applying the conductivity enhancement process. For improved visualization, the color scale in α is displayed for an inclination value of ±1.6°, although peaks reaching up to 6° in the head-to-head direction can be found. The three CSHG views display three times the same DW viewed under three different azimuthal angles that are rotated by 120° each. These views thus illustrate the full circumference of the as-poled and enhanced DW.*

Source: Adapted with permission from Godau et al. 2017. Copyright (2017) American Chemical Society.

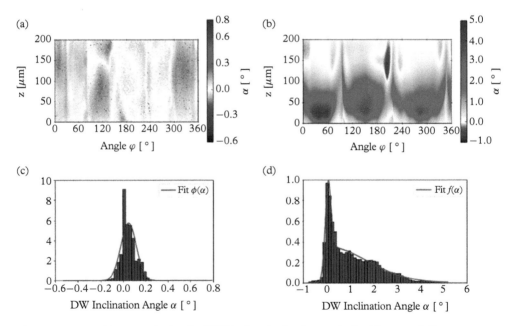

Figure 7.7 *"World map projections" of DW inclination landscapes in the (a) pre- and (b) post-enhancement states. The DWs were unfolded onto the φ (azimuth), z plane. Corresponding histograms of the inclination angle distributions of the data displayed in (a) and (b) are shown in panels (c) and (d), respectively.*

Source: Reprinted with permission from Wolba et al. 2018.

Spatially resolved inclination maps as recorded by CSHG microscopy have opened the door to a detailed analysis of the electronic transport along DWs (Godau et al. 2017; Wolba et al. 2018), which previously had to rely on only one globally averaged inclination value estimated from PFM images taken on both sides of the crystal.

7.2.3 Non-Ising Ferroelectric Walls Revealed by Collinear Backward SHG Polarimetry

The domain wall structure is determined by the minimization of the energy terms at play in the given ferroic system (see Chapters 1 and 4). In magnetic materials, the modulus of the magnetic moment is considered as a constant, leaving only a continuous rotation from one direction to another as a possible transition between domains. The transition from one domain orientation to another is governed by the competition between two energy terms: (1) the anisotropy energy, which tends to keep the magnetization oriented along specific orientations, thus favoring an abrupt transition between two different directions; and (2) the exchange energy, which tends to avoid inhomogeneous magnetic structures and thus favors smooth transitions between adjacent moments. The balance between these two terms results in a characteristic length along which the magnetization direction gradually rotates within the spatial region on the order of several tens of nanometers, thereby forming an extended wall. Depending on whether the spin rotation is perpendicular to the plane of the wall or parallel to it, two fundamental types of one-dimensional magnetic domain walls are distinguished: the Bloch type (Bloch 1932; Landau and Lifshitz 1935) and the Néel type (Néel 1955).

Unlike magnetic materials, in which the magnetization vector is allowed to rotate, e.g. across a wall, the polarization in ferroelectric systems is strongly coupled to the crystal lattice. This coupling results in a significant increase in energy for any misalignment between the polarization and the symmetry-allowed directions in the lattice (Zhirnov 1959). Therefore, DWs in ferroelectrics are typically of the Ising type (Pöykkö and Chadi 1999; Meyer and Vanderbilt 2002) (see Figure 7.8). In these model ferroelectric DWs, the polarization amplitude abruptly changes (within a few atomic cells) at the domain boundary region without rotation. Nevertheless, and even if rotation is limited in ferroelectrics, a swirling of the polarization was recently reported in various systems. Flux closure domains, resembling the Landau pattern commonly observed in magnetic systems, have been shown to exist in ferroelectric films close to electrodes (Aguado-Puente and Junquera 2008; Jia et al. 2011; Kim et al. 2014), in mesoscopic structures (McQuaid et al. 2011), and in ferroelectric/dielectric heterostructures (Tang et al. 2015; Lu et al. 2018). By controlling strain and polarization charge screening in ferroelectric thin films and multilayers, exotic topological structures like vortices (Peters et al. 2016; Yadav et al. 2016; Hong et al. 2017) and bubble domains (Yadav et al. 2016) were stabilized. Chiral ferroelectric structures have been lately discovered (Das et al. 2018; Shafer et al. 2018) and skyrmionic structures were predicted to exist in ferroelectrics, too (Nahas et al. 2015; Gonçalves et al. 2019).

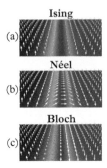

Figure 7.8 *Topology of fundamental domain walls showing (a) the conventional Ising-type walls as well as (b) Néel and (c) Bloch configurations. The arrows indicate the direction and the magnitude of the order parameter, in this case the polarization.*

Furthermore, recent theoretical work suggests that much more complex internal structures can be expected (Lee et al 2009, 2010; Angoshtari and Yavari 2010; Marton et al. 2010; Behera et al. 2011; Stepkova et al. 2012; Taherinejad et al. 2012; Yudin et al. 2012; Gu et al. 2014; Wojdeł and Íñiguez 2014) even in the simplest case of 180° DWs in uniaxial ferroelectrics, which were hitherto assumed to present a locally centrosymmetric Ising-like structure. Both continuum theory as well as atomistic models show a possible rotation of the polarization across ferroelectric DWs. In the case of nominally uncharged 180° DWs, this implies the existence of an in-plane component of the polarization oriented either perpendicular to the wall (Néel-like configuration) or parallel to it (Bloch-like configuration), as shown in Figure 7.8. These walls were first named by the Gopalan group according to the corresponding fundamental types of magnetic domain walls (Lee et al. 2009).

Motivated by theoretical predictions, intensive efforts have focused on the experimental observation of non-Ising walls. Transmission electron microscopy (TEM) has quickly emerged as the method of choice for the observation of domain wall regions (Jia et al. 2008; Li et al. 2013; Gonnissen et al. 2016; Wei et al. 2016; Lee et al. 2017) owing to its incomparable lateral resolution (see also Chapter 10). Yet, this technique is difficult to apply in a routine fashion. The sample preparation required for TEM is in general arduous and highly invasive. Also, the interaction with the intense electron beam in a TEM may alter the domain structure in certain systems (Scott and Kumar 2014).

Recently, the possibility of investigating the local structure of ferroelectric domain boundaries by means of SHG microscopy was successfully applied to study thermotropic phase boundaries in $BaTiO_3$ and $KNbO_3$ (Lummen et al. 2014), and to reveal Néel-type walls in tetragonal $Pb(Zr_{0.2},Ti_{0.8})O_3$ thin films with small a-domain incursions (De Luca et al. 2017) as well as in a pure c-domain phase (Cherifi-Hertel et al. 2017). A c-domain phase is considered a region where the polarization is oriented out-of-plane, while a-domains contain an in-plane polarization. The existence of chiral Bloch walls and the

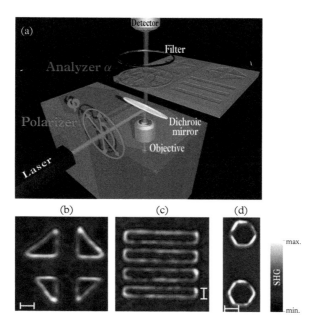

Figure 7.9 *(a) Schematic representation of the SHG experimental setup. The measurements are conducted using an inverted confocal microscope in the back-reflection geometry. The laser is focused at normal incidence with respect to the surface of the sample, and the SHG signal is collected in the collinear geometry. C-domains (in which the polarization is out-of-plane, pointing either upward or downward) of different shapes: (b) right triangles, (c) rectangles, and (d) hexagonal shape domains are written in $Pb(Zr_{0.2}, Ti_{0.8})O_3$ thin films (b,c) and in $LiTaO_3$ bulk crystals (d). Localized intensity is observed at the domain boundary regions. The bars in images (b,c) correspond to a scale of 2 μm and 15 μm in image (d).*

Source: Adapted from Cherifi-Hertel et al. 2017.

related Bloch lines was recently observed in periodically poled $LiTaO_3$ crystals by means of 3D confocal SHG microscopy (Cherifi-Hertel et al. 2017). The detailed methodology allowing the investigation of such non-Ising Néel and Bloch walls is described in the following.

7.2.3.1   *Method*

This method relies on polarimetry analysis of local SHG emission and a comparison of the experimental result to the numerical simulations in which the optical response is modeled using symmetry arguments and taking into account the wall orientation and topology.

SHG microscopy measurements are conducted by means of a scanning confocal microscope in the back reflection geometry (see Figure 7.9). This geometry is particularly interesting for the study of nontransparent materials and thin films since it does not require specific sample preparation prior to the experiment. This ensures the overall noninvasive character of the method, and allows a better lateral resolution with respect to the transmission geometry. Typically, a pulsed fs laser (100 fs pulses with a repetition rate of 80 MHz) is used with a fundamental wavelength in the range 750 nm to 1000 nm. The laser beam is directed at normal incidence to the sample, and focused through an objective lens (typical numerical aperture (NA) = 0.66). The SHG images are obtained by scanning the sample with respect to the incoming beam using computer-controlled stepping motors. The output intensity is spectrally filtered and transferred to a photomultiplier box for imaging. When a spectral analysis of the emission signal is necessary, a mirror is inserted to shift the emitted signal toward a spectrometer. Polarimetry measurements are performed by recording SHG images at different polarizer and analyzer angles.

The SHG signal is described by the fundamental equations of the second-order nonlinear process recalled earlier. Equation (7.2) describes the SHG signal obtained at a given polarization ϕ of the incident laser beam, and the polarization analysis of SHG emission is performed by applying the Jones formalism. This allows for a complete description of the SHG intensity $I^{SHG}(\varphi, \alpha) \approx \left| \boldsymbol{P}^{2\omega}(\varphi, \alpha) \right|^2$, if the symmetry properties of the material are known (i.e. the d_{ij} elements of the susceptibility tensor), and vice versa. By choosing a set of polarizations (φ and α angles) that is properly aligned with respect to the crystal orientation, the elements of the $\chi^{(2)}$ tensor can be directly accessed through polarimetry measurements. Quantitative evaluations tend, however, to be difficult as they require comparative measurements with respect to a reference system (e.g. quartz or ammonium dihydrogen phosphate) and a specific measurement geometry. Nevertheless, the qualitative probe of the $\chi^{(2)}$ elements provides valuable information on the symmetry and on the ferroic domain structure. This standard approach was adapted to the study of the internal structure of polar walls, taking advantage of the local character of the SHG emission at the wall regions as shown in Figure 7.9. The susceptibility tensor at the DWs can be gradually "constructed" through SHG measurements at selected (φ, α) pair values. The tensor that is thereby obtained is then analyzed with respect to the symmetry of the parent material while taking into account the expected polarization orientation. In practice, image series are recorded at each (φ, α) step. Integration of the intensity over a selected area of the image sequence yields the local SHG variation as a function of the analyzer angle. The obtained results are represented in polar plots and fitted using the analytic model given by Equation (7.2) assuming various DW types. The agreement or disagreement between the experimental data and the data obtained by assuming different analytic models allows us to infer the structure of the observed DWs.

The optical tensor is usually known for the case of regular symmetry of the parent material in which the polarization is oriented along the directions allowed by the symmetry (e.g. c-domains, in which the polarization is along the z-axis in tetragonal ferroelectrics). Modeling the SHG signal for Néel- or Bloch-type walls requires knowledge of the related nonlinear optical susceptibility tensor (i.e. the d_{ij} elements of the SHG tensor). Assuming that the optical tensor is isomorphic to the piezoelectric tensor, the same transformation operations that hold for the piezoelectric tensor at a given symmetry also apply to the SHG tensor. Any modification of the tensor corresponding to an arbitrary coordinate system with a specific ferroelectric polarization of the underlying material can thus be deduced using the rotation transformation relations as they are commonly applied to piezoelectric tensors (Newnham 2005; Kalinin et al. 2006). For this, it is sufficient to know the reference SHG tensor d^0 defined in the crystallographic reference frame of the parent phase. The matrix transformation method is used for the derivation of the SHG tensors at non-Ising DWs in tetragonal $Pb(Zr_{0.2},Ti_{0.8})O_3$ thin films and trigonal $LiTaO_3$. As a starting point, we consider the tensor corresponding to the ideal Ising-type DW structure d_{ij}^0, with the c-polarization axis along z. Any deviation from the model Ising case is characterized by three rotation angles ϕ, θ, ψ and the corresponding new tensor $d_{ij}^{New} = A_{ik}d_{kl}^0\alpha_{lj}^{-1}$, where A is the rotation matrix (the product of three individual rotations about the Euler angles ϕ, θ, ψ), and α^{-1} is the transformation matrix (its elements are functions of the directional cosines of the rotation matrix).

7.2.3.2 Néel Walls Revealed in Tetragonal $Pb(Zr_{0.2},Ti_{0.8})O_3$

Thin tetragonal $Pb(Zr_{0.2},Ti_{0.8})O_3$ films epitaxially grown on buffered $SrTiO_3(001)$ (Gariglio et al. 2007) are investigated by means of SHG microscopy and polarimetry to experimentally probe the internal structure of the walls. Domains of opposite polarization separated by nominally neutral 180° DWs are prepared by applying a bias voltage (± 8–10 V) between a conductive atomic force microscopy tip and the conductive buffer layer. The tip is scanned over a selected area to pattern the desired domains. The corresponding SHG images show a localized emission at the domain boundary regions (Figure 7.1(a), Figures 7.9(b,c)).

The spectral analysis of the optical signal reveals an unambiguous SHG process by demonstrating a frequency doubling process combined with a quadratic dependence of the intensity with the power of the fundamental wave (Figure 7.1). This result is particularly surprising since no SHG signal should be generated from c-domains in tetragonal systems at this measurement geometry with normal incidence. Furthermore, Ising-type DWs are centrosymmetric, and they exhibit a vanishing polarization at the center of the wall. Therefore, no SHG signal is expected at the domain wall regions either.

Further insight into the internal structure of the walls in tetragonal $Pb(Zr_{0.2},Ti_{0.8})O_3$ is obtained through an analysis of SHG polarimetry data measured at horizontal (HDWs) and vertical walls (VDWs) (Figures 7.10(a,b)). The polar plots arising from HDWs and VDWs fit well with the analytic model, accounting for an average in-plane polarization oriented perpendicular to the walls corresponding to a Néel-type configuration.

More precise 2D simulations of the SHG images are obtained by subdividing the DWs into discrete regions, in which the ferroelectric polarization is allowed to rotate. The $\chi^{(2)}$ SHG tensor is then calculated at each rotation angle. This allows calculation of the SHG intensity at any position (pixel) in the domains and in the domain wall regions for given polarizer and analyzer angles. The result is displayed in Figures 7.10(c,d) for both Néel and Bloch DW structures. The results obtained in the two cases are complementary. A comparison of these simulations with the experimental results presented in panel (a) of

EXPERIMENT

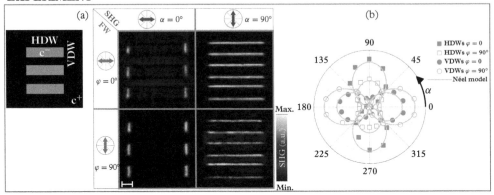

SIMULATIONS

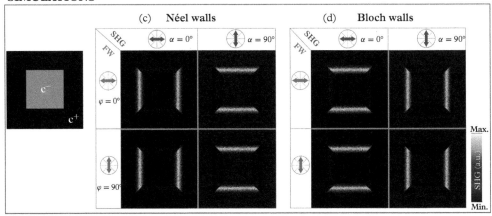

Figure 7.10 *Experimental study of the internal structure of 180° DWs in Pb(Zr$_{0.2}$,Ti$_{0.8}$)O$_3$ thin films showing (a) polarimetry SHG images recorded at different combinations of analyzer and polarizer angles. The complete polar plot recorded at both the horizontal (HDWs) and the vertical DWs (VDWs) of a rectangular shape c-domain is displayed in panel (b). 2D numerical simulations of the SHG polarimetry images are shown in (c) for a Néel-type internal structure, and in (d) for a Bloch-type internal structure of 180° ferroelectric walls in tetragonal Pb(Zr$_{0.2}$,Ti$_{0.8}$)O$_3$. The bar in image (a) corresponds to a scale of 2 μm.*

Source: Adapted from Cherifi-Hertel et al. 2017.

Figure 7.10 shows that the internal structure of the walls in tetragonal Pb(Zr$_{0.2}$,Ti$_{0.8}$)O$_3$ is of the Néel type (Cherifi-Hertel et al. 2017). An investigation of the same system in transmission geometry has led to the same conclusions (De Luca et al. 2017). The origin of the Néel character of the domain walls observed by De Luca et al (De Luca et al. 2017) was ascribed to the domain wall tilt across the film thickness. Theory predicts a Bloch DW structure in tetragonal PbTiO$_3$ (Wojdeł and Íñiguez 2014) at low temperature.

Since tetragonal $Pb(Zr_{0.2},Ti_{0.8})O_3$ is often used as a stand-in system for pure $PbTiO_3$, a similar temperature-driven DW transformation can be expected in $Pb(Zr_{0.2},Ti_{0.8})O_3$.

7.2.3.3 *Observation of Chiral Bloch Walls and Bloch Lines in Stoichiometric LiTaO₃*

As an independent test of this approach, a bulk crystal is selected instead of a thin film to exclude surface-related contributions that may affect the SHG polarimetry response (Ducuing and Bloembergen 1963; Shen 1999). A periodically poled nearly stoichiometric $LiTaO_3$ bulk crystal with trigonal point group symmetry is chosen as a test sample. To suppress spurious surface contributions to the SHG polarimetry signal, the measurements were performed at a focal distance of 100 to 300 μm below the surface. Figure 7.11(a) displays a clear SHG signal at the boundary regions of the alternated stripe domains. As opposed to $Pb(Zr_{0.2},Ti_{0.8})O_3$, the SHG polarimetry analysis shows maximum SHG polarization along the walls in $LiTaO_3$. Nevertheless, a small SHG signal is still measurable when the analyzer angle of the SHG polarization is perpendicular to the walls. This result is consistent with a dominant Bloch-like configuration and a weaker Néel component. This finding agrees with the theoretical predictions reported by Scrymgeour et al. showing the coexistence of longitudinal and transverse polarization components at 180° DWs in trigonal lithium niobate (Scrymgeour et al. 2005).

Bloch-type walls may exist in two variants since the ferroelectric polarization between c-domains of opposite sign can rotate either clockwise or counterclockwise across the wall. The absence of mirror symmetry makes Bloch walls chiral. This property became particularly interesting if the chirality changes not only from one wall to another, but also within the same wall, as in magnetic systems (Slonczewski 1974). The transition regions separating Bloch walls with opposite chirality form point defects forming lines in the bulk of magnetic crystals, and therefore they are called Bloch lines. These topological nanostructures have inspired the design of original solid state magnetic memory devices (Konishi 1983; Zimmermann et al. 1991) and have stimulated numerous studies among the magnetism community (Thiaville et al. 1991; Thiaville et al. 2001; Jourdan et al. 2009; Borich et al. 2016). It has been recently predicted that such structures should also exist in ferroelectric (Stepkova et al. 2015) and ferroelastic (Salje and Scott 2014; Salje and Carpenter 2015) systems.

Bloch lines at chiral Bloch-type walls are topological nanostructures occurring in the sample volume. Confocal microscopy is particularly suited for mapping the SHG response in 3D (Uesu et al. 2004; Kawado et al. 2008). Although SHG does not allow for the discrimination between DW segments with antiparallel internal polarization (because of its quadratic dependence on the susceptibility), a transverse transition (Néel-like) between two segments with opposite chirality (see the schematic representation of Bloch lines in Figure 7.11(b)) may result in the SHG image as a region with a different contrast (Cherifi-Hertel et al. 2017). The strong modulation of the SHG signal in the 3D profile of the wall (see Figure 7.11(c)) suggests an even more complex 3D wall structure involving both the wall distortion and a chirality transition. This structure is reminiscent of topological features known from micromagnetism.

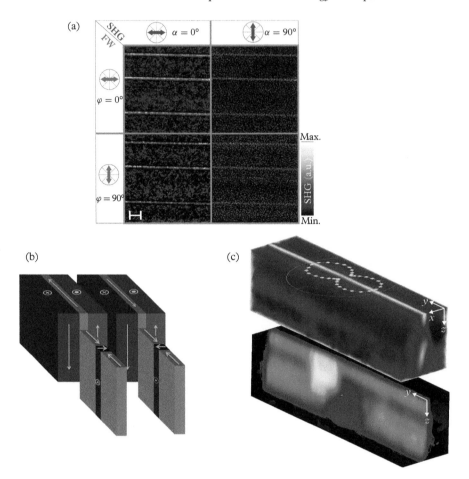

Figure 7.11 *(a) SHG Polarimetry analysis of 180° DWs in periodically poled LiTaO₃. The bar corresponds to a scale of 5 μm. (b) Schematic representation showing the two variants of Bloch walls. Bloch walls are chiral, and chirality reversal through transition regions forming Bloch lines is expected across the crystal. (b) 3D SHG showing the profile of a chiral Bloch wall in LiTaO₃ crystal through a depth of 100 μm.*

Source: Adapted from Cherifi-Hertel et al. 2017.

7.2.3.4 *Accuracy of SHG Polarimetry Measurements in Backward Geometry*

The quantitative interpretation of the SHG signal in back reflection geometry requires specific precautions. The SHG response can be particularly affected by phase matching effects. The back-reflected forward SHG process is characterized by a wavevector mismatch $\Delta \boldsymbol{k}_r = \boldsymbol{k}(2\omega) - 2\boldsymbol{k}(\omega)$, where $\boldsymbol{k}(\omega)$ and $\boldsymbol{k}(2\omega)$ are the wavenumbers of second-harmonic and fundamental waves, respectively. Both are ordinarily polarized in collinear SHG at the normal incidence angle of the fundamental wave. The very same configuration can also support backward SHG with a wavevector mismatch $\Delta \boldsymbol{k}_i = \boldsymbol{k}(2\omega) + 2\boldsymbol{k}(\omega)$, where $\boldsymbol{k}(2\omega)$ and $\boldsymbol{k}(\omega)$ are the same quantities defined above. In principle, both forward and backward signals contribute to the overall SHG intensity, although the former might be expected to be the dominant effect, in light of its larger coherence length (1.55 μm as opposed to a value of $\approx$50 nm for backward SHG). Therefore, the phase mismatch affects the total SHG efficiency. However, when the electromagnetic wave propagates along the z-axis, both the second-harmonic and the fundamental waves are ordinarily polarized (i.e. polarized in the xy-plane). In this case, the values of Δk_r, Δk_i, and of the propagation distance (given by the crystal thickness) are constants. Their contribution affects the SHG intensity, but it should not alter the shape of the SHG polar plots on which the local symmetry analysis is based. Note that this would not be the case if the experiments were done with waves propagating along x or y.

Additional effects such as surface contributions (Ducuing and Bloembergen 1963; Shen 1999) due to the strong discontinuity of the electric field induced by the symmetry breaking at the surface, birefringence (Brasselet et al. 2010), and focusing effects (Hsieh et al. 2010) can also induce distortions in SHG polarimetry measurements. Note that these effects are not specific to the backward detection geometry. Therefore, extreme care is recommended in general for the analysis of symmetry based on SHG polarimetry measurements.

7.2.4 Collinear Forward SHG at Zigzag Walls

Typical DWs in LiNbO$_3$ as prepared by electrical poling or manipulated through conductivity enhancement exhibit only small or moderate inclinations (up to several degrees), and their imaging in nonlinear microscopy is mainly accomplished through Cherenkov-type phase matching. However, it is also possible to create rather exotic domain configurations in LiNbO$_3$ by following a special heat treatment protocol (Kugel and Rosenman 1993). The resulting DWs exhibit a full head-to-head inclination of $\alpha = +90°$ (h2h) in the macroscopic average. Such DWs have been studied by scanning electron microscopy (Aristov et al. 1984; Kokhanchik and Irzhak 2007), (Kokhanchik 2009), conductive AFM (Schröder et al. 2012), and X-PEEM (Pawlik et al. 2017). Surprisingly, they turned out to be highly corrugated and topologically nontrivial on the microscale, exhibiting locally the full range of $\pm90°$ of inclination angles.

As shown in Figure 7.12, when such DWs in LiNbO$_3$ crystals are investigated by nonlinear optical microscopy through x and y surfaces, again a strong confinement of the SHG intensity to the DWs becomes apparent, which occasionally is superimposed by a pattern of interference fringes (Kämpfe et al. 2015). Their explanation involved a deeper analysis of the relevant phase matching processes, beyond the provision of reciprocal lattice vectors perpendicular to a single DW in the focus spot. In fact, it is possible that the axial length of the focus (primarily determined by the Rayleigh length, but also including aberration distortions from the high refractive index of LiNbO$_3$) is

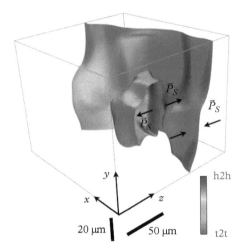

Figure 7.12 *Domain wall configuration in an undoped congruent LiNbO₃ crystal, obtained by a heating treatment (about 10 h above 1000 °C). The larger DW in the background runs through the whole sample with a macroscopically averaged inclination of +90° (full head-to-head, depicted in red). In the foreground, a detached segment is visible, which exhibits the full range of possible inclinations, even −90° (depicted in blue).*

large enough to include two DWs, or two separate spots on the same DW, which act as interfering coherent emitters. As SHG is also generated at the sample surface ("surface SHG"), interference can also occur when a single DW and the surface are both within the focal volume. In all these situations, collinear phase matching becomes possible, i.e. the direct emission of an SHG beam in the forward direction (see Figure 7.13). The two DW positions, or DW and crystal surface, form in this case a rudimentary lattice with reciprocal lattice vectors of various magnitudes aligned parallel to k_0. The interference fringes are due to the spatially varying distance between the two DWs (or DW and surface), which lead to either constructive or destructive interference at the remote detector (see Figure 7.14).

The benefit of this effect for DW characterization is that it allows extrapolation of the recorded DW position from a single 2D scan into the third dimension (i.e. perpendicular to the scanning plane). Hence, 3D information is accessible through a 2D scan, which increases the acquisition speed and may allow for dynamic characterizations. The simplest configuration in which this idea can be exploited is the interference between SHGs from a DW and a planar crystal surface, with the scanning plane parallel to this surface. In this case, the periodicity of the interference fringes is directly related to the DW inclination with respect to the scanning plane, as explained in detail in Kämpfe et al. (2015). Correspondingly, the analysis of the interference signal from two points on the same DW will yield a spatial autocorrelation of the DW interface shape.

(a)

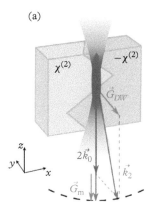

Figure 7.13 *Differentiation between collinear and noncollinear (Cherenkov) phase matching in second-harmonic generation. In the collinear case, two positions of the same DW within the focus volume temporally form a rudimentary lattice, providing the needed reciprocal vector G_m aligned parallel to the optical axis. In contrast, the Cherenkov process relies on the G_{DW} from a single location on the DW and perpendicular to it.*

Source: Reprinted from Kämpfe et al. 2015, with the permission of AIP Publishing.

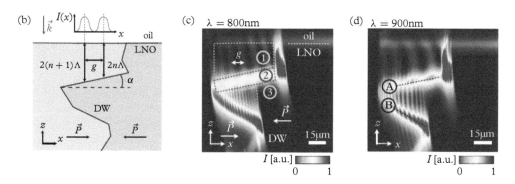

Figure 7.14 *Interference fringes in the recorded SHG signal from collinear phase matching. (b) Sketch of the geometrical relation between the DW and the crystal surface, with the latter acting as the second SHG emission source within the focus volume. For a constant inclination toward the surface, the distance between the surface and the DW increases linearly with lateral distance, leading to a constant width g of the fringes. (c) and (d) x, z scans at $\lambda = 800$ nm and 900 nm, respectively. The coordinates are named according to the laboratory system, with z denoting the optical axis of the microscope and x the polar axis of the crystal. As expected, all fringe distances scale proportional to λ. Two different fringe systems can be distinguished. The one with the larger periodicity length in the upper part of the images is generated by DW-surface interference, while the other with narrower stripes and higher intensity is due to DW-DW interference.*

Source: Reprinted from Kämpfe et al. 2015, with the permission of AIP Publishing.

7.3   Linear Optical Coherence Tomography at DWs

DWs are not only capable of enhancing the SHG response in $LiNbO_3$ by providing phase matching in their vicinity, but also scatter light in the more conventional regime of linear optics. An elegant method of observing this effect in backscattering geometry is spectral domain optical coherence tomography (SD-OCT), a technique widely employed in biology and medicine for studying the optical properties of various tissues such as the retina. The sample is inserted in one arm of a Michelson interferometer which is operated with a coherent broad band source. 3D images of the backscattering efficiency throughout the sample volume can be obtained by scanning in 2D only, as the third (axial) dimension is spectrally encoded and can be reconstructed by a Fourier transform of the spectrometer signal (see Figure 7.15). If a sample consists of a stack of weakly reflecting (and otherwise transparent) layers, each layer depth (= constant path length difference between the interferometer arms) corresponds to a characteristic wavenumber oscillation period in the recorded intensity, stemming from the interference of the light backscattered from the sample with the light from the reference arm. The spectral encoding of one dimension means that the method is inherently fast and thus well suited for the investigation of DW dynamics. Moreover, quantitative results for the backscattered intensity (in dB) are routinely obtained.

Periodically poled $LiNbO_3$ (PPLN) constitutes a prototype sample for OCT measurements, as a large number of aligned DWs is optically accessible through an almost perfectly transparent bulk. While early measurements could only trace the first few DWs below the crystal surface (Pei et al. 2011), the preparation of a slanted entry face (to remove surface reflection from the detection path; see Figure 7.16 for the setup) and application of image reconstruction algorithms for the correction of the nonlinear dispersion allow the backscattering signal to be recorded from more than 100 DWs without loss in spatial resolution (Haußmann et al. 2017; Kirsten et al. 2017). The typical acceptance angle lies in the range of about $\pm 1°$; i.e. the scattered signal from DWs with a higher inclination is not recorded by the interferometer.

OCT is a technique which neatly bridges the two objectives of applying optical techniques to DWs: (1) 3D geometrical information can be obtained reliably and quickly, and (2) the internal physical properties of DWs which manifest themselves in the optical scattering signal are directly accessible. Concerning point (1), the versatility of OCT is most impressively shown by resolving the evolution of DWs under external fields above the coercive threshold in the x, y-plane at typical video frame rate speeds (e.g. 16 fps). Figures 7.17(a,b) show two single frames from such a video observation which were taken within 4 s, while Figure 7.17(c) displays their respective difference. In the difference image, black features correspond to DWs that vanished over time, while white features represent DWs that popped up in the later image, either by being shifted to a new location or by a realignment of their orientation which allowed for a better match between the scattering direction and the acceptance angle range. The sample used in these experiments was a 5% Mg-doped PPLN single crystal (supplier Crystal Technology, Inc.) with a typical DW-DW distance of $\approx 3.5 \, \mu m$. For the measurements, a laboratory

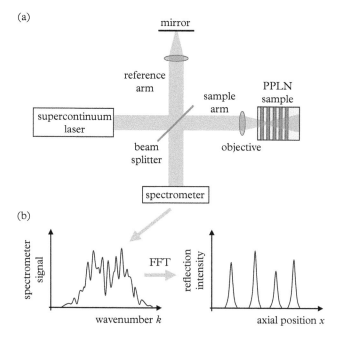

Figure 7.15 *Principle of spectral domain optical coherence tomography (SD-OCT): (a) A supercontinuum laser is fed into a Michelson interferometer. The light in the sample arm is focused with a low NA into a ferroelectric sample containing aligned DWs, e.g. periodically poled LiNbO₃. Back-reflections from DWs at various depths are collected by the same objective and superposed with the reference signal on the detector. (b) Interference with the light from the reference arm results in a characteristic spectral oscillation that allows reconstructing the spatial (axial) position through the fast Fourier transformation (FFT). Note that the beam position on the sample is varied in the optical x, y frame through galvoscanners.*

Source: *Reprinted with permission from Haußmann et al. 2017.*

setup consisting of a supercontinuum laser (NKT Photonics) in the spectral range from 541 nm to 927 nm, together with a 4096 pixel spectrometer detector allowing for a maximal readout rate of 140 kHz (Basler Sprint spL4096-140km) was used. This system allowed for an axial resolution of 0.28 nm in LiNbO₃, with a lateral resolution of ≈4 μm.

Regarding internal DW properties (2), the backscattering signal can be analyzed with respect to various incident (optical) polarizations, and under external electrical fields of variable strengths. In our experiments on 5% Mg-doped PPLN, we noted strong backscattering signals up to –57 dB in extraordinary polarization, i.e. with the optical electric field vector parallel to the polar (z) axis (see Figure 7.18). In the perpendicular (ordinary) polarization, the backscattering signal is drastically reduced, and even falls below the noise level of the detector (about –87 dB). This behavior turned out to be perfectly stable under the application of static external fields, yielding identical OCT images of the DW stack up to the coercive field, at which DW movements set in.

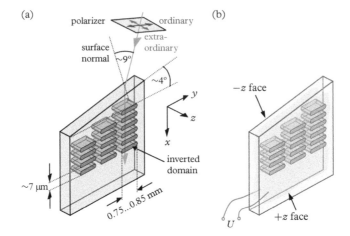

Figure 7.16 *(a) Illumination and domain structure geometry of a periodically poled LiNbO₃ sample in an optical coherence tomography measurement. Coordinate directions correspond to the conventional crystallographic axes in LiNbO₃. The incident light passes a polarizer that is adjusted to probe either the extraordinary (green) or ordinary (blue) light polarization. (b) Geometry of the applied silver paint electrodes. Note the safety rim of ≈0.4 mm along the borders that prevents sparking or electrical breakdown.*

Source: Reprinted with permission from Haußmann et al. 2017.

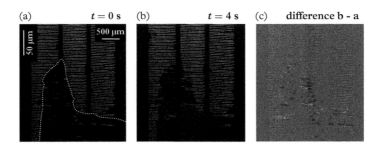

Figure 7.17 *DW movements induced by an external electric field above the switching threshold. (a) and (b) SD-OCT images of the x, y-plane at the central z position for t = 0 s and t = 4 s in 5% Mg:PPLN. At t = 0 s the electrical field of 3.2 kV/mm was abruptly switched on. The periodic DW structure had, however, been already damaged under the electrode during previous experiments. The dotted line corresponds approximately to the edge of the hand-painted silver electrode. (c) difference image of (b) and (a). Deviations from neutral grey colour correspond to shifted, erased or reappearing DWs. Note that the vertical scaling (OCT axial direction, crystallographic x axis) is different from the horizontal (y) axis.*

Source: Reprinted with permission from Haußmann et al. 2017.

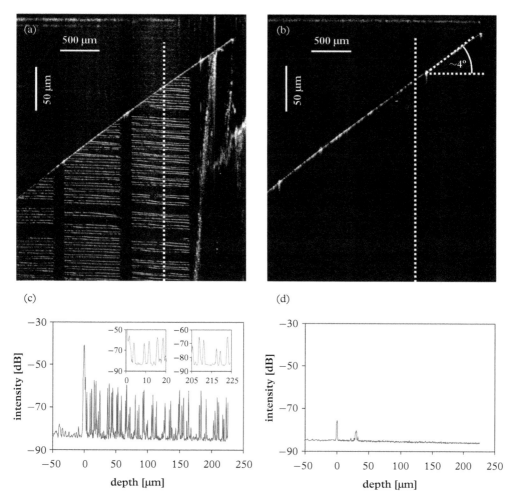

Figure 7.18 *Polarization dependence of the OCT contrast in 5% Mg:PPLN (central x, y plane). (a) Polarizer set along the extraordinary polarization direction, and (b) along the ordinary direction. The slanted bright line in both images corresponds to the sample surface. The spurious signals on the right side of (a) are not related to the domain structure and might be due to reflections from the samples edges. (c) and (d) Signal profile sections along the dotted lines in (a) and (b). Zero depth corresponds to the surface. Insets in (c) show that the vertical resolution close to the sample surface does not deteriorate for depths as large as 200 μm. As seen from (d), no DW contrast above the noise level could be resolved for ordinary polarization. The peak near 30 μm is probably caused by spurious reflections from the focusing optics and/or polarizer.*

Source: Reprinted with permission from Haußmann et al. 2017.

Even then, the signal contrast itself did not change, allowing recording of the previously discussed dynamic behavior.

The surprising fact is that an influence of external fields is expected, as the electro-optic effect common to most ferroelectrics leads to a refractive index difference between the bulk domains on both sides of the DW, thereby causing a nonzero reflectance according to Fresnel's formulas (suitably modified for anisotropic media). Moreover, this effect is not expected to show a strong polarization dependence. $LiNbO_3$ is known to exhibit an internal field, which causes electro-optic index steps between domains even without an external field. The behavior of the reflectance studied in other experiments using optical Bragg scattering under almost grazing incidence did indeed match the expected field dependence of the electro-optic effect, including the compensation of the internal field (Kösters et al. 2006). The fact that this behavior was not noticeable in OCT clearly indicates that a genuine optical property of the DW was observed here, caused either by refractive index spikes or kinks localized at the DW or by the metallic reflectivity of the electron gas itself.

Mechanical stress at DWs can induce such index variations by the elasto-optic effect or indirectly via the combined action of the piezoelectric and electro-optic effects. A first step toward an experimental quantification of the index profile in $LiNbO_3$ (congruent, undoped) was taken by Kim and Gopalan, using an aperture (fiber) scanning near field optical microscope (SNOM) in collection mode. From the recorded signal, the index profile was reconstructed (Kim 2003; Kim and Gopalan 2005b). However, it has to be noted that the refractive index structure can be more complicated than in the uniaxially birefringent bulk, due to the possibility of inducing biaxial birefringence at DWs (Yang and Mohideen 1998). Biaxiality at DWs was also predicted by a microscopic model based on the quantum mechanical orbital approximation for ions in the crystal lattice combined with dipole–dipole interaction for 180° DWs in tetragonal $BaTiO_3$ (Chaib et al. 2002).

More interesting from the viewpoint of DW conductivity is the possibility that the charge carrier accumulation at DWs, assumed to behave as a metallic 2D electron gas in the ideal case, can also give rise to a reflectivity due to a localized increase in the imaginary part of the refractive index. Within the Drude model framework, an interface between a metallic bulk and vacuum gradually loses its high reflectivity when the so-called *plasma frequency* is exceeded. The situation of a narrow layer of charge carriers embedded in a semiconductor does, however, require a more detailed analysis. The optical behavior of a Drude 2D electron gas under these conditions has already been studied in the 1980s, albeit without any connection to ferroelectric DWs (Kuijk and Vounckx 1989). Important parameters that define the optical behavior of the electron gas layer are the charge carrier concentration n_e, their effective mass m_{eff} and mobility μ, as well as the background permittivity ε_b of the semiconductor. For a completely screened charged DW with the inclination α, the carrier concentration can be estimated from the polarization charge and the thickness of the screening layer d_{DW} to

$$n_e \approx \frac{2 \cdot P_S \cdot \sin\alpha}{e \cdot d_{DW}} \qquad (7.6)$$

For LiNbO$_3$, such data (apart from the spontaneous polarization $P_S = 0.7$ C/m^2, and the inclination angles which can be measured by CSHG and OCT) are only scarcely available and suffer from great uncertainties. Rough estimations using values of $d_{DW} \approx 20$ nm, $m_{eff} \approx 0.05\,m_e$ (Eliseev et al. 2011), $\mu \approx 0.8 \dots 830$cm^2/Vs (Ionita and Jaque 1998), and $\varepsilon_b = 5$ result in an increase in the DW reflectance over the visible range into the mid-IR (neglecting phononic resonances that drastically influence the spectral behavior of ε_b). There is, however, the possibility of increasing the carrier density, and, accordingly, the plasma frequency, by flooding the conductive sheet with electrons through forced injection using sufficient voltages. As a result, plasmonic effects at DWs might be accessible even in the visible wavelength range. Moreover, if the strong polarization contrast of the OCT signal is indeed due to the electron gas, this would imply a strong anisotropy in material parameters such as the effective mass and mobility.

7.4 Future Perspectives

In spite of the great progress recently made in the investigation of non-Ising domains walls, several questions remain regarding the origin of the deviation from the ideal Ising configuration as well as the connection between the wall conductivity, its deformation, orientation with respect to the crystal axes, and its internal structure. The combination of advanced theoretical studies with experiments, including innovative DW engineering methods and optical observations supported by the modeling of the optical signal, is expected to shed new light on the mechanisms at play in non-Ising DWs. The results presented in this chapter show that the convolution of a focused laser beam with extremely narrow ferroelectric DWs results in an artificially broadened emission at the domain wall regions replicating the size of the fundamental beam. Although this convolution process makes local SHG polarimetry analysis easier, the SHG profile at DW regions is not relevant for obtaining reliable information on the DW size. Future development of proper deconvolution methods will allow a precise study of the profile of non-Ising walls and its evolution during propagation.

· ·

REFERENCES

Aguado-Puente P., Junquera, J., "Ferromagneticlike closure domains in ferroelectric ultrathin films: first-principles simulations," *Phys. Rev. Lett.* 100(17), 177601 (2008).
Angoshtari A., Yavari A., "Atomic structure of steps on 180° ferroelectric domain walls in PbTiO$_3$," *J. Appl. Phys.* 108(8), 084112 (2010).
Aristov V. V., Kokhanchik L. S., Voronovskii Y. I., "Voltage contrast of ferroelectric domains of lithium niobate in SEM," *Phys. Status Solidi (A)* 86(1), 133–141 (1984).
Ayoub M., Futterlieb H., Imbrock J., Denz C., "3D imaging of ferroelectric kinetics during electrically driven switching," *Adv. Mater.* 29(5), 1603325 (2017).
Béa H., Paruch P. "A way forward along domain walls," *Nat. Mater.* 8, 168–169 (2009).

Behera R. K., Lee C.-W., Lee D., Morozovska A. N., Sinnott S. B., Asthagiri A., Gopalan V., Phillpot S. R., "Structure and energetics of $180°$ domain walls in $PbTiO_3$ by density functional theory," *J. Phys.: Condens. Mat.* 23(17), 175902 (2011).

Berth G., Quiring V., Sohler W., Zrenner A., "Depth-resolved analysis of ferroelectric domain structures in Ti:PPLN waveguides by nonlinear confocal laser scanning microscopy," *Ferroelectrics* 352(1), 78–85 (2007).

Bloch F., "Zur Theorie des Austauschproblems und der Remanenzerscheinung der Ferromagnetika," *Z. Phys.* 74(5–6), 295–335 (1932).

Bloembergen N., *Nonlinear Optics*, 4 edn. (World Scientific, Singapore, 1996).

Borich M. A., Tankeev A. P., Smagin V. V., "Micromagnetic structure of the domain wall with Bloch lines in an electric field," *Phys. Solid State* 58(7), 1375–1383 (2016).

Bozhevolnyi S. I., Hvam J. M., "Second-harmonic imaging of ferroelectric domain walls," *Appl. Phys. Lett.* 73(13), 1814–1816 (1998).

Bozhevolnyi S. I., Pedersen K., Skettrup T., Zhang X., Belmonte M., "Far- and near-field second-harmonic imaging of ferroelectric domain walls," *Opt. Commun.* 152(4–6), 221–224 (1998).

Brasselet S., Aït-Belkacem D., Gasecka A., Munhoz F., Brustlein S., Brasselet, S., "Influence of birefringence on polarization resolved nonlinear microscopy and collagen SHG structural imaging," *Opt. Express* 18(14), 14859–14870 (2010).

Campbell M. P., McConville J. P. V., McQuaid R. G. P., Prabhakaran D., Kumar A., Gregg J. M., "Hall effect in charged conducting ferroelectric domain walls," *Nat. Commun.* 7, 1–6 (2016).

Catalan G., Seidel J., Ramesh R., Scott J., "Domain wall nanoelectronics," *Rev. Mod. Phys.* 84(1), 119–156 (2012).

Chaib H., Otto T., Eng L. M., "Theoretical study of ferroelectric and optical properties in the $180°$ ferroelectric domain wall of tetragonal $BaTiO_3$," *Phys. Status Solidi (B)* 233(2), 250–262 (2002).

Chauleau J.-Y., Haltz E., Carrétéro C., Fusil S., Viret M., "Multistimuli manipulation of antiferromagnetic domains assessed by second-harmonic imaging," *Nat. Mater.* 16(8), 803–807 (2017).

Cherifi-Hertel S., Bulou H., Hertel R., Taupier G., Dorkenoo K. D. H., Andreas C., Guyonnet J., Gaponenko I., Gallo K., Paruch P., "Non-Ising and chiral ferroelectric domain walls revealed by nonlinear optical microscopy," *Nat. Commun.* 8, 75794 (2017).

Coppola G., Ferraro P., Iodice M., De Nicola S., Grilli S., Mazzotti D., De Natale P., "Visualization of optical deflection and switching operations by a domain- engineered-based $LiNbO_3$ electro-optic device," *Opt. Express* 11(10), 1212–1222 (2003).

Das S., Ghosh A., McCarter M. R., Hsu S. L., Tang Y. L., Damodaran A. R., Ramesh R., Martin L. W. "Perspective: emergent topologies in oxide superlattices," *APL Mater.* 6(10), 100901 (2018).

De Luca G., Rossell M. D., Schaab J., Viart N., Fiebig M., Trassin M., "Domain wall architecture in tetragonal ferroelectric thin films," *Adv. Mater.* 29(7), 1–5 (2017).

Denev S. A., Lummen T. T. A., Barnes E., Kumar A., Gopalan V., "Probing ferroelectrics using optical second harmonic generation," *J. Am. Ceram. Soc.* 94(9), 2699–2727 (2011).

Deng X., Chen X., "Domain wall characterization in ferroelectrics by using localized nonlinearities," *Opt. Express*, 18(15), 15597 (2010).

Ducuing J., Bloembergen N., "Observation of reflected light harmonics at the boundary of piezoelectric crystals," *Phys. Rev. Lett.* 10(11), 474 (1963).

Eliseev E. A., Morozovska A. N., Svechnikov G. S., Gopalan V., Shur V. Y. "Static conductivity of charged domain walls in uniaxial ferroelectric semiconductors," *Phys. Rev. B* 83(23), 235313 (2011).

Ferraro P., Grilli S., De Natale P. (Eds.), *Ferroelectric Crystals for Photonic Applications*, 1 edn., Volume 91 (Springer Series in Materials Science, Springer Berlin Heidelberg, Berlin, Heidelberg, 2009).

Fiebig M., Fröhlich D., Lottermoser T., Maat M., "Probing of ferroelectric surface and bulk domains in $RMnO_3$ ($R=$ Y, Ho) by second harmonic generation," *Phys. Rev. B* 66(14), 144102 (2002a).

Fiebig M., Lottermoser T., Fröhlich D., Goltsev A. V., Pisarev R. V., "Observation of coupled magnetic and electric domains," *Nature* 419(6909), 818 (2002b).

Flörsheimer M., Paschotta R., Kubitscheck U., Brillert C., Hofmann D., Heuer L., Schreiber G., Verbeek C., Sohler W., Fuchs H., "Second-harmonic imaging of ferroelectric domains in $LiNbO_3$ with micron resolution in lateral and axial directions," *Appl. Phys. B* 67(5), 593–599 (1998).

Fragemann A., Pasiskevicius V., Laurell F., "Second-order nonlinearities in the domain walls of periodically poled $KTiOPO_4$," *Appl. Phys. Lett.* 85(3), 375–377 (2004).

Gahagan K. T., Gopalan V., Robinson J. M., Jia Q. X, Mitchell T. E., Kawas M. J., Schlesinger T. E., Stancil D. D., "Integrated electro-optic lens/scanner in a $LiTaO_3$ single crystal," *Appl. Opt.* 38(7), 1186 (1999).

Gariglio S., Stucki N., Triscone J.-M., Triscone G., "Strain relaxation and critical temperature in epitaxial ferroelectric $Pb(Zr_{0.20}Ti_{0.80})O_3$ thin films," *Appl. Phys. Lett.* 90(20), 202905 (2007).

Godau C., Kämpfe T., Thiessen A., Eng L. M., Haumann A., "Enhancing the domain wall conductivity in lithium niobate single crystals," *ACS Nano* 11(5), 4816–4824 (2017).

Gonçalves M. A., Pereira, E., Sayalero C., Garca-Fernández P., Junquera J., Íñiguez J., "Theoretical guidelines to create and tune electric skyrmion bubbles," *Sci. Adv.* 5(2), eaau7023 (2019).

Gonnissen J., Batuk D., Nataf G. F., Jones L., Abakumov A. M., Van Aert S., Schryvers D., Salje E. K. H., "Direct observation of ferroelectric domain walls in $LiNbO_3$: wall-meanders, kinks, and local electric charges," *Adv. Funct. Mater.* 26(42), 7599–7604 (2016).

Gopalan V., Dierolf V., Scrymgeour D. A., "Defect domain wall interactions in trigonal ferro-electrics," *Annu. Rev. Mater. Res.* 37(1), 449–489 (2007).

Gu Y., Li M., Morozovska A. N., Wang Y., Eliseev E. A., Gopalan V., Chen L.-Q., "Flexoelectricity and ferroelectric domain wall structures: phase-field modeling and DFT calculations.," *Phys. Rev. B* 89(17), 174111 (2014).

Guyonnet J., Gaponenko I., Gariglio S., Paruch P., "Conduction at domain walls in insulating $Pb(Zr_{0.2}Ti_{0.8})O_3$ thin films," *Adv. Mater.* 23(45), 5377–5382 (2011).

Haußmann A., Kirsten L., Schmidt S., Cimalla P., Wehmeier L., Koch E., Eng L. M., "Three-dimensional, time- resolved profiling of ferroelectric domain wall dynamics by spectral-domain optical coherence tomography," *Ann. Phys.* 529(8), 1700139 (2017).

Hong Z., Damodaran A. R., Xue F., Hsu S.-L., Britson J., Yadav A. K., Nelson C. T., Wang J.-J., Scott J. F., Martin L. W., Ramesh R., Chen L.-Q., "Stability of polar vortex lattice in ferroelectric superlattices," *Nano Lett.* 17(4), 2246–2252 (2017).

Houchmandzadeh B., Lajzerowicz J., Salje E., "Order parameter coupling and chirality of domain walls," *J. Phys.: Condens. Mat.* 3(27), 5163–5169 (1991).

Hsieh C.-L., Pu Y., Grange R., Psaltis D., "Second harmonic generation from nanocrystals under linearly and circularly polarized excitations," *Opt. Express* 18(11), 11917–11932 (2010).

Hu X. P., Xu P., Zhu S. N., "Engineered quasi-phase-matching for laser techniques [Invited]," *Photon. Res.* 1(4), 171–185 (2013).

Ionita I., Jaque, F., "Photoconductivity and electron mobility in $LiNbO_3$ co-doped with Cr^{3+} and MgO," *Opt. Mater.* 10(2), 171–173 (1998).

Janovec V., Přívratská J., "Domain structures," in: *International Tables for Crystallography* (International Union of Crystallography, Chester, England, 2006, pp. 449–505).

Jia C.-L., Mi S.-B., Urban K., Vrejoiu I., Alexe M., Hesse D., "Atomic-scale study of electric dipoles near charged and uncharged domain walls in ferroelectric films," *Nat. Mater.* 7(1), 57–61 (2008).

Jia C.-L., Urban K. W, Alexe M., Hesse D., Vrejoiu, I., "Direct observation of continuous electric dipole rotation in flux-closure domains in ferroelectric Pb(Zr,Ti)O$_3$," *Science* 331(6023), 1420–1423 (2011).

Jourdan T., Masseboeuf A., Laon F., Bayle-Guillemaud P., Marty A. "Magnetic bubbles in FePd thin films near saturation," *J. Appl. Phys.* 106(7), 0–8 (2009).

Kalinin S. V., Rodriguez B. J., Jesse S., Shin J., Baddorf A. P., Gupta P., Jain H., Williams D. B., Gruver-man A., "Vector Piezoresponse Force Microscopy," *Microsc. Microanal.* 12(03), 206–220 (2006).

Kämpfe T., *Charged Domain Walls in Ferroelectric Single Crystals*, Ph.D. thesis (Technische Universität Dresden, 2016).

Kämpfe T., Reichenbach P., Haußmann A., Woike T., Soergel E., Eng L. M., "Real-time three-dimensional profiling of ferroelectric domain walls," *Appl. Phys. Lett.* 107(15), 152905 (2015).

Kämpfe T., Reichenbach P., Schröder M., Haußmann A., Eng L. M., Woike T., Soergel E., "Optical three-dimensional profiling of charged domain walls in ferroelectrics by Cherenkov second-harmonic generation," *Phys. Rev. B* 89(3), 035314 (2014).

Kawado S., Yokota H., Kaneshiro J., Uesu Y., "SHG tomography for domain structure observations—an application to 3D visualizations of periodically poled domains," *Ferroelectrics* 368(1), 23–27 (2008).

Kim S., *Optical, Electrical and Elastic Properties of Ferroelectric Domain Walls in LiNbO$_3$ and LiTaO$_3$*, Ph.D. thesis (Pennsylvania State University, 2003).

Kim S., Gopalan V., "Optical index profile at an antiparallel ferroelectric domain wall in lithium niobate," *Mater. Sci. Eng.: B* 120(1–3), 91–94 (2005a).

Kim S., Gopalan V., "Optical index profile at an antiparallel ferroelectric domain wall in lithium niobate," *Mater. Sci. Eng.: B* 120(1), 91–94 (2005b).

Kim Y.-M., Morozovska A., Eliseev E., Oxley M. P., Mishra R., Selbach S. M., Grande T., Pantelides S. T., Kalinin S. V., Borisevich A. Y., "Direct observation of ferroelectric field effect and vacancy-controlled screening at the BiFeO$_3$/La$_x$Sr$_{1-x}$MnO$_3$ interface," *Nat. Mater.* 13(11), 1019–1025 (2014).

Kim Y., Alexe M., Salje E. K. H., "Nanoscale properties of thin twin walls and surface layers in piezoelectric WO$_{3-x}$," *Appl. Phys. Lett.* 96(3), 032904 (2010).

Kirsten L., Haußmann A., Schnabel C., Schmidt S., Cimalla P., Eng L. M., Koch E., "Advanced analysis of domain walls in Mg doped LiNbO$_3$ crystals with high resolution OCT," *Opt. Express* 25(13), 14871–14882 (2017).

Kokhanchik L. S., "The use of surface charging in the SEM for lithium niobate domain structure investigation," *Micron* 40(1), 41–45 (2009).

Kokhanchik L., Irzhak D., "Investigation of periodic domain structures in lithium niobate crystals," *Ferroelectrics* 352(1), 134–142 (2007).

Konishi S., "A new-ultra-density solid state memory: Bloch line memory," *IEEE Trans. Magnetics* 19(5), 1838–1840 (1983).

Kösters M., Hartwig U., Woike T., Buse K., Sturman B., "Quantitative characterization of periodically poled lithium niobate by electrically induced Bragg diffraction," *Appl. Phys. Lett.* 88(18), 182910 (2006).

Kugel V. D., Rosenman G., "Domain inversion in heat-treated LiNbO$_3$ crystals," *Appl. Phys. Lett.* 62(23), 2902–2904 (1993).

Kuijk M., Vounckx R., "Optical plasma resonance in semiconductors: novel concepts for modulating far-infrared light," *J. Appl. Phys.* 66(4), 1544–1548 (1989).

Landau L., Lifshits E., "On the theory of the dispersion of ferromagnetic bodies," *Phys. Zeitsch. der Sow.* 8(14), 153–169 (1935).

Lee D., Behera R. K., Wu P., Xu H., Li Y. L., Sinnott S. B., Phillpot S. R., Chen L. Q., Gopalan V., "Mixed Bloch-Néel-Ising character of 180° ferroelectric domain walls," *Phys. Rev. B* 80(6), 060102 (2009).

Lee D., Xu H., Dierolf V., Gopalan V., Phillpot S. R., "Structure and energetics of ferroelectric domain walls in LiNbO$_3$ from atomic-level simulations," *Phys. Rev. B* 82(1), 014104 (2010).

Lee M. H., Chang C.-P., Huang F.-T., Guo G. Y., Gao B., Chen C. H., Cheong S.-W., Chu M.-W., "Hidden antipolar order parameter and entangled Néel-type charged domain walls in hybrid improper ferroelectrics," *Phys. Rev. Lett.* 119(15), 157601 (2017).

Li L., Gao P., Nelson C. T., Jokisaari J. R, Zhang Y., Kim S. J., Melville A., Adamo C., Schlom D. G., Pan X., "Atomic scale structure changes induced by charged domain walls in ferroelectric materials," *Nano Lette.* 13(11), 5218–5223 (2013).

Lu L., Nahas Y., Liu M., Du H., Jiang Z., Ren S., Wang D., Jin L., Prokhorenko S., Jia C. L., Bellaiche L., "Topological defects with distinct dipole configurations in PbTiO$_3$/SrTiO$_3$ multilayer films," *Phys. Rev. Lett.* 120(17), 1–6 (2018).

Lummen T. T. A., Gu Y., Wang J., Lei S., Xue F., Kumar A., Barnes A. T., Barnes E., Denev S., Belianinov A., Holt M., Morozovska A. N., Kalinin S. V., Chen L.-Q., Gopalan V., "Thermotropic phase boundaries in classic ferroelectrics," *Nat. Commun.* 5, 3172 (2014).

Marton P., Rychetsky I., Hlinka J., "Domain walls of ferroelectric BaTiO$_3$ within the Ginzburg-Landau-Devonshire phenomenological model," *Phys. Rev. B* 81(14), 144125 (2010).

McQuaid R. G. P., McGilly L. J., Sharma P., Gruverman A., Gregg J. M., "Mesoscale flux-closure domain formation in single-crystal BaTiO$_3$," *Nat. Commun.* 2(1), 404 (2011).

Meier D., Maringer M., Lottermoser T., Becker P., Bohaty L., Fiebig M., "Observation and coupling of domains in a spin-spiral multiferroic," *Phys. Rev. Lett.* 102(10), 107202 (2009).

Meyer B., Vanderbilt, D., "Ab initio study of ferroelectric domain walls in PbTiO$_3$," *Phys. Rev. B* 65(10), 104111 (2002).

Mundy J. A., Schaab J., Kumagai Y., Cano A., Stengel M., Krug I. P., Gottlob D. M., Doğanay H., Holtz M. E., Held R., Yan Z., Bourret E., Schneider C. M., Schlom D. G., Muller D. A., Ramesh R., Spaldin N. A., Meier D., "Functional electronic inversion layers at ferroelectric domain walls," *Nat. Mater.* 16(6), 622–627 (2017).

Nahas Y., Prokhorenko S., Louis L., Gui Z., Kornev I., Bellaiche L., "Discovery of stable skyrmionic state in ferroelectric nanocomposites," *Nat. Commun.* 6(1), 8542 (2015).

Nataf G. F., Guennou M., Haußmann A., Barrett N., Kreisel J., "Evolution of defect signatures at ferroelectric domain walls in Mg-doped LiNbO$_3$," *Phys. Status Solidi—Rapid Res. Lett.* 10(3), 222–226 (2016).

Néel L., "Energie des parois de Bloch dans les couches minces," *Comptes rendus hebdomadaires des séances de l'Académie des sciences* 241(35), 535–536 (1955).

Newnham R. E., *Properties of Materials*, 1 edn (Oxford University Press, Oxford, 2005).

Nordlander J., De Luca G., Strkalj N., Fiebig M., Trassin M., "Probing ferroic states in oxide thin films using optical second harmonic generation," *Appl. Sci.* 8(4), 570 (2018).

Pawlik A.-S., Kämpfe T., Haußmann A., Woike T., Treske U., Knupfer M., Büchner B., Soergel E., Streubel R., Koitzsch A., Eng L. M., "Polarization driven conductance variations at charged ferroelectric domain walls," *Nanoscale* 9(30), 10933–10939 (2017).

Pei S.-C., Ho T.-S., Tsai C.-C., Chen T.-H., Ho Y., Huang P.-L., Kung A. H., Huang S.-L., "Non-invasive characterization of the domain boundary and structure properties of periodically poled ferroelectrics," *Opt. Express* 19(8), 7153–7160 (2011).

Peters J. J. P., Apachitei G., Beanland R., Alexe M., Sanchez A. M. "Polarization curling and flux closures in multiferroic tunnel junctions," *Nat. Commun.* 7(1), 13484 (2016).

Pöykkö S., Chadi D. J., "Ab initio study of 180° domain wall energy and structure in PbTiO₃," *Appl. Phys. Lett.* 75(18), 2830 (1999).

Přívratská, J., "Possible appearance of spontaneous polarization and/or magnetization in domain walls associated with non-magnetic and non-ferroelectric domain pairs," *Ferroelectrics*, 353(1), 116–123 (2007).

Rüsing M., Neufeld S., Brockmeier J., Eigner C., Mackwitz P., Spychala K., Silberhorn C., Schmidt W. G., Berth G., Zrenner A., Sanna S., "Imaging of 180° ferroelectric domain walls in uniaxial ferroelectrics by confocal Raman spectroscopy: unraveling the contrast mechanism," *Phys. Rev. Mater.* 2(10), 103801 (2018).

Salje E. K. H., "Multiferroic domain boundaries as active memory devices: trajectories towards domain boundary engineering," *Chemphyschem.* 11(5), 940–950 (2010).

Salje E. K. H., Carpenter M. A., "Domain glasses: twin planes, Bloch lines, and Bloch points," *Phys. Status Solidi (B)* 252(12), 26392648 (2015).

Salje E. K. H., Scott J. F., "Ferroelectric Bloch-line switching: a paradigm for memory devices?," *Appl. Phys. Lett.* 105(25), 252904 (2014).

Schaab J., Skjærvø S. H., Krohns S., Dai X., Holtz M. E., Cano A., Lilienblum M., Yan Z., Bourret E., Muller D. A., Fiebig M., Selbach S. M., Meier D., "Electrical half-wave rectification at ferroelectric domain walls," *Nat. Nanotechnol.* 13(11), 1028–1034 (2018).

Schröder M., Haußmann A., Thiessen A., Soergel E., Woike T., Eng L. M., "Conducting domain walls in lithium niobate single crystals," *Adv. Funct. Mater.* 22(18), 3936–3944 (2012).

Scott J. F., Kumar A., "Faceting oscillations in nano-ferroelectrics," *Appl. Phys. Lett.* 105(5), 052902 (2014).

Scrymgeour D. A., Gopalan V., Itagi A., Saxena A., Swart P. J., "Phenomenological theory of a single domain wall in uniaxial trigonal ferroelectrics: lithium niobate and lithium tantalate," *Phys. Rev. B* 71(18), 184110 (2005).

Seidel J., Martin L. W., He Q., Zhan Q., Chu Y.-H., Rother A., Hawkridge M. E., Maksymovych P., Yu P., Gajek M., Balke N., Kalinin S. V. Gemming S., Wang F., Catalan G., Scott J. F., Spaldin N. A., Orenstein J., Ramesh R., "Conduction at domain walls in oxide multiferroics," *Nat. Mater.* 8, 229–234 (2009).

Shafer P., García-Fernández P., Aguado-Puente P., Damodaran A. R., Yadav A. K., Nelson C. T., Hsu S.-L., Wojdeł J. C., Íñiguez J., Martin L. W., Arenholz E., Junquera J., Ramesh R., "Emergent chirality in the electric polar ization texture of titanate superlattices," *Proc. Natl. Acad. Sci.* 115(5), 915–920 (2018).

Shen Y. R. "Surface contribution versus bulk contribution in surface nonlinear optical spectroscopy," *Appl. Phys. B* 68(3), 295–300 (1999).

Shen Y. R., *The Principles of Nonlinear Optics (Wiley Classics Library)* (Wiley-Interscience, New York, 1984).

Sheng Y., Best A., Butt H.-J., Krolikowski W., Arie A., Koynov K., "Three-dimensional ferroelectric domain visualization by Čerenkov-type second harmonic generation" *Opt. Express* 18(16), 16539–16545 (2010).

Slonczewski J. C., "Theory of Bloch-line and Bloch-wall motion," *J. Appl. Phys.* 45(6), 2705–2715 (1974).

Stepkova V., Marton P., Hlinka J., "Stress-induced phase transition in ferroelectric domain walls of BaTiO$_3$," *J. Phys.: Condens. Mat.* 24(21), 212201 (2012).

Stepkova V., Marton P., Hlinka J., "Ising lines: natural topological defects within ferroelectric Bloch walls," *Phys. Rev. B* 92(9), 094106 (2015).

Taherinejad M., Vanderbilt D., Marton P., Stepkova V., Hlinka J., "Bloch-type domain walls in rhombohedral BaTiO$_3$," *Phys. Rev. B* 86(15), 155138 (2012).

Tang Y. L., Zhu Y. L., Ma X. L., Borisevich A. Y., Morozovska A. N., Eliseev E. A., Wang W. Y., Wang Y. J., Xu Y. B., Zhang Z. D., Pennycook S. J., "Observation of a periodic array of flux-closure quadrants in strained ferroelectric PbTiO$_3$ films," *Science* 348(6234), 547–551 (2015).

Thiaville A., Ben Youssef J., Nakatani Y., Miltat, J., "On the influence of wall microdeformations on Bloch line visibility in bubble garnets (invited)," *J. Appl. Phys.* 69(8), 6090–6095 (1991).

Thiaville A., Miltat J., Ben Youssef, J., "Dynamics of vertical Bloch lines in bubble garnets: experiments and theory," *Eur. Phys. J. B* 23(1), 37–47 (2001).

Tolédano P., Guennou M., Kreisel J., "Order-parameter symmetries of domain walls in ferro-electrics and ferroelastics," *Phys. Rev. B: Condens. Mat. Mater. Phys.* 89(13), 1–5 (2014).

Uesu Y., Kurimura S., Yamamoto Y., "Optical second harmonic images of 90° domain structure in BaTiO$_3$ and periodically inverted antiparallel domains in LiTaO$_3$," *Appl. Phys. Lett.* 66(17), 2165–2167 (1995).

Uesu Y., Shibata H., Suzuki S., Shimada S., "3D images of inverted domain structure in LiNbO$_3$ using SHG interference microscope," *Ferroelectrics* 304(1), 99–103 (2004).

Valev V. K., Clercq B. D., Zheng X., Denkova D., Osley E. J., Vandendriessche S., Silhanek A. V., Volskiy V., Warburton P. A., Vandenbosch G. A. E., Ameloot M., Moshchalkov V. V., Verbiest T., "The role of chiral local field enhancements below the resolution limit of Second Harmonic Generation microscopy," *Opt. Express* 20(1), 256–264 (2012).

Wehmeier L., Kämpfe T., Haußmann A., Eng L. M., "In situ 3D observation of the domain wall dynamics in a triglycine sulfate single crystal upon ferroelectric phase transition," *Phys. Status Solidi (RRL) Rapid Res. Lett.* 11(11), 1700267 (2017).

Wei X.-K., Jia C.-L., Sluka T., Wang B.-X., Ye Z.-G., Setter N., "Néel-like domain walls in ferroelectric Pb(Zr,Ti)O$_3$ single crystals," *Nat. Commun.* 7, 12385 (2016).

Wojdeł J. C., Íñiguez J., "Ferroelectric transitions at ferroelectric domain walls found from first principles," *Phys. Rev. Lett.* 112(24), 247603 (2014).

Wolba B., Seidel J., Cazorla C., Godau C., Haußmann A., Eng L. M., "Resistor network modeling of conductive domain walls in lithium niobate," *Adv. Electron. Mater.* 4(1), 1700242 (2018).

Xiao S. Y., Kämpfe T., Jin Y. M., Haußmann A., Lu X. M., Eng L. M., "Dipole–tunneling model from asymmetric domain–wall conductivity in LiNbO$_3$ single crystals," *Phys. Rev. Appl.* 10(3), 1–9 (2018).

Yadav A. K., Nelson C. T., Hsu S. L., Hong Z., Clarkson J. D., Schlepütz C. M., Damodaran A. R., Shafer P., Arenholz E., Dedon L. R., Chen D., Vishwanath A., Minor A. M., Chen L. Q., Scott J. F., Martin L. W., Ramesh R., "Observation of polar vortices in oxide superlattices," *Nature* 530(7589), 198–201 (2016).

Yamada M., Saitoh M., Ooki H., "Electricfield induced cylindrical lens, switching and deflection devices composed of the inverted domains in LiNbO$_3$ crystals," *Appl. Phys. Lett.* 69(24), 3659–3661 (1996).

Yang M.-M., Bhatnagar A., Luo Z.-D., Alexe M., "Enhancement of local photovoltaic current at ferroelectric domain walls in BiFeO$_3$," *Sci. Rep.* 7(1), 43070 (2017).

Yang S. Y., Seidel J, Byrnes S J, Shafer P, Yang C.-H., Rossell M. D., Yu P, Chu Y.-H., Scott J. F., Ager J. W., Martin L. W., Ramesh R., "Above–bandgap voltages from ferroelectric photovoltaic devices," *Nat. Nanotechnol.* 5(2), 143–147 (2010).

Yang T. J., Mohideen U., "Nanoscale measurement of ferroelectric domain wall strain and energy by near-field scanning optical microscopy," *Phys. Lett. A* 250(1), 205–210 (1998).

Yu N. E., Kurimura S., Nomura Y., Kitamura K., "Stable high-power green light generation with a periodically poled stoichiometric lithium tantalate," *Mater. Sci. Eng. B: Solid-State Mater. Adv. Technol.* 120(1–3), 146–149 (2005).

Yudin P. V., Tagantsev A. K., Eliseev E. A., Morozovska A. N., Setter N., "Bichiral structure of ferroelectric domain walls driven by flexoelectricity," *Phys. Rev. B* 86(13), 134102 (2012).

Zhirnov V., "A contribution to the theory of domain walls in ferroelectrics," *Soviet Phys. Jetp-Ussr* 8(5), 822–825 (1959).

Zimmermann L., Miltat J., Pougnet P., "Stability of information in Bloch line memories," *IEEE Trans. Magnetics* 27(6), 5508–5510 (1991).

8

Turing Patterns in Ferroelectric Domains: Nonlinear Instabilities

J. F. Scott

Schools of Chemistry and of Physics
University of St. Andrews
St. Andrews, Fife, United Kingdom

8.1   Introduction to Instabilities and Turing Patterns

The properties of ferroelectrics are of fundamental interest to academia and industry. This chapter outlines how ferroelectric domain patterns link to various fundamental cross-disciplinary instabilities. First, the startling resemblance between Turing patterns and the 'labyrinthine' polar nanoregions associated with ferroelectric relaxors is highlighted. Next, patterns arising from the Landau–Ginzburg approach are examined and linked to experimentally observed domain patterns. The link between Turing patterns and Landau–Ginzburg patterns is particularly noteworthy as although the terms and considerations of the two approaches differ, both can describe ferroic domain configurations in lead-based ferroelectrics in different boundary conditions. Zhabotinskii–Belousov patterns and Richtmyer–Meshkov instabilities are considered, before the evolution of these patterns with increasing strain using the Wang–Zhao model is examined. The chapter concludes by discussing the dimensionality of $PbTiO_3$. We note, that because many of these processes require diffusion, they should be absent (or qualitatively different) near quantum critical points.

Alan Turing is perhaps best known for his work in solving the Enigma code in the 1940s and for his pioneering work on artificial intelligence, but he was also a superb theoretical biologist. Among his important contributions was the development of reaction-diffusion models (Turing 1952), which were later used to explain patterning in chemical reactions, including the famous Zhabotinskii–Belousov reaction (Belousov 1959; Zhabotinskii 1964).

The idea that diffusion would destabilize a uniform distribution of atoms or molecules in nature is somewhat paradoxical and counterintuitive, because diffusion usually stabilizes things. However, simple reaction-diffusion equations predict many observed patterns in chemistry and biology (Kondo and Miura 2010), and qualitative similarities

J. F. Scott, *Turing Patterns in Ferroelectric Domains: Nonlinear Instabilities* In: *Domain Walls: From Fundamental Properties to Nanotechnology Concepts.* Edited by: Dennis Meier, Jan Seidel, Marty Gregg, and Ramamoorthy Ramesh, Oxford University Press (2020). © J. F. Scott.
DOI: 10.1093/oso/9780198862499.003.0008

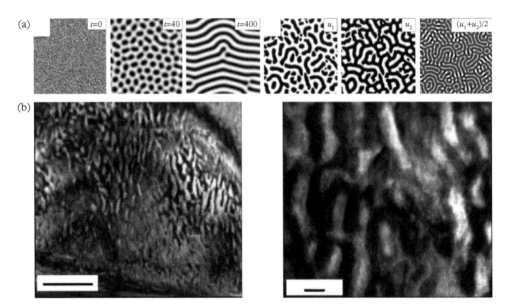

Figure 8.1 *(a) Some Turing-like patterns under different conditions: critically Turing patterns will arise out of a uniform homogeneous medium. (b) A typical relaxor-like ferroelectric domain pattern, in this case one induced by a magnetic field at room temperature in a single crystal slice of multiferroic lead zirconium titanate-lead iron tantalate. Scale bars are 500 nm and 50 nm, respectively.*

Source: (a) Figure reprinted with permission from L. Yang and I. R. Epstein, "Symmetric, asymmetric, and antiphase: Turing patterns in a model system with two identical coupled layers," Phys. Rev. E: Stat. Nonlinear, Soft Matter Phys. 69(22), 26211 (2004, Feb). Copyright (2004) by the American Physical Society. (b) Image from Yang and Epstein 2004 and reproduced under Creative Commons license.

even exist between these Turing patterns and polar nanoregions that are typical of relaxor ferroelectrics (Figure 8.1).

To illustrate the basic concept, starting with an equation of the form

$$\frac{\partial U}{\partial t} = f(U) + D\nabla^2 U \tag{8.1}$$

for two components:

$$\frac{\partial U_1}{\partial t} = f_1(U_1, U_2) + D_1\nabla^2 U_1 \tag{8.2}$$

$$\frac{\partial U_2}{\partial t} = f_1(U_1, U_2) + D_2\nabla^2 U_2 \tag{8.3}$$

it can be shown that there is a threshold for a static pattern instability for a certain ratio $D_1 > D_2$. Here, one component, u_1, is termed an activator and the other, u_2, an inhibitor. In fact, the static pattern predicted by this approach—termed type I-s (s for static) (Cross and Hohenberg 1993; Hohenberg and Krekhov 2015)—is rarely observed, because most chemical reactions involve diffusion coefficients D_1 and D_2 that are close in magnitude; however, specific reagents have been studied in the successful attempt to produce such patterns. These static patterns are not common in nature, or at least, when they are—they are not obviously a diffusion-based process. The non-static oscillatory patterns, however, are common. These are often labeled as type III-o (o for oscillatory) in the Turing scheme, which separates instabilities into those which are static in both space and time (s), those which are static in space but oscillate in time (I-o), and those which oscillate in both space and time (III-o). For example, the spiral domain instabilities, like the Zhabotinskii–Belousov instability, oscillate in both space and time.

8.2 Ferroelectric Domain Instabilities from Landau–Ginzburg

As illustrated by Hohenberg and Krekhov (2015), from the Landau–Ginzburg model there are a total of five patterns predicted for domain instabilities (Figure 8.2). All five, needles, chevrons, zippers, bulls-eyes, and spirals, have recently been illustrated (Scott et al. 2017) (Figure 8.3). Note that the chevron pattern, Figure 8.2(b), can maintain the

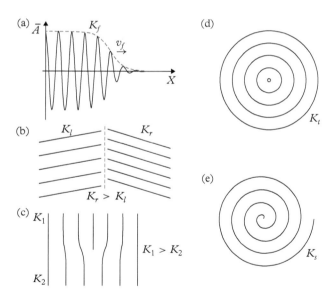

Figure 8.2 *Theoretical domain instability patterns for the Ginzburg–Landau equation. (a) A front in one dimension. (b) A domain boundary, with different periodicities on either side of the interface. (c) A dislocation. (d) A target, or "bulls-eye", pattern. (a)–(d) are for the real Ginzburg–Landau equation. (e) Two-dimensional spiral pattern for complex Ginzburg–Landau equation.*

Source: *All panels reprinted from Hohenberg and Krekhov 2015, copyright (2015), with permission from Elsevier.*

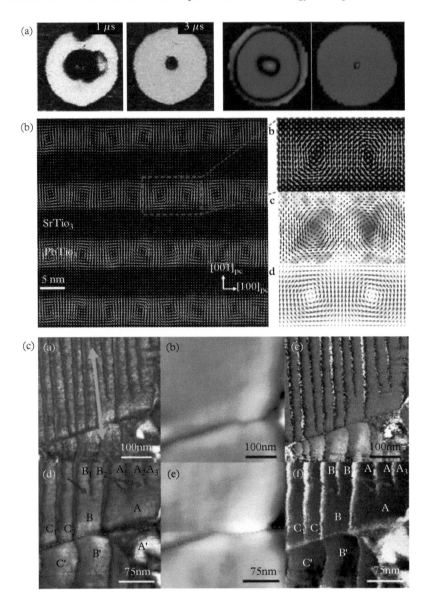

Figure 8.3 *(a) The two images on the left are experimentally measured piezoresponse force microscopy (PFM) images of bull's-eye patterns on 1-μm-diameter circular capacitors of lead zirconate titanate (PZT) taken during switching. The two images on the right are corresponding simulations of the switching. (b) Exquisite image of a spiral domain pattern, showing vortex–antivortex pairs. Taken with high-resolution scanning tunneling electron microscope (HR-STEM) taken on a cross section of a $(SrTiO_3)_{10}/(PbTiO_3)_{10}$ superlattice with the polar displacement vector added with yellow arrows. (c) Dislocation and twinning instabilities in PZT across a domain boundary. Taken using enhanced piezoresponse force microscopy (E-PFM), showing amplitude, topography, and phase respectively.*

Source: (a) Image from Gruverman et al. 2008. (b) Image taken from Yadav et al. 2016. (c) Image from Ivry et al. 2011 and reproduced with permission from John Wiley and Sons.

domain width on both sides of the reflection plane or change the width; this width change can be an integral ratio (doubling, tripling, ...) but it can also be a non-integral ratio, resulting in incommensurate domain structures. Of these five patterns, the chevron and dislocation patters are described by Hohenberg as arising from Turing static instabilities, meaning they are static in both space and time; however, recent unpublished work by A. Hershovitz and Y. Ivry measure the zipping and unzipping velocity of the dislocation mechanism, so in that sense they are not truly static. The bull's-eye pattern (Figure 8.2(d)) is analogous to the Taylor–Couette instability with a rotating central cylinder; and the spiral pattern (Figure 8.2(e)) is related to the Zhabotinskii–Belousov instability; but oddly enough, the bull's-eye and spiral patterns are not related to each other.

In addition to the five nonlinear distortions shown in theoretical detail in Hohenberg and Krekhov (2015), and shown in Figures 8.2 and 8.3, there are several other relevant nonlinear distortions. It is known that from conventional hydrodynamics that the Taylor–Couette instability with a rotating central cylinder exhibits five different instabilities (Metaxas et al. 2007; Altmeyer et al. 2011; Leclercq et al. 2014). Only one of these is related to the "bulls-eye" in Figure 8.2(d). The other four have not been identified yet in ferroelectric domains. However, one of these is vortex dominated, and it seems likely that this is closely related to the vortices in ferroelectrics; very recently S.-W. Cheol has claimed that vortex domains are universal in ferroelectrics (EMRS, Warsaw, September 27, 2018), but in our view they are only one of nine unrelated possibilities.

Note that the oscillatory instabilities can occur in the context of faceting of ferroelectric nanocrystals (Ahluwalia et al. 2013; Baudry et al. 2014, 2015, 2017; Scott and Kumar 2014; Ng et al. 2015). The most important point is that unlike the theories of Landau and Lifshitz (1935), or Kittel (1946), for ferroic domains these instabilities do not have straight rectilinear sides, as experimentally illustrated by Figure 8.4 and theoretically calculated in Eliseev et al. (2018) and Morozovska et al. (2018).

8.3   Nonlinear Geometric Distortions

The basic assumption of the present work is that domains have elastic moduli μ that are slightly larger than those of the underlying substrates. If the reverse is true, nothing very interesting happens: ferroelectric domains simply float around like islands on top of the substrate. But if the domain moduli exceed the substrate moduli, then under tensile or especially compressional stress, the domains undergo wrinkling, folding, bending, and other geometric nonlinear distortions. This is shown in detail in Wang and Zhao (2015) and depicted in Figure 8.5.

Several questions remain unanswered: Although most investigators agree that there is a well-defined wrinkling-to-folding phase transition (Metaxas et al. 2007; Wang and Zhao 2015), some recent researchers have inferred the existence of a preliminary precursor ripple-to-wrinkle transition.

Other kinds of nonlinear domain instability exist, outside the families considered by Hohenberg, notable including the Richtmyer–Meshkov instability, which is a well-known instability in fluids and in plasmas, including astrophysics (Meshkov 1969;

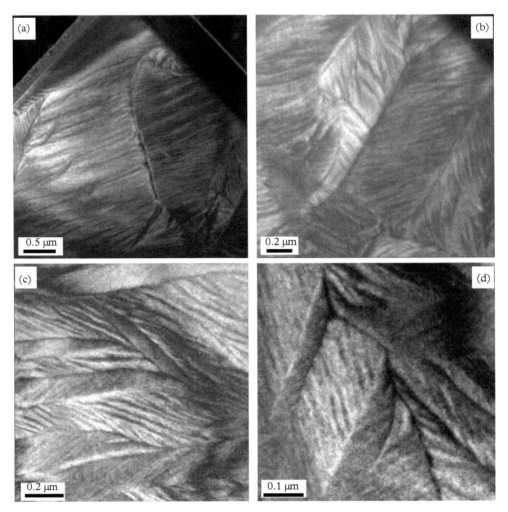

Figure 8.4 *Transmission electron micrograph (TEM) of the nonlinear domain pattern of a single crystal slice of lead zirconium titanate–lead iron tantalate. It is a noteworthy material as it is one of the few single crystal multiferroics that can have its ferroelectric domains switched with the application of a magnetic field at room temperature (Evans et al. 2013, 2014, 2015).*
Source: Image from Scott et al. (2017).

Meshkov et al. 1997; Brouillette 2002). In a configuration of two fluids with the more dense fluid above the other, a mechanical force from below will create nanodomains of the less dense fluid propagating upward through the denser liquid. If the denser fluid is at the bottom, this phenomenon does not occur. A similar situation has been observed for lead germanate, $Pb_5Ge_3O_{11}$. Here the large white + domain wall first develops a wave or wrinkle analogous to a capillary wave, followed by ejection upward in the presence of

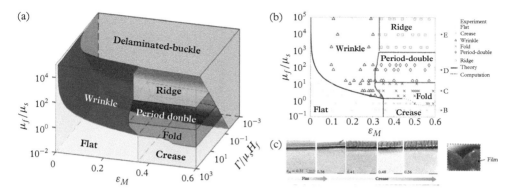

Figure 8.5 *Calculated 3-D phase diagram of surface instabilities induced by mismatch strains. The instability pattern is determined by three non-dimensional parameters: mismatch strain ε_M, modulus ratio μ_f/μ_s, and normalized adhesion energy $\Gamma/(\mu_s H_f)$. (b) Experimental validation of the calculated phase diagram (a). (c) Experimental images of the formation of crease.*

Source: All image from Wang and Zhao (2015) and reproduced under the Creative Commons Attribution 4.0 International Public License.

+E field from below, Figure 8.6. If the field is reversed, these nanodomain instabilities do not occur, and needle-like domains form instead (A. Gruverman, PhD thesis, University of Ekaterinburg 1990; Gruverman et al. 2008).

8.4   Dimensionality Effects in PbTiO$_3$

We use PbTiO$_3$ as an example of the dimensionality effects in PbTiO$_3$. Very recently the fabrication of ultra-tetragonal PbTiO$_3$ has been reported (Zhang et al. 2018), using epitaxial growth on PbO, which produced a switchable out-of-plane polarization of approximately 236 μC/cm^2, nearly 200% of the previous record polarization in a ferroelectric. In this chapter, we report the structure of the local ferroelectric domain state, imaged with piezoresponse force microscopy (PFM). Surprisingly, two variants are produced by the same processing. This is somewhat surprising because earlier measurements on similar ultra-tetragonal PbTiO$_3$ give structures that were either initially planar (non-switching polarizations in planar fourfold closure domains), or otherwise gave an out-of-plane polarization that was apparently metastable and disappeared totally with time (Catalan et al. 2006). In this chapter, we report a similar phenomenon in which specimens prepared in the same way as those with large out-of-plane polarizations display only in-plane fourfold nanodomains ca. 50 nm in diameter; we suggest that the domain interface might be thicker ((3D) rather than (2D)). Such fourfold vertices are rare in perovskites (McQuaid et al. 2014), and mechanisms are suggested.

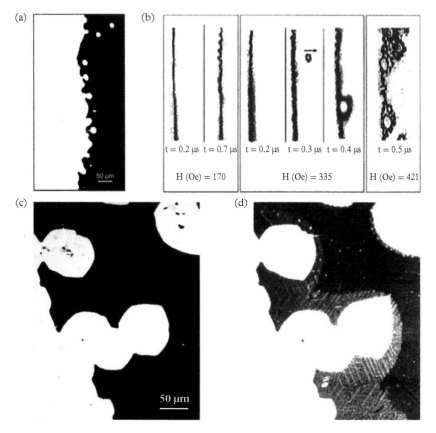

Figure 8.6 *Emission of ferroelectric nanodomains in lead germanate, under the application of an electric field. Similar spherical ejection is observed for the Richtmyer–Meshkov instabilities.*
Source: Original from A. Gruverman, Ph.D. thesis, Ural State University, Ekaterinburg, USSR (1990).

8.5   Fourfold Vertices in PbTiO$_3$

In addition to the previously published transmission electron microscopy (TEM), to demonstrate the ferroelectric properties of these specimens independent of hysteretic switching, we show (Figure 8.7) the PFM of domains and domain walls. Note the fourfold vertices' in agreement with the model of Srolovitz and Scott (1986). This four-state planar Potts model is often termed the Ashkin–Teller model after the scientists who first published an equivalent interaction model (Ashkin and Teller 1943). It is still studied in the case of a large applied external field (Berche et al. 2013), which is the case here with a very large interfacial lattice mismatch strain. We emphasize that in the absence of such a large field, threefold vertices are generally dominant.

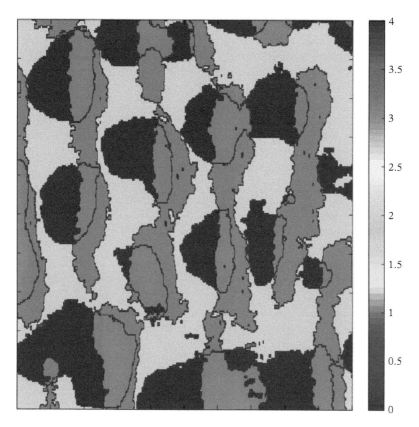

Figure 8.7 *PFM pattern of ultra-tetragonal PbTiO₃. Average domain diameter: 50 nm. Colors designate polarizations (left, down, right, up), showing perfect fourfold closure structures.*

Source: Original work by Amit Kumar, Joseph G. M. Guy, Linxing Zhang, Jun Chen, J. Marty Gregg, and James F. Scott (unpublished).

Here we consider another possibility: that the thickness of the domain, and whether it approximates a 2D or 3D structure is the dominant effect (see the third dimension in Figure 8.5 and Wang and Zhao 2015). In the domain vertex model of Srolovitz and Scott (1986), either fourfold domains are unstable and rapidly separate into closely spaced pairs of Y-shaped threefold vertices, or the pairs of threefold domains are unstable and rapidly coalesce into a single fourfold X-shaped vertex; this depends upon parameters in the Potts model used. To be specific, a four-state Potts model is required to describe the domain wall vertices in any case. However, to explain pairs of adjacent threefold vertices, the Potts model is a scalar model (a simple extension of an Ising model); in this model, any fourfold vertex quickly decomposes to adjacent pairs. But a vector Potts model ("clock model") is required to describe the fourfold vertices. In this case, pairs of adjacent threefold vertices quickly coalesce to X-shaped fourfold vertices. Hence, a given sample can exhibit threefold or fourfold vertices but not both. A few systems illustrate both cases, however. This does not violate the Srolovitz–Scott model, but it implies that the interaction parameter is anisotropic, with out-of-plane polarized domains satisfying a scalar $n = 4$ Potts model and in-plane polarized domains satisfying a clock model. This has not been reported before in any ferroelectric, with the exception of lead strontium titanate (PST) (Nesterov 2015), where pairs of threefold vertices are found and—"very rarely"—a few fourfold vertices (it is tempting to suggest that these very few vertices may have been interrupted in a transient state after nucleation). $SrBi_2Ta_2O_9$ (SBT) also exhibits both threefold and fourfold vertex geometries (Zhu et al. 2001). This may give some insight into the more complex circular domains reported recently (Janolin et al. 2008; Essoumhi et al. 2014). Note in particular that applying an external electric field to rotate the polarizations will necessarily create and destroy vertex structures, which had not previously been known.

We can discuss further a physical reason for the out-of-plane domains satisfying a scalar Potts model but in-plane domains conforming to the vector Potts prediction. In this respect, we can point out that these domains are not topologically equivalent, with out-of-plane polarizations allowed to be only up or down. In recent studies by McQuaid et al., no intact fourfold closure vertices were observed; all were split by strain (Chang et al. 2013; McQuaid et al. 2014). The collapse of fourfold vertices into adjacent pairs of threefold vertices is further considered by Opperman (1975), who emphasizes qualitative changes in dynamics due to axial anisotropy (Oppermann 1974).

We note parenthetically that $n = 4$ Potts systems cannot produce Berezinskii–Kosterlitz–Thouless melting in two dimensions (although $n = 6$ can); the $n = 4$ systems do not form hexatic two-dimensional states (Zhao et al. 2013). It is also worth noting that the two-dimensional Potts model gives second-order phase transitions for $n < 4$ but first-order phase transitions for $n = 4$ or greater, so the $n = 4$ state is special.

8.6 Voronoi Partitions and n-Fold Vertices in $PbTiO_3$

It is also useful to point out that the general question in nature of whether fourfold or threefold vertices are stable is rather well known. In two dimensions, Voronoi partitions (domain walls in the present context) subdivide space into four quarters only when the objects of interest (nucleation sites at defects in the present examples) are symmetrically placed on the vertices of a rectangle. Otherwise, any other configuration generates closely spaced pairs of threefold vertices. This is a result of domains growing from initial nucleation sites to fill two-dimensional Euclidean space. Weaire, Aste, and Rivier describe this (Weaire and Rivier 1984; Aste and Weaire 2008) as a two-dimensional law: "Only three states meet in a common vertex," or "vertex connectivity = 3." However, in three dimensions, the opposite is true, and fourfold vertices are stable; we find 100% fourfold vertices in ultra-tetragonal $PbTiO_3$ on PbO. Therefore the change in vertex geometry from fourfold for in-plane PZT to threefold for out-of-plane polarization may arise from the effective dimensionality in the two cases: Are our thin films truly two-dimensional? The observation that these rules hold in general for Voronoi partitions implies that the phenomenon may be independent of the nature of interaction ("jamming") between or among ferroelectric domains. Voronoi partitions are sometimes referred to as Dirichlet tessellations.

We prefer the strong axial field model of Berche et al. (2013) as an explanation ("Teller–Ashkin model") for our observed fourfold vertices but include the Voronoi discussion for completeness. Parenthetically, since the number of domains intersecting a vertex must be n > 4 for Kosterlitz–Thouless melting to occur in two dimensions, such melting is unlikely for most ferroelectric domain walls.

8.7 Summary

Here we have shown how fundamental pattern formations can be manifested in ferroelectrics—notably Turing patterns, which arise from diffusion considerations, and the five domain patters arising from the Landau–Ginzburg model. We went on to highlight other instabilities, such as the potential manifestation of Richtmyer–Meshkov instabilities in lead germanate, before considering how the model of Wang and Zhao (2015) links different non-equilibrium—time-dependent—configurations. We finished by looking at the dimensionality of ultra-tetragonal $PbTiO_3$.

∙∙

REFERENCES

Ahluwalia R., Ng N., Schilling A., McQuaid R. G. P., Evans D. M., Gregg J. M., Srolovitz D. J., Scott J. F., "Manipulating ferroelectric domains in nanostructures under electron beams," *Phys. Rev. Lett.* 111(16), 165702 (2013).

Altmeyer S., Hoffmann C., Lücke M., "Islands of instability for growth of spiral vortices in the Taylor-Couette system with and without axial through flow," *Phys. Rev. E* 84(4), 46308 (2011).

Ashkin J., Teller E., "Statistics of two-dimensional lattices with four components," *Phys. Rev.* 64(5–6), 178–184 (1943).

Aste T., Weaire D., *The Pursuit of Perfect Packing*, 2nd edn. (CRC Press, Boca Raton, FL, 2008).

Baudry L., Luk'yanchuk I. A., Razumnaya A., "Dynamics of field-induced polarization reversal in thin strained perovskite ferroelectric films with c-oriented polarization," *Phys. Rev. B* 91(14), 144110 (2015).

Baudry L., Lukyanchuk I., Vinokur V. M., "Ferroelectric symmetry-protected multibit memory cell," *Sci. Rep.* 7, 42196 (2017).

Baudry L., Sené A., Luk'yanchuk I. A., Lahoche L., Scott J. F., "Polarization vortex domains induced by switching electric field in ferroelectric films with circular electrodes," *Phys. Rev. B* 90(2), 24102 (2014).

Belousov B. P., "Периодически действующая реакцияи ее механизм [Periodically acting re-action and its mechanism]," Сборник рефератов по радиационной медицине in: [*Abstracts Radiat. Med.*]145–147 (1959).

Berche B., Butera P., Shchur L. N., "The two-dimensional 4-state Potts model in a magnetic field," *J. Phys. A Math. Theor.* 46(9), 95001 (2013).

Brouillette M., "The Richtmyer-Meshkov instability," *Annu. Rev. Fluid Mech.* 34(1), 445–468 (2002).

Catalan G., Janssens A., Rispens G., Csiszar S., Seeck O., Rijnders G., Blank D. H., Noheda B., "Polar domains in lead titanate films under tensile strain," *Phys. Rev. Lett.* 96(12), 127602 (2006).

Chang L.-W., Nagarajan V., Scott J. F., Gregg J. M., "Self-similar nested flux closure structures in a tetragonal ferroelectric," *Nano Lett.* 13(6), 2553–2557 (2013).

Cross M. C., Hohenberg P. C., "Pattern formation outside of equilibrium," *Rev. Mod. Phys.* 65(3), 851–1112 (1993).

Eliseev E. A., Fomichov Y. M., Kalinin S. V., Vysochanskii Y. M., Maksymovich P., Morozovska A. N., "Labyrinthine domains in ferroelectric nanoparticles: manifestation of a gradient-induced morphological transition," *Phys. Rev. B* 98(5), 54101 (2018).

Essoumhi A., El Kazzouli S., Bousmina M., "Review on Palladium-containing Perovskites: synthesis, physico-chemical properties and applications in catalysis," *J. Nanosci. Nanotechnol.* 14(2), 2012–2023 (2014).

Evans D. M., Alexe M., Schilling A., Kumar A., Sanchez D., Ortega N., Katiyar R. S., Scott J. F., Gregg J. M., "The nature of magnetoelectric coupling in Pb(Zr,Ti)O$_3$-Pb(Fe,Ta)O$_3$," *Adv. Mater.* 27(39), 6068–6073 (2015).

Evans D. M., Schilling A., Kumar A., Sanchez D., Ortega N., Arredondo M., Katiyar R. S., Gregg J. M., Scott J. F., "Magnetic switching of ferroelectric domains at room temperature in multiferroic PZTFT," *Nat. Commun.* 4, 1534 (2013).

Evans D. M., Schilling A., Kumar A., Sanchez D., Ortega N., Katiyar R. S., Scott J. F., Gregg J. M., "Switching ferroelectric domain configurations using both electric and magnetic fields in Pb(Zr,Ti)O$_3$-Pb(Fe,Ta)O$_3$ single-crystal lamellae," *Philos. Trans. R. Soc. A Math. Phys. Eng. Sci.* 372(2009), 20120450 (2014).

Gruverman A., Wu D., Fan H.-J., Vrejoiu I., Alexe M., Harrison R. J., Scott J. F., "Vortex ferroelectric domains," *J. Phys. Condens. Mat.* 20, 342201 (2008).

Hohenberg P. C., Krekhov A. P., "An introduction to the Ginzburg–Landau theory of phase transitions and nonequilibrium patterns," *Phys. Rep.* 572, 1–42 (2015).

Ivry Y., Chu D., Scott J. F., Durkan C., "Domains beyond the grain boundary," *Adv. Funct. Mater.* 21(10), 1827–1832 (2011).

Janolin P. E., Bouvier P., Kreisel J., Thomas P. A., Kornev I. A., Bellaiche L., Crichton W., Hanfland M., Dkhil B., "High-Pressure Effect on PbTiO$_3$: an Investigation by Raman and X-Ray Scattering up to 63 GPa," *Phys. Rev. Lett.* 101(23), 237601 (2008).

Kittel C., "Theory of the structure of ferromagnetic domains in films and small particles," *Phys. Rev.* 70(11–12), 965–971 (1946).

Kondo S., Miura T., "Reaction-diffusion model as a framework for understanding biological pattern formation," *Science* 329(5999), 1616–1620 (2010).

Landau L. D., Lifshitz E., "On the theory of the dispersion of magnetic permeability in ferromagnetic bodies," *Phys. Zeitschrift der Sowjetunion* 8, 153 (1935).

Leclercq C., Pier B., Scott J. F., "Absolute instabilities in eccentric Taylor–Couette–Poiseuille flow," *J. Fluid Mech.* 741, 543–566 (2014).

McQuaid R. G. P., Gruverman A., Scott J. F., Gregg J. M., "Exploring vertex interactions in ferroelectric flux-closure domains," *Nano Lett.* 14(8), 4230–4237 (2014).

Meshkov E. E., "Instability of the interface of two gases accelerated by a shock wave," *Fluid Dyn.* 4(5), 101–104 (1969).

Meshkov E. E., Nevmerzhitsky V. V., Zmushko N. V., "On possibilities of investigating hydrodynamic instabilities and turbulent mixing development in spherical geometry," in: *The Physics of Compressible Turbulent Mixing* (G. Jourdan and L. Houas, Eds.), pp. 343–347 (Caractère Imprimeur, Marseille, 1997).

Metaxas P. J., Jamet J. P., Mougin A., Cormier M., Ferre J., Baltz V., Rodmacq B., Dieny B., Stamps R. L., "Creep and flow regimes of magnetic domain-wall motion in ultrathin Pt/Co/Pt films with perpendicular anisotropy," *Phys. Rev. Lett.* 99(21), 217208 (2007).

Morozovska A. N., Fomichov Y. M., Maksymovych P., Vysochanskii Y. M., Eliseev E. A., "Analytical description of domain morphology and phase diagrams of ferroelectric nanoparticles," *Acta Mater.* 160, 109–120 (2018).

Nesterov O., *Control of Periodic Ferroelastic Domains in Ferroelectric $Pb_{1-x}Sr_xTiO_3$ Thin Films for Nano-Scaled Memory Devices*, Thesis, University of Groningen (2015).

Ng N., Ahluwalia R., Kumar A., Srolovitz D. J., Chandra P., Scott J. F., "Electron-beam driven relaxation oscillations in ferroelectric nanodisks," *Appl. Phys. Lett.* 107(15), 152902 (2015).

Oppermann R., "Change of the order of a phase transition induced by axial anisotropy," *J. Phys. C Solid State Phys.* 7(20), L366–369 (1974).

Oppermann R., "Wilson expansions for an extended Potts model," *J. Phys. A. Math. Gen.* 8(4), L43–46 (1975).

Scott J. F., Evans D. M., Katiyar R. S., McQuaid R. G. P., Gregg J. M., "Nonequilibrium ferroelectric-ferroelastic 10 nm nanodomains: wrinkles, period-doubling, and power-law relaxation," *J. Phys. Condens. Mat.* 29(30), 304001 (2017).

Scott J. F., Hershkovitz A., Ivry Y., Lu H., Gruverman A., Gregg J. M., "Superdomain dynamics in ferroelectric-ferroelastic films: switching, jamming, and relaxation," *Appl. Phys. Rev.* 4(4), 41104 (2017).

Scott J. F., Kumar A., "Faceting oscillations in nano-ferroelectrics," *Appl. Phys. Lett.* 105(5), 52902 (2014).

Srolovitz D. J., Scott J. F., "Clock-model description of incommensurate ferroelectric films and of nematic-liquid-crystal films," *Phys. Rev. B* 34(3), 1815–1819 (1986).

Turing A. M., "The chemical basis of morphogenesis," *Philos. Trans. R. Soc. Lond. B. Biol. Sci.* 237(641), 37–72 (1952).

Wang Q., Zhao X., "A three-dimensional phase diagram of growth-induced surface instabilities," *Sci. Rep.* 5, 8887 (2015).

Weaire D., Rivier N., "Soap, cells and statistics—random patterns in two dimensions," *Contemp. Phys.* 25(1), 59–99 (1984).

Yadav A. K., Nelson C. T., Hsu S. L., Hong Z., Clarkson J. D., Schlepuetz C. M., Damodaran A. R., Shafer P., Arenholz E., Dedon L. R., Chen D., Vishwanath A., Minor A. M., Chen L. Q., Scott J. F., Martin L. W., Ramesh R., "Observation of polar vortices in oxide superlattices," *Nature* 530, 198 (2016).

Yang L., Epstein I. R., "Symmetric, asymmetric, and antiphase Turing patterns in a model system with two identical coupled layers," *Phys. Rev. E—Stat. Nonlinear, Soft Mat. Phys.* 69(22), 26211 (2004).

Zhabotinskii A. M., Периодический процесс окисления малоновой кислоты растворе [Periodical process of oxidation of malonic acid solution]," Биофизика *[Biofizika]* 9, 306–311 (1964).

Zhang L., Chen J., Fan L., Diéguez O., Cao J., Pan Z., Wang Y., Wang J., Kim M., Deng S., Wang J., Wang H., Deng J., Yu R., Scott J. F., Xing X., "Giant polarization in super-tetragonal thin films through interphase strain," *Science* 361(6401), 494–497 (2018).

Zhao Y., Li W., Xi B., Zhang Z., Yan X., Ran S.-J., Liu T., Su G., "Kosterlitz–Thouless phase transition and re-entrance in an anisotropic three-state Potts model on the generalized kagome lattice," *Phys. Rev. E* 87(3), 32151 (2013).

Zhu X., Zhu J., Zhou S., Li Q., Liu Z., Ming N., "Domain structures and planar defects in $SrBi_2Ta_2O_9$ single crystals observed by transmission electron microscopy," *Appl. Phys. Lett.* 78(6), 799–801 (2001).

9

Photoelectric Effects at Domain Walls

M.-M. Yang and M. Alexe

Department of Physics, the University of Warwick, Coventry, CV4 7AL, UK

9.1 Introduction

Over the last decade, the field of the ferroelectric photovoltaic effect has been experiencing a significant revival triggered by the observation of an above-bandgap photovoltage in the $BiFeO_3$ thin films. Ferroelectric materials have been long studied as a potential alternative to conventional semiconductor-based materials for photovoltaic applications, not only because of their multiple charge separation mechanisms but also due to the intriguing coupling between light and ferroelectric properties. Ferroelectrics can spontaneously generate a short-circuit current under uniform illumination owing to the asymmetric momentum distribution of the non-equilibrium photo-excited carriers in the k-space, termed the bulk photovoltaic effect (Sturman and Fridkin 1992). In contrast to the conventional photovoltaic effect based on a gradient of the chemical potential, the ferroelectric photovoltaic effect exhibits distinctive features, such as light-polarization-dependent and switchable short-circuit current, as well as the anomalous photovoltage exceeding the value of the bandgap. Although solar energy harvesting based on ferroelectric materials suffers a low-power conversion efficiency due to their poor light absorption in the visible range and high resistance, the progress in oxide thin film growth has significantly improved the efficiency of ferroelectric solar cells in recent years (Nechache et al. 2014; Spanier et al. 2016). Meanwhile, the coupling of light with intrinsic degrees of freedom, such as strain, electrical polarization, and magnetism in (multi-)ferroelectrics, offers a fertile and rich playground to explore the new functionalities of (multi-)ferroelectrics and to develop related applications. In this regard, light-induced reversible ferroelectric switching and domain wall motion have been recently achieved (Rubio-Marcos et al. 2017; Li et al. 2018; Yang and Alexe 2018). In this chapter, we will review the history and the state-of-the-art research of the ferroelectric photovoltaic effect, in particular, the role of domain walls.

M.-M. Yang and M. Alexe, *Photoelectric Effects at Domain Walls* In: *Domain Walls: From Fundamental Properties to Nanotechnology Concepts.*
Edited by: Dennis Meier, Jan Seidel, Marty Gregg, and Ramamoorthy Ramesh, Oxford University Press (2020). © M.-M. Yang and M. Alexe.
DOI: 10.1093/oso/9780198862499.003.0009

9.2 Bulk Photovoltaic Effect

Two years after the invention of modern prototype solar cells, Chynoweth at Bell Labs found that the prototypical ferroelectric material, $BaTiO_3$, under constant illumination generated a steady-state DC photocurrent, i.e. photovoltaic effect, distinctive from any transient pyroelectric current (Chynoweth 1956). It was only in the mid-1970s that the field of the ferroelectric photovoltaic effect took off. Glass et al. reported an anomalous high photovoltage over 2000 V with a saturated field of 10^5 V/cm in doped $LiNbO_3$ single crystals under 10 mW/cm² uniform illumination (Figure 9.1(a); see Glass et al. 1974, 1975). As the separation of non-equilibrium photo-induced charges occurs in the bulk of the chemically homogenous $LiNbO_3$ single crystals, this photovoltaic effect is termed the bulk photovoltaic (BPV) effect. Later, another intriguing feature of the BPV effect was revealed. As shown in Figure 9.1(b,c), the short-circuit current generated by the BPV effect depends on the polarization of the incident light (Belinicher et al. 1977; Fridkin and Magomadov 1979; Festl et al. 1982). Similar phenomena, i.e. anomalous photovoltage and light-polarization-dependent short-circuit current, were also reported for other ferroelectric materials, such as $BaTiO_3$, $LiTaO_3$, and $KNbO_3$ (Koch et al. 1975, 1976; Gunter 1978; Krätzig and Orlowski 1978). Later, it was demonstrated that the BPV effect is a universal property of all the materials lacking centrosymmetry (Fridkin 1984; Sturman and Fridkin 1992).

Phenomenologically, the BPV effect in non-centrosymmetric crystals is described by a third rank tensor β_{ijk} (Fridkin and Magomadov 1979). The short-circuit current density along the i_{th} crystallographic direction ($i, j = x, y, z$) generated by illumination with a constant light intensity I_0 is

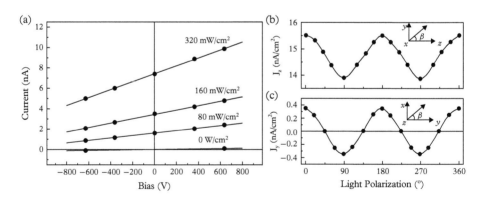

Figure 9.1 *(a) Current-voltage characteristics of Fe:LiNbO₃ at 300 K illuminated with various intensities of 472.7 nm light. Light polarization dependence of the short-circuit photocurrent. (b) $\mathcal{J}_Z$ along the Z-direction. (c) $\mathcal{J}_Y$ along the Y-direction in Fe:LiNbO₃ crystal. The light illumination direction is indicated in the insets.*

Source: (a) Adapted with permission from Springer Nature, Glass et al. (1975). (b,c) Adapted with permission from JETP, Fridkin and Magomadov (1979).

$$\mathfrak{J}_i = I_0 \beta_{ijk} e_j e_k^* + i I_0 \gamma_{ij} \left(e_j \times e_k^* \right)_j \qquad (9.1)$$

where e_j and e_k are the projections of the light polarization vector. The first term of Equation (9.1) is termed the linear bulk photovoltaic effect and is the main topic of this chapter. The second term represents the circular photovoltaic/photogalvanic (CPV) effect where γ_{ij} is the gyrotropic tensor, which arises from the spin-orbit coupling effect. The CPV effect only occurs under the illumination of elliptically polarized light and changes its direction when the helicity of the incident light changes. This indicates that the CPV effect vanishes under the illumination of linearly polarized light. For more details on the CPV effect, refer to works by Sturman and Fridkin (1992) and Ganichev and Prettle (2003) and the references therein. For the sake of simplicity, the BPV effect discussed hereafter refers only to the linear BPV effect.

Experimentally the BPV current density $\mathfrak{J}_i$ can also be expressed in terms of the Glass coefficient G:

$$\mathfrak{J}_i = G \alpha I_0 \qquad (9.2)$$

where α is the absorption coefficient. Thus, the Glass coefficient matrix is given by $G_{ijk} = \beta_{ijk}/\alpha$. The value of the Glass coefficients for ferroelectric materials is generally on the order of 10^{-11} m·V^{-1} (Sturman and Fridkin 1992). The open-circuit photovoltage V_{oc} is an extrinsic property, being proportional to the device length L but inversely proportional to the sum of dark and photoconductivity:

$$V_{oc} = \frac{\mathfrak{J}_i}{\sigma_d + \sigma_{ph}} L \qquad (9.3)$$

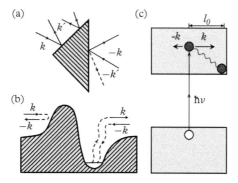

Figure 9.2 *(a) Asymmetry of elastic scattering by a wedge. (b) Asymmetry of photoexcitation and recombination from a non-central potential well. (c) Asymmetric distribution of non-thermalized carriers in a non-centrosymmetric crystal. The white circle denotes the hole, and the black circle denotes the light-excited electron.*

Source: Adapted with permission from Springer Nature, Fridkin (2001).

where σ_d and σ_{ph} are the dark conductivity and photoconductivity, respectively. In the case of wide-bandgap ferroelectric materials, photoconductivity is generally much larger than their dark conductivity. Thus, the photovoltage can be simplified as

$$V_{oc} = \frac{\mathfrak{F}_i}{\sigma_{ph}} L \qquad (9.4)$$

The corresponding light-induced electrical field E_{PV} can be expressed as

$$E_{PV} = \mathfrak{F}_i \bigg/ \sigma_{ph} \qquad (9.5)$$

In regard to the microscopic mechanism, the BPV current is associated with the violation of the detailed balancing for the photoexcited non-equilibrium carriers owing to the breaking of inversion symmetry. For example, Figure 9.2(a) shows the possible scattering processes at a wedge, simulating a non-central potential wherein the transition probability $W_{kk'}$ of the electron transition from the state with momentum k' to the state with momentum k is not equal to the reverse transition: $W_{kk'} \neq W_{k'k}$. This transition asymmetry also applies to the excitation and recombination processes in the non-centrosymmetric crystals (see Figure 9.2(b)), giving rise to an asymmetric momentum distribution of the non-equilibrium carriers in k space. This results in a spontaneous flow of non-equilibrium carriers toward a preferred crystallographic direction in a homogeneous non-centrosymmetric crystal under uniform illumination (Fridkin 2001). Among the three elementary processes, general arguments and model formulas imply that the contribution related to the photoexcitation asymmetry usually predominates in the BPV current. As depicted in Figure 9.2(c), the photoexcited hot electrons in the non-centrosymmetric crystals lose their energy during the thermalization process with a time scale of $\tau_{nt} \approx 10^{-12} - 10^{-14}$ s, resulting in a shift of l_0 toward a certain direction. The associated current can be expressed as

$$\mathfrak{F}_i = e\alpha I_0 l_0 \Phi \xi^{ex} (\hbar v)^{-1} \qquad (9.6)$$

where e is the electron charge, l_0 is the mean free path for the photoexcited hot carriers, Φ is the quantum yield, ξ^{ex} is the photoexcitation asymmetry parameter, and $\hbar v$ is the photon energy. The value of l_0 was estimated to be 10–100 nm; e.g. the l_0 for $BaTiO_3$ was measured to be 40 nm (Sturman and Fridkin 1992; Gu et al. 2017). ξ^{ex} is generally smaller than 0.1, which depends on the crystallographic structure of non-centrosymmetric crystals as well as the incident light wavelength. Also, the photoconductivity can be expressed as

$$\sigma_{ph} = e\alpha I_0 \Phi (\hbar v)^{-1} (\mu \tau) \qquad (9.7)$$

where μ and τ are the mobility and lifetime of non-equilibrium carriers. The electric field generated by the BPV effect can be rewritten as

$$E_{PV} = \frac{l_0 \xi^{ex}}{\mu \tau} \tag{9.8}$$

The maximum power conversion efficiency η_{max} can be estimated as

$$\eta_{max} = G E_{PV}/4 \tag{9.9}$$

Accordingly, non-centrosymmetric materials with a low $\mu\tau$ product will exhibit a large light-induced electric field as well as photovoltage. The contribution to the photoconductivity consists of two parts, i.e. hot carriers during the thermalization process and the thermalized carriers. Following thermalization, the hot carriers become polarons, which do not contribute to the BPV current (Schirmer et al. 2011). The mobility and lifetime product of hot carriers $(\mu\tau)_{nt}$ are much smaller than those of the thermalized carriers $(\mu\tau)_t$. Thus, if one can collect the hot carriers during their mean free path l_0, both the light-induced electrical field E_{PV} and conversion efficiency η_{max} can be dramatically enhanced. It has been experimentally proved that E_{PV} increases by four orders of magnitude and η_{max} reaches 1% when the sample thickness is decreased from 0.4 cm to about 40 nm (Zenkevich et al. 2014; Gu et al. 2017).

The microscopic model described above is termed the ballistic current on the basis that the BPV current is produced by freely moving carriers. On the other hand, an alternative mechanism termed shift current also contributes to the BPV current, which is related to the density matrix elements that are non-diagonal with respect to the numbers of bands (Sturman and Fridkin 1992). The shift current is caused by the shift of non-equilibrium carriers in real lattice space at the virtual electron transition from band n' to band n'' (von Baltz and Kraut 1981; Belinicher et al. 1982). The corresponding BPV tensor β_{ijk} can be expressed as the product of the transition intensity $I_{ij}(n', n'', k, \omega)$ between the i-th and j-th directions and the shift vector $R_l(n', n'', k)$ (Young et al. 2012):

$$\beta_{ijl} = e \sum_{n',n''} \int dk I_{ij}\left(n',n'',k,\omega\right) R_l\left(n',n'',k\right) \tag{9.10}$$

The shift current theory was first elaborated in the early 1980s and has been recently developed to calculate the BPV current in ferroelectric materials such as $BaTiO_3$ and $BiFeO_3$, the results of which are consistent with experimental results (Young et al. 2012; Cook et al. 2017; Rangel et al. 2017). It offers insights into some entangled fundamentals behind the BPV effect and provides implications for exploring the BPV effect in exotic materials, such as topological insulators and atomic layer materials (Tan and Rappe 2016; Fregoso et al. 2017; Gong et al. 2018). For example, the shift current theory reveals that the BPV current is independent of the material's polarization except that

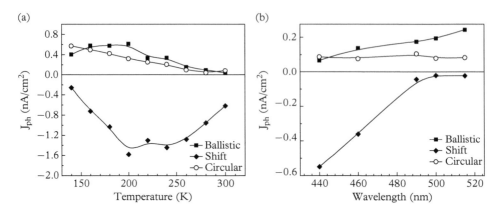

Figure 9.3 *(a) Temperature and (b) light wavelength dependence of the calculated values of the ballistic, circular, and shift photovoltaic current in Bi$_{12}$GeO$_{20}$ crystal.*
Source: Adapted with permission from Taylor & Francis, Astafiev et al. (1988).

they both originate from the same non-centrosymmetric structure (Young et al. 2012; Yang et al., 2017).

The possibility of the existence of both contributions, i.e. the ballistic current and the shift current, which are comparable in magnitude but different in a physical sense, raises the problem of their separation (Belinicher and Sturman 1988). Three features of the ballistic current and shift current can be used to distinguish one contribution from another (Astafiev et al. 1988). First, the BPV current due to the impurity-band transition, such as in the case of Fe:LiNbO$_3$ singe crystal under visible light illumination, primarily originates from the ballistic current. Second, the shift current, which is not connected with the free carriers in allowed bands, does not affect the Hall current/voltage in a magnetic field. Lastly, the circular photovoltaic effect is pure ballistic current. By measuring the circular photovoltaic effect and the linear BPV effect in a magnetic field, i.e. the magneto-photovoltaic effect, one can draw a distinction between the shift current and the ballistic current (Ivchenko et al. 1988). For instance, Figure 9.3 shows the temperature-resolved and wavelength dependent contributions of the shift and ballistic current in Bi$_{12}$GeO$_{20}$, indicating that the shift and ballistic current compensate each other in this case (Astafiev et al. 1988; Burger et al. 2019). Also, unlike the shift current effect, the theoretical calculation of the ballistic current remains, however, blank.

9.3 Tip-Enhanced Photovoltaic Effect

The ferroelectric photovoltaic effect normally suffers from a low quantum efficiency and a low power conversion coefficient of less than 10^{-4}. These disadvantages may be overcome by simply reducing the electrode size to a nanoscale level, e.g. by replacing a macroscopic-sized electrode by a conductive atomic force microscopy (AFM) tip (Alexe and Hesse 2011; Spanier et al. 2016). This intriguing phenomenon is termed the tip-enhanced photovoltaic (TEPV) effect. Alexe et al. showed that the photocurrent and implicitly external quantum efficiency of BiFeO$_3$ crystals can be enhanced by seven orders of magnitude by using a point-contact geometry in comparison with the device using macroscopic coplanar electrodes. As shown in Figure 9.4, despite its reduced dimensions, the AFM tip is able to collect a similar amount of photocurrent compared to the coplanar electrodes under the same illumination conditions, leading to a dramatically enhanced current density of 10–100 A/cm^2 at the tip contact area. Thereafter, it is found that TEPV also occurs in the BiFeO$_3$ thin film and the photocurrent collected by the AFM tip exhibits a sinusoidal dependence on the incident light polarization, demonstrating that the mechanism behind this effect is the BPV effect (Yang et al. 2017b). It also shows that the photocurrent collected by the AFM tip as well as the above-bandgap photovoltage is due to the photoelectric process over the whole illuminated area rather than just at the tip contact area.

Meanwhile, Spanier et al. reported the TEPV effect in BaTiO$_3$ single crystals with a photocurrent of 12 pA and a V_{OC} of 1.2 V under 1 sun AM 1.5 G illumination. By assuming the radius of the effective current collection area by the AFM tip as the thermalization distance of the photoexcited non-equilibrium carriers, i.e. the mean free path l_0 of hot carriers ($l_0 \sim 100$ nm for BaTiO$_3$), a corresponding current density of 19.1 mA/cm^2 and a power conversion efficiency of 4.8% are obtained, which exceeds the Shockley–Queisser (S-Q) limit for a material of 3.2 eV bandgap by $\approx$50% (Shockley and Queisser 1961). Moreover, the magnitude of the photocurrent observed in the TEPV effect increases with the number of the tips contacting the ferroelectric surface, making it feasible to boost the conversion efficiency of ferroelectric solar cells and apply them

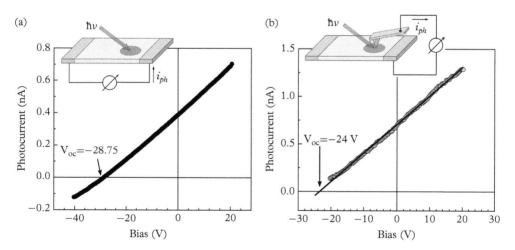

Figure 9.4 *Photocurrent-voltage characteristics of the BiFeO$_3$ single crystal (a) measured macroscopically across the entire crystal by illumination in the region marked in the lower right inset, and (b) measured by probing with the AFM tip in the middle of the illuminated area.*

Source: *Adapted with permission from Alexe and Hesse 2011 under the terms of the CC-BY 4.0 license.*

in the solar industry (Alexe and Hesse 2011; Spanier et al. 2016). In addition, because of the dramatically enhanced current density, the associated electric field underneath the AFM tip is also greatly enhanced, which makes it possible to switch ferroelectric domains, induce defects migration and domain wall motion, etc. (Yang and Alexe 2018; Luo et al. 2019; Vats et al. 2019). This will be discussed in detail in the following section.

9.4 The Role of Domain Walls in the Ferroelectric Photovoltaic Effect

The field of ferroelectric photovoltaic effect remained at a standstill for over two decades until its recent revival by the report of switchable diode behavior in $BiFeO_3$ single crystals and anomalous photovoltage in $BiFeO_3$ thin films possessing periodic arrays of 71° domain walls (Figure 9.5; Choi et al. 2009; Yang and Ramesh 2010; also see Chapter 10 for more details on the domains and domain walls in $BiFeO_3$). Using scandate single crystal substrates with proper thermal/chemical treatment, $BiFeO_3$ thin films grow epitaxially with stripe domains separated by 71° domain walls. By illuminating these films with a 285 mW/cm^2 white light, the devices with coplanar electrodes running parallel to the domain walls, i.e. perpendicular to in-plane ferroelectric polarization, exhibit an above-bandgap photovoltage ($V_{OC} \approx 16$ V) and an in-plane short-circuit current

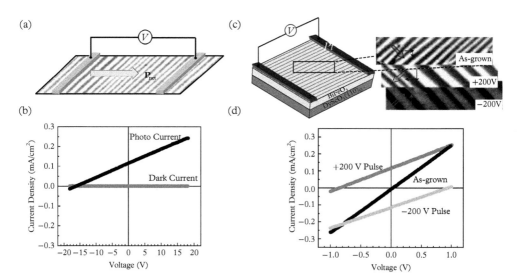

Figure 9.5 *(a) Schematic of the domain structure and perpendicular geometry of coplanar electrodes in the $BiFeO_3$/$DyScO_3$ thin film. (b) Corresponding current density–voltage characteristic. (c) Schematic with PFM images show the device domain structures of as-grown (top), +200 V poled (middle), and -200 V poled (bottom). (d) Current density–voltage characteristics corresponding to different ferroelectric polarization configurations.*

Source: Adapted with permission from Springer Nature, Yang and Ramesh (2010).

density $\mathcal{J}_{SC} \approx 120 \ mA/cm^2$ flowing along the net in-plane ferroelectric polarization. The photovoltage V_{OC} was found to increase linearly in magnitude with increasing electrode spacing, and the short-circuit current was reported to be insensitive to the polarization of incident light (Yang and Ramesh 2010; Seidel et al. 2011). The photovoltage sign of the ferroelectric device can be reversibly switched by applying an external bias to switch the ferroelectric polarization. Based on these phenomena, Yang et al. proposed the crucial role of domain walls in the photovoltaic effect of BiFeO$_3$ thin films. Specifically, owing to the polarization divergence in the direction perpendicular to the domain wall, the associated charge density $\rho = -\nabla \cdot P$ forms an electric dipole and leads to an electrical field within the walls and a potential step from one side to the other (Figure 9.6; see Meyer and Vanderbilt 2002; Seidel et al. 2009). This potential step enables efficient separation of the photoexcited electron–hole pairs and develops a potential drop over the domain wall. Due to the absence of such a potential drop in the domain bulk, photoexcited electron–hole pairs are expected to quickly recombine, leading to no net photovoltaic effect. Once domain walls run parallel to each other over the whole device, as in the case of BiFeO$_3$ thin film with pure 71° domain walls, the accumulation of all the potential drops in domain walls gives rise to an anomalous photovoltage exceeding the value of the bandgap.

However, this model has stirred up considerable debate and been challenged in many ways. First, subsequent studies revealed substantial photocurrent I_{SC}, and anomalous photovoltage was also observed in the ferroelectric photo-cell where coplanar electrodes run perpendicular to the domain walls (Figure 9.7; see Bhatnagar et al. 2013; Yang et al. 2017b). A simple way to gain a large photovoltage in BiFeO$_3$ solar cells is to reduce its conduction under illumination, e.g. by reducing the sample tempera-

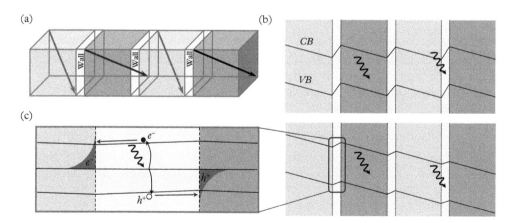

Figure 9.6 *(a) Schematic of four domains (three domain walls) in an order array of 71° domain walls. (b) Corresponding band diagram showing the valence band (VB) and conduction band (CB) across these domains and domain walls in the dark. (c) Evolution of band structure upon illumination of the domain wall array.*

Source: *Adapted with permission from Springer Nature, Yang and Ramesh (2010).*

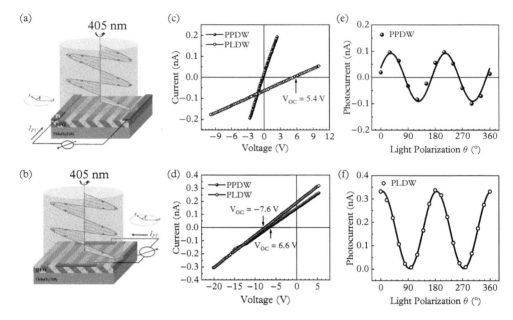

Figure 9.7 *Schematics show coplanar electrodes on the surface of BiFeO₃ thin film running (a) parallel (PLDW) and (b) perpendicular (PPDW) to the stripe domain walls. Current–voltage characteristics under illumination with monochromatic light (λ = 405 nm) of BiFeO₃ thin films comprising (c) 109° and (d) 71° periodic stripe domains. (e), (f) Light polarization dependence of the short-circuit photocurrent in the BiFeO₃ thin films with pure 71° domain walls. The curves are the fitting with Equation (9.1).*

Source: Adapted with permission from Yang et al. 2017b under the terms of the CC-BY 4.0 license.

ture, which is independent of the number of domain walls between the electrodes. Second, the photocurrent in BiFeO₃ thin films under illumination can flow in the direction perpendicular to the ferroelectric polarization (Ji et al. 2011; Yang et al. 2017a). In particular, a similar anomalous photovoltaic effect also occurs in the monodomain BiFeO₃ thin film without any domain walls (Matsuo et al. 2016; Yang et al. 2017a). Third, in contrast to the domain-wall-based model proposed by Yang, the spatially resolved distribution of photocurrent in BiFeO₃ single crystals characterized by the photoelectric AFM reveals a similar amount of photocurrent over the domains and the domain walls (Alexe and Hesse 2011). This indicates that the photoexcited electron–hole pairs are not largely recombined in the domain bulk. Most importantly, both the magnitude and direction of the photocurrent observed in BiFeO₃ show a distinctive dependence on the incident light polarization (Figures 9.7(e,f); see Ji et al. 2011; Bhatnagar et al. 2013; Matsuo et al. 2016; Yang et al. 2017a). This peculiar light polarization dependence of the photocurrent clearly differentiates the anomalous photovoltaic effect of BiFeO₃ from the conventional photovoltaic effect as in the classic *p-n* junction. In a conventional photovoltaic effect in which separation of the non-equilibrium photo-generated carriers is based on a gradient of the chemical potential, the photocurrent does not depend on light polarization.

All these intriguing features in the anomalous photovoltaic effect of BiFeO$_3$ can be readily elucidated in the theory of the BPV effect. The photocurrent in BiFeO$_3$ can be described by Equation (9.1) with the BPV tensor given as (Young et al. 2012; Bhatnagar et al. 2013)

$$\beta_{ij} = \begin{bmatrix} 0 & 0 & 0 & 0 & \beta_{15} & -\beta_{22} \\ -\beta_{22} & \beta_{22} & 0 & \beta_{15} & 0 & 0 \\ \beta_{31} & \beta_{31} & \beta_{33} & 0 & 0 & 0 \end{bmatrix} \tag{9.11}$$

V_{OC} can be described by Equation (9.3). Based on the above phenomena, the role of domain walls in the anomalous photovoltage can instead be deciphered by Figure 9.8 (Bhatnagar et al. 2013). In the geometry with electrodes running parallel to the domain walls, the latter serve as an additional resistance to the whole device. However, in the geometry with electrodes perpendicular to the domain walls, the latter act as shunts due to their enhanced conductivity compared to the domain bulk. This results in enhanced conduction of the ferroelectric device and a consequent decrease in the magnitude of V_{OC} according to Equation (9.3) (Seidel et al. 2009, 2010; Farokhipoor and Noheda 2011; Sluka et al. 2013). The effect of domain walls on the photocurrent conduction was also studied by utilizing photoelectric AFM wherein the conductive AFM tip acts as a mobile electrode that is able to detect the spatial distribution of the photocurrent as well

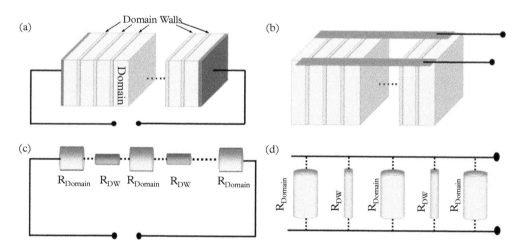

Figure 9.8 *Schematic showing the periodic arrangement of domains and domain wall in BiFeO$_3$ thin films for the (a) parallel (PLDW) and (b) perpendicular (PPDW) geometry with respect to the domain walls. An equivalent circuit considering that the domain bulk and domain wall have different resistances, R$_{Domain}$ and R$_{DW}$, is shown for both the geometries in (c) PLDW and (d) PPDW.*

Source: *Adapted with permission from Bhatnagar et al. 2013 under the terms of the CC-BY 4.0 license.*

as photoconductivity with a nanometer resolution (Yang et al. 2017b). This shows that domain walls could effectively facilitate the transport and collection of photocurrents generated in the bulk of ferroelectrics, owing to their enhanced photoconductivity.

9.5 Light-Induced Polarization Switching and Domain Wall Motion

Manipulation of ferroic domain and domain walls by light at room temperature is a fascinating topic in modern solid-state physics due to cross-fertilization in research fields that are largely decoupled and have potential for the establishment of new technological paradigms. Full optical control of the magnetic domain has been achieved in ferromagnetic, ferrimagnetic, and antiferromagnetic materials via heating and the inverse Faraday effect (Van der Ziel et al. 1965; Duong et al. 2004; Stanciu et al. 2007; Lambert et al. 2014; Manz et al. 2016). Despite being in its infancy, progress is currently being made in developing effective methodologies capable of full optical control of the ferroelectric and/or ferroelastic domains (see Chapters 10 to 12 for electric field control of ferroelectric domains/domain walls).

Rubio-Marcos et al. reported light-induced ferroelectric domain wall motion and reversible control of macroscopic polarization in $BaTiO_3$ single crystals (Rubio-Marcos et al. 2015, 2017). As schematically illustrated in Figure 9.9(a), the studied $BaTiO_3$ crystal mainly consists of two types of domains with ferroelectric polarization aligning

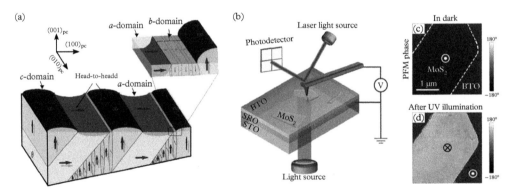

Figure 9.9 *(a) Scheme shows a domain structure composed of a-domain and c-domain with a head-to-head configuration. The head-to-head configuration maximizes the internal stress close to the domain wall, leading to the formation of b-domain. The inset of the b-domain structure shows how internal stress at the domain wall is minimized by a bundle of alternate a-domain and c-domain. (b) A sketch of the experiment geometry to study the light-induced polarization switching in $MoS_2/BaTiO_3/SrRuO_3$ heterostructure with an AFM system. Domain structure characterized in the dark before (c) and after (d) UV illumination. STO: $SrTiO_3$; SRO: $SrRuO_3$; BTO: $BaTiO_3$.*

Source: (a) Adapted with permission from Rubio-Marcos et al. 2015 under the terms of the CC-BY 4.0 license. (b–d) Adapted with permission from Li et al. 2018 under the terms of the CC-BY 4.0 license.

orthogonal to each other. The domain with ferroelectric polarization running along the surface normal direction is referred as the *c*-domain and the one with polarization in the in-plane direction is termed the *a*-domain. The boundary separating the *a*- and *c*-domains is called the *a-c* domain wall. Due to the large internal compressive stress between the *a*- and *c*- domains, subtle *b*-domains form at the *a-c* domain walls. Illumination on the $BaTiO_3$ single crystal with a wavelength of 532 nm induces a movement of the *a-c* domain wall toward the *a*-domain, thereby transforming *a*-domains into *c*- and *b*-domains. This is further demonstrated by high-resolution X-ray diffraction (XRD). Illumination on the $BaTiO_3$ crystal gives rise to an increase in the diffraction intensity of both the *c*- and *b*-domains. After turning off the light, the domain wall moves back to the position before the illumination. This effect also leads to a reversible optical control of the macroscopic polarization and its related properties, such as permittivity. With respect to the underlying mechanism responsible for this effect, the authors ascribe it to the interaction between the alternating current (AC) electric field of light and the compensation charges at the *a-c* domain walls. However, the intriguing light polarization dependence of the domain wall motion under polarized light hints at the possible role of the bulk photovoltaic effect. The combination of the bulk photovoltaic effect and the converse piezoelectric effect would induce a photostriction effect in the $BaTiO_3$ crystal, thus modulating the mechanical energy at the ferroelastic domain wall and affecting its geometry.

Ferroelectric domain structures are also determined by the boundary conditions, especially those related to the screening of ferroelectric polarization (Fridkin 1980). Light, by generating non-equilibrium carriers in ferroelectric materials as well as semiconductor electrodes, can effectively tailor the electrical boundary condition, thus affecting the domain structures as well as other ferroelectric properties. One of the clear-cut examples is that light can effectively lower the ferroelectric Curie temperature and decrease the temperature hysteresis of the phase transition (Fridkin 1980). Also, light can reduce the coercive field of lithium niobate crystals by an order of magnitude and modify the shape of polarization-electric field hysteresis loop (Gruverman et al. 2002; Sones et al. 2005). Along this path, Li et al. demonstrated optical control of ferroelectric polarization in the $BaTiO_3$ thin film capacitor using an ultraviolet (UV) light ($\lambda = 377 \pm 25$ nm) as the illumination source (Li et al. 2018). As illustrated in Figure 9.9(b), the $BaTiO_3$ thin film grown on a conductive $SrRuO_3$ layer buffered $SrTiO_3$ crystal substrate forms a capacitor with a two-dimensional narrow-gap semiconductor MoS_2 acting as the top electrode. UV illumination switches the pristine outward polarization to the downward direction only within the area covered by the MoS_2 layer (Figures 9.9(c,d)). This effect is attributed to the redistribution of the photo-generated carriers and screening charges at the MoS_2–$BaTiO_3$ interface. It is a two-step process which involves the formation of intra-layer excitons during light absorption followed by their decay into inter-layer excitons. This results in positive charge accumulation at the interface forcing the polarization reversal from the upward to the downward direction.

To achieve full optical control of ferroelectric domain walls, one needs to switch the ferroelectric polarization in a reversible and non-volatile manner. This is achieved in the $BiFeO_3$ thin films by the mediation of the tip-enhanced photovoltaic effect (see Yang and Alexe 2018). The enhanced short-circuit photocurrent density at the tip contact area generates a local electric field that well exceeds the coercive field, thus enabling ferroelectric polarization switching. Reversible optical control of ferroelectric polarization can be achieved via a unique feature of the BPV effect, i.e. the dependence on light polarization of the short-circuit current. Figure 9.10(a) illustrates the device geometry consisting of $BiFeO_3$ thin film grown on a $TbScO_3$ substrate with a patterned $SrRuO_3$ conductive layer located on the center of the substrate. The conductive AFM tip contacts the $BiFeO_3$ films on top of the $SrRuO_3$ layer, and stripe platinum electrodes are patterned on the surface of the $BiFeO_3$ film grown on bare $TbScO_3$ substrate. During illumination of the $BiFeO_3$ film, the conductive tip senses a current without applying any external voltage, which originates in the BPV effect of the $BiFeO_3$ film. More importantly, both the magnitude and direction of the photocurrent depend on the light polarization direction (see Figure 9.10(b)). Consequently, the light-generated

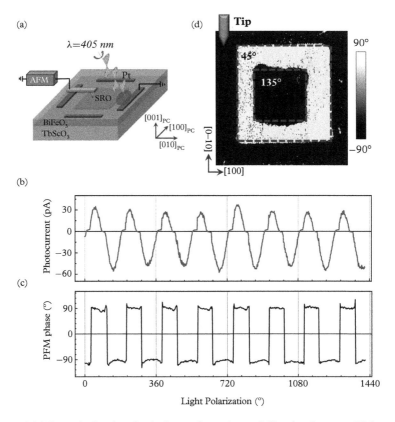

Figure 9.10 *(a) Schematic showing the device configuration and illumination area. Light polarization dependence of (b) short-circuit photocurrent, (c) out-of-plane PFM phase signal. (d) Light-polarization-dependent optically written out-of-plane domain structures. The AFM cantilever is along the $BiFeO_3$ $[010]_{pc}$ crystallographic direction as shown in (d).*

Source: Adapted with permission from John Wiley & Son, Yang and Alexe (2018).

electric field underneath the AFM tip can be controlled by the incident light polarization via a half-wave plate. As a result, the phase signal detected by the piezoresponse force microscope (PFM) on the $BiFeO_3$ film, which indicates the polarization direction underneath the tip, can be reversibly tuned in a non-volatile way by simply rotating the light polarization (Figure 9.10(c)). By scanning the AFM tip on the $BiFeO_3$ surface with a certain light polarization, e.g., 45° or 135°, ferroelectric polarization over a large area can be reversibly switched by light (Figure 9.10(d)). Note that the light-polarization-induced variation of the photocurrent direction is not a general phenomenon in the bulk photovoltaic effect. It is determined by the electrode configuration, direction of light incidence, crystallographic structure, and the fine electronic features of the non-centrosymmetric materials (Yang et al. 2017a,b).

9.6   Summary

We conclude that the BPV effect remains very intriguing. On the one hand it has a microscopic generation mechanism that needs a better fundamental understanding, while on the other, due to the very large photo-generated voltages, it has a certain potential for the development of novel photoelectric devices. The BPV effect generates light-polarization-dependent photocurrent and anomalous photovoltage in non-centrosymmetric materials. In principle, the conversion efficiency is independent of the Shockley–Queisser limit. In conjunction with the tip-enhanced effect, the photocurrent density in ferroelectric solar cells can be enhanced by reducing the size of the contact electrode. The role of the domain walls is not as fundamental as was initially believed, but nevertheless, light is able to the move the domain walls and reversibly switch the ferroelectric polarization. The interaction between light and ferroelectric materials not only provides an alternative approach to converting solar energy to electricity, but also offers a playground to explore the coupling between photoelectric effects and other (multi-)ferroic or correlated-electron properties

Acknowledgments

M.A. acknowledges the Wolfson Research Merit and Theo Murphy Blue-sky Awards of Royal Society. The work was partly supported by the EPSRC (UK) through grants no. EP/M022706/1, EP/P031544/1 and EP/P025803/1.

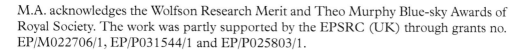

REFERENCES

Alexe M., Hesse D., "Tip-enhanced photovoltaic effects in bismuth ferrite," *Nat. Commun.* 2, 256 (2011).
Astafiev S., Fridkin V. M., Lazarev V. G., Shlensky A. L., "Magnetophotovoltaic effect in crystals without a center of symmetry," *Ferroelectrics* 83, 3–18 (1988).
Belinicher V., Malinovskii V., Sturman B., "Photogalvanic effect in a crystal with polar axis," *Soviet J. Exp. Theor. Phys.* 46, 362 (1977).

Belinicher V. I., Ivchenko E. L., Sturman B. I., "Kinetic theory of the displacement photovoltaic effect in piezoelectrics," *Zh. Eksp. Teor. Fiz.* 83, 649–661 (1982).

Belinicher V. I., Sturman B. I., "The relation between shift and ballistic currents in the theory of photogalvanic effect," *Ferroelectrics* 83, 29–34 (1988).

Bhatnagar A., Roy Chaudhuri A., Heon Kim Y., Hesse D., Alexe M., "Role of domain walls in the abnormal photovoltaic effect in $BiFeO_3$," *Nat. Commun.* 4, 2835 (2013).

Burger A. M., Agarwal R., Aprele A., Schruba E., Gutierrez-Perez A., Fridkin V. M., Spanier J. E., "Direction observation of shift and ballistic photovoltaic currents," *Sci. Adv.* 5, eaau5588 (2019).

Choi T., Lee S., Choi Y. J., Kiryukhin V., Cheong S.-W., "Switchable ferroelectric diode and photovoltaic effect in $BiFeO_3$," *Science* 324(5923), 63–66 (2009).

Chynoweth A. G., "Surface space-charge layers in barium titanate," *Phys. Rev.* 102, 705–714 (1956).

Cook A. M., Fregoso B. M., de Juan F., Coh S., Moore J. E., "Design principles for shift current photovoltaics," *Nat. Commun.* 8, 14176 (2017).

Duong N., Satoh T., Fiebig M., "Ultrafast manipulation of antiferromagnetism of NiO," *Phys. Rev. Lett.* 93, 117402 (2004).

Farokhipoor S., Noheda B., "Conduction through 71° domain walls in $BiFeO_3$ thin films," *Phys. Rev. Lett.* 107, 127601 (2011).

Festl H. G., Hertel P., Kratzig E., von Baltz R., "Investigations of the photovoltaic tensor in doped $LiNbO_3$," *Phys. Status. Solidi. B* 118, 157–164 (1982).

Fregoso B. M., Morimoto T., Moore J. E., "Quantitative relationship between polarization differences and the zone-averaged shift photocurrent," *Phys. Rev. B* 96, 075421 (2017).

Fridkin V. M., *Ferroelectric Semiconductors* (Consultants Bureau, New York, 1980).

Fridkin V. M., "Review of recent work on the bulk photovoltaic effect in ferro and piezoelectrics," *Ferroelectrics* 53, 169–187 (1984).

Fridkin V. M., "Bulk photovoltaic effect in noncentrosymmetric crystals," *Crystallogr. Rep.* 46, 654 (2001).

Fridkin V. M., Magomadov R. M., "Anomalous photovoltaic effect in $LiNbO_3$: Fe in polarized light," *Pis'ma Zh. Eksp. Teor Fiz* 30, 4 (1979).

Ganichev S. D., Prettle W., "Spin photocurrent in quantum wells," *J. Phys.: Condens. Mat.* 15, R935–R983 (2003).

Glass A. M., von der Linde D., Auston D. H., Negran T. J., "Excited state polarisation, bulk photovoltaic effect and the photorefractive effect in electrically polarized media," *J. Electron. Mater.* 4, 915–943 (1975).

Glass A. M., von der Linde D., Negran T. J., "High-voltage bulk photovoltaic effect and the photorefractive process in $LiNbO_3$," *Appl. Phys. Lett.* 25, 233–235 (1974).

Gong S. J., Zheng F., Rappe A. M., "Phonon influence on bulk photovoltaic effect in the ferroelectric semiconductor GeTe," *Phys. Rev. Lett.* 121, 017402 (2018).

Gruverman A., Rodriguez B. J., Nemanich R., Kingon A., "Nanoscale observation of photoinduced domain pinning and investigation of imprint behaviour in ferroelectric thin films," *J. Appl. Phys.* 92, 2734–2739 (2002).

Gu Z., Imbrenda D., Bennett-Jackson A. L., Falmbigl M., Podpirka A., Parker T. C., Shreiber D., Ivill M. P., Fridkin V. M., Spanier J. E., "Mesoscopic free path of nonthermalized photogenerated carriers in a ferroelectric insulator," *Phys. Rev. Lett.* 118, 096601 (2017).

Gunter P., "Photovoltages, photocurrents and photorefractive effects in $KNbO_3$: Fe," *Ferroelectrics* 22, 671–674 (1978).

Ivchenko E. L., Lyanda-Geller Y. B., Pikus G. E., "Magneto-photogalvanic effects in noncentrosymmetric crystals," *Ferroelectrics* 83, 19–27 (1988).

Ji W., Yao K., Liang Y. C., "Evidence of bulk photovoltaic effect and large tensor coefficient in ferroelectric $BiFeO_3$ thin films," *Phys. Rev. B* 84, 094115 (2011).

Koch W. T. H., "Anomalous photovoltage in $BaTiO_3$," *Ferroelectrics* 13, 305–307 (1976).

Koch W. T. H., Munser R., Ruppel W., Würfel P., "Bulk photovoltaic effect in $BaTiO_3$," *Solid State Commun.* 17, 147–150 (1975).

Krätzig E., Orlowski R., "1976 $LiTaO_3$ as holographic storage material," *Appl. Phys.* 15, 133–139 (1978).

Lambert C.-H., Mangin S., Varaprasad B. S., Takahashi Y. K., Hehn M., Cinchetti M., Malinowski G., Hono K., Fainman Y., Aeschlimann M., Fullerton E. E., "All-optical control of ferromagnetic thin films and nanostructures," *Science* 345, 1337–1340 (2014).

Li T., Lipatov A., Lu H., Lee H., Lee J. W., Torun E., Wirtz L., Eom C. B., Íñiguez J., Sinitskii A., Gruverman A., "Optical control of polarization in ferroelectric heterostructures," *Nat. Commun.* 9, 3344 (2018).

Luo Z. D., Park D. S., Yang M. M., Alexe M., "Light-controlled nanoscopic writing of electronic memories using the tip-enhanced bulk photovoltaic effect," *ACS Appl. Mater. Interf.* 11, 8276–8283 (2019).

Manz S., Matsubara M., Lottermoser T., Büchi J., Iyama A., Kimura T., Meier D., Fiebig M., "Reversible optical switching of antiferromagnetism in $TbMnO_3$," *Nat. Photon.* 10, 653 (2016).

Matsuo H., Kitanaka Y., Inoue R., Noguchi Y., Miyayama M., Kiguchi T., Konno T. J., "Bulk and domain-wall effects in ferroelectric photovoltaics," *Phys. Rev. B* 94, 214111 (2016).

Meyer B., Vanderbilt D., "Ab initio study of ferroelectric domain walls in $PbTiO_3$," *Phys. Rev. B* 65, 104111 (2002).

Nechache R., Harnagea C., Li S., Cardenas L., Huang W., Chakrabartty J., Rosei F., "Bandgap tuning of multiferroic oxide solar cells," *Nat. Photon.* 9, 61–67 (2014).

Rangel T., Fregoso B. M., Mendoza B. S., Morimoto T., Moore J. E., Neaton J. B., "Large bulk photovoltaic effect and spontaneous polarization of single-layer monochalcogenides," *Phys. Rev. Lett.* 119, 067402 (2017).

Rubio-Marcos F., Campo A. D., Marchet P., Fernández J. F., "Ferroelectric domain wall motion induced by polarized light," *Nat. Commun.* 6, 6594 (2015).

Rubio-Marcos F., Ochoa D. A., Campo A. D., García M. A., Castro G. R., Fernández J. F., García J. E., "Reversible optical control of macroscopic polarization in ferroelectrics," *Nat. Photon.* 12, 29–32 (2017).

Schirmer O. F., Imlau M., Merschjann C., "Bulk photovoltaic effect of $LiNbO_3$:Fe and its small-polaron-based microscopic interpretation," *Phys. Rev. B* 83, 165106 (2011).

Seidel J., Martin L. W., He Q., Zhan Q., Chu Y. -H., Rother A., Hawkridge M. E., Maksymovych P., Yu P., Gajek M., Balke N., Kalinin S. V., Gemming S., Wang F., Catalan G., Scott J. F., Spaldin N. A., Orenstein J., Ramesh R., "Conduction at domain walls in oxide multiferroics," *Nat. Mater.* 8, 229 (2009).

Seidel J., Maksymovych P., Batra Y., Katan A. J., Yang S., He Q. G., Baddorf A. P., Kalinin S. V., Yang C.-H., Yang J.-C., Chu Y.-H., Salje E. K. H., Wormeester H., Salmeron M. B., Ramesh R., "Domain wall conductivity in La-doped $BiFeO_3$," *Phys. Rev. Lett.* 105, 197603 (2010).

Seidel J., Fu D., Yang S. Y., Alarcón-Lladó E., Wu J., Ramesh R., Ager J. W., "Efficient photovoltaic current generation at ferroelectric domain walls," *Phys. Rev. Lett.* 107(12), 126805 (2011).

Shockley W., Queisser H. J., "Detailed balance limit of efficiency of p-n junction solar cells," *J. Appl. Phys.* 32, 510–519 (1961).

Sluka T., Tagantsev A. K., Bednyakov P., Setter N., "Free-electron gas at charged domain walls in insulating BaTiO$_3$," *Nat. Commun.* 4, 1808 (2013).

Sones C. L., Wengler M. C., Valdivia C. E., Mailis S., Eason R. W., Buse K., "Light-induced order of magnitude decrease in the electric field for domain nucleation in MgO-doped lithium niobite crystals," *Appl. Phys. Lett.* 86, 212901 (2005).

Spanier J. E., Fridkin V. M., Rappe A. M., Akbashev A. R., Polemi A., Qi Y., Gu Z., Young S. M., Hawley C. J., Imbrenda D., Xiao G., Bennett-Jackson A. L., Johnson C., "Power conversion efficiency exceeding the Shockley–Queisser limit in a ferroelectric insulator," *Nat. Photon.* 10, 611–616 (2016).

Stanciu C., Hansteen F., Kimel A. V., Kirilyuk A., Tsukamoto A., Itoh A., Rasing T., "All-optical magnetic recording with circularly polarized light," *Phys. Rev. Lett.* 99, 047601 (2007).

Sturman B. I., Fridkin V. M., *Photovoltaic and Photo-Refractive Effects in Noncentrosymmetric Materials*, Volume 8 (CRC Press, Paris, 1992).

Tan L. Z., Rappe A. M., "Enhancement of the bulk photovoltaic effect in topological insulators," *Phys. Rev. Lett.* 116, 237402 (2016).

Van der Ziel J., Pershan P., Malmstrom L., "Optically-induced magnetization resulting from the inverse Faraday effect," *Phys. Rev. Lett.* 15, 190 (1965).

Vats G., Bai Y., Zhang D., Juuti J., Seidel J., "Optical control of ferroelectric domains: nanoscale insight into macroscopic observations," *Adv. Opt. Mater.*, 1800858 (2019).

von Baltz R., Kraut W., "Theory of bulk photovoltaic effect in pure crystals," *Phys. Rev. B* 23, 5590–5596 (1981).

Yang M.-M., Luo Z.-D., Kim D. J., Alexe M., "Bulk photovoltaic effect in monodomain BiFeO$_3$ thin films," *Appl. Phys. Lett.* 110, 183902 (2017a).

Yang M.-M., Bhatnagar A., Luo Z.-D., Alexe M., "Enhancement of local photovoltaic current at ferroelectric domain walls in BiFeO$_3$," *Sci. Rep.* 7, 43070 (2017b).

Yang M. M., Alexe M., "Light-induced reversible control of ferroelectric polarization in BiFeO$_3$," *Adv. Mat.* 30, 1704908 (2018).

Yang S. Y., Ramesh R., "Above-bandgap voltages from ferroelectric photovoltaic devices," *Nat. Nanotechnol.* 5, 143 (2010).

Young S. M., Rappe A. M., "First principles calculation of the shift current photovoltaic effect in ferroelectrics," *Phys. Rev. Lett.* 109, 116601 (2012).

Young S. M., Zheng F., Rappe A. M., "First-principles calculation of the bulk photovoltaic effect in bismuth ferrite," *Phys. Rev. Lett.* 109, 236601 (2012).

Zenkevich A., Matveyev Y., Maksimova K., Gaynutdinov R., Tolstikhina A., Fridkin V., "Giant bulk photovoltaic effect in thin ferroelectric BaTiO$_3$ films," *Phys. Rev. B* 90, 1614089(R) (2014).

10

Transmission Electron Microscopy Study of Ferroelectric Domain Walls in BiFeO₃ Thin Films: Structures and Switching Dynamics

L. Li[1] and X. Pan[1,2]

[1]*Department of Materials Science and Engineering, University of California–Irvine, California 92697, United States*
[2]*Department of Physics and Astronomy, University of California–Irvine, California 92697, United States*

10.1 Introduction

A ferroelectric domain wall (DW) is a quasi-two-dimensional (quasi-2D) boundary that separates two ferroelectric domains that differ in the orientation of their spontaneous polarization, as introduced in Chapter 1. It can be created, erased, and reconfigured within the same physical volume of the ferroelectric matrix by an external electric field (see also Chapters 11 and 13). Over the past few years, there has been increasing evidence showing that ferroelectric DWs can possess functionalities that are absent in the bulk of the domains. Examples include enhanced ferroelectricity (Privratska and Janovec 1999; Wei et al. 2014), increased conductivity (Chapter 6, 12 and 15) and photovoltages (Chapter 9) (Seidel et al. 2009; Yang et al. 2010), and magneto transport properties (He et al. 2012), all of which make ferroelectric DWs attractive for applications as building blocks for next-generation reconfigurable nanodevices. Understanding the static structures and dynamic behaviors of ferroelectric DWs is important for the development of practical applications. A wide range of advanced imaging and spectroscopic techniques based on aberration-corrected transmission electron microscopy (TEM) or scanning transmission electron microscopy (STEM), as illustrated in Figure 10.1, have become very powerful methods for the characterization of ferroelectric DWs or related structures. For example, by using atomic-resolution STEM energy-dispersive spectroscopy (EDS) spectrum imaging, the chemical composition across an atomically sharp interface between a ferroelectric thin film and a substrate can be mapped out

L. Li and X. Pan, *Transmission Electron Microscopy Study of Ferroelectric Domain Walls in BiFeO₃ Thin Films: Structures and Switching Dynamics* In: *Domain Walls: From Fundamental Properties to Nanotechnology Concepts.* Edited by: Dennis Meier, Jan Seidel, Marty Gregg, and Ramamoorthy Ramesh, Oxford University Press (2020). © L. Li and X. Pan.
DOI: 10.1093/oso/9780198862499.003.0010

(Figure 10.1(a)). By using dark-field diffraction-contrast TEM imaging, patterns of ferroelectric DWs can be observed from the cross section of a nanoscale thin film (Figure 10.1(b)). With electron energy-loss spectroscopy (EELS), a variation in the oxidation states of transition metals that are changed with the alternative polarizations across a ferroelectric DW at a ferroelectric/substrate interface can be revealed (Figure 10.1(c)). With in situ TEM, real-time observation of DW motion under an applied electrical bias can be achieved (Figure 10.1(d)). By using STEM imaging with sub-angstrom resolution, atomic structures across DWs can be resolved (Figure 10.1(e)). Based on such STEM imaging, the spatial distribution of polarization vectors can be determined unambiguously at the atomic scale (Figure 10.1(f)). Combining all these techniques within a single TEM instrument, a direct correlation can be established between the properties of ferroelectric DWs and the underlying physical microstructures.

In this chapter, we present a review on the recent progress in TEM studies of ferroelectric DWs in one of the most widely studied ferroelectric systems: $BiFeO_3$ thin films. This system has been chosen as representative of a much wider range of ferroelectric perovskites with functional DWs, due to its strong spontaneous polarization; coexistence of ferroelectricity, ferroelasticity, and antiferromagnetism; and numerous functionalities at the DWs. Several review articles addressing the same system or similar ones have been published in recent years (Kalinin et al. 2010; Wu et al. 2016; Li et al. 2019). Here, we first briefly introduce the instrumentation, experimental procedures, imaging mechanisms, and analytical methods of the state-of-the-art TEM-based techniques. The application of these techniques to the study of DW structures and switching behaviors is demonstrated, with particular emphasis on the critical roles of interfaces and defects, and the interplay between different types of DWs. The phenomena and mechanisms discovered in the $BiFeO_3$ model system are also applicable to many other ferroelectric materials with similar DW structures. The results not only advance our fundamental understanding of static and dynamic properties of ferroelectric DWs, but also form the basis for designing practical ferroelectric-DW-based devices.

10.2   Characterization of DW Structures by TEM

10.2.1   Diffraction-Contrast TEM

The "diffraction-contrast" imaging in TEM is also known as "dark-field" or "bright-field" imaging. The principle underlying such imaging as compared to normal TEM imaging is illustrated by the ray diagrams in Figures 10.2(a,b). In normal TEM imaging, the image contrast consists of signals from all the transmitted and diffracted beams, which are shown by the ray traces in Figure 10.2(a). By contrast, a dark-field image is generated by using only selected diffracted beams by carefully tilting the TEM specimen and placing an appropriate aperture at the back focal plane. In Figure 10.2(b), for example, the selected diffracted beams are shown by the ray traces diffracted to the right-side spot at the focal plane, which transmit through the aperture. Alternatively, the centered ray traces at the focal plane, i.e. the transmitted beams, can be selected to form a bright-field

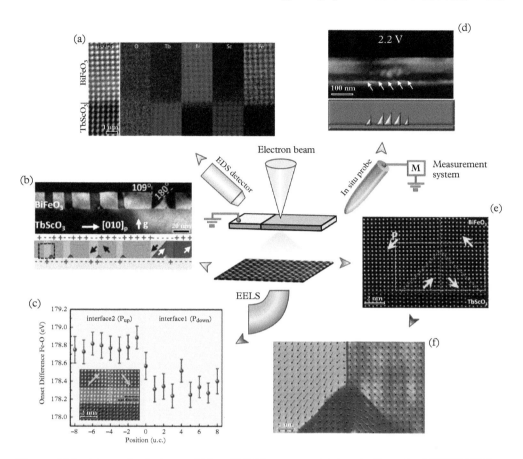

Figure 10.1 *Advanced techniques within a TEM instrument. (a) Atomically resolved STEM EDS spectrum image of a $BiFeO_3/TbScO_3$ interface. (b) Cross-sectional dark-field TEM image of periodic 109° DW patterns in a $BiFeO_3$ film grown on a $TbScO_3$ substrate. (c) High-resolution EELS line scan taken at a $BiFeO_3/TbScO_3$ interface across a 109° DW showing the energy onset difference between O-K and Fe-L_3 edges across the 109° DW, which indicates a shift in the Fe oxidation state, which changes with the alternative polarizations across the 109° DW at the $BiFeO_3/TbScO_3$ interface. The inset shows a HAADF STEM image obtained at the interface, where the polarization directions are marked by green arrows. (d) In situ TEM observation of nucleation of triangular domains with inclined charged DWs in a $BiFeO_3$ film when a positive voltage of 2.2 V is applied at the tip. (e) HAADF STEM image of the intersection of a 109° DW with the $BiFeO_3/TbScO_3$ interface. Yellow arrows indicate the polarization direction, and yellow dashed lines indicate the position of the DWs. (f) Polarization vector map of the vortex domain structure in $BiFeO_3$ as highlighted by the green rectangle in (e). Note: EDS, energy-dispersive spectroscopy; EELS, electron energy-loss spectroscopy; HAADF, high-angle annular dark-field; P, polarization.*

Source: (a,c) Reproduced with permission from Zhang et al. 2018. Copyright 2018 Nature Publishing Group. (b,e,f) Reproduced with permission from Nelson et al. 2011. Copyright 2011 American Chemical Society. (d) Reproduced with permission from Nelson et al. 2011. Copyright 2011 The American Association for the Advancement of Science.

image. In such diffraction-contrast TEM images, DWs separating ferroelectric domains can be readily identified because of a strong bright-dark contrast present in different domains. For example, in Figure 10.2(c), by using the $(1\bar{1}0)_O$ diffraction for planar-view imaging and the $(110)_O$ diffraction for cross-sectional imaging, ordered patterns of 109° DWs are observed in a 100 nm $BiFeO_3$ film grown on (110)O $TbScO_3$ (O is used to represent orthorhombic index). In addition to the domain contrast, the diffraction-contrast imaging also imparts a distinct contrast to defects such as dislocations or impurities (not present in Figure 10.2(c)), which is useful for observing their interactions with DWs.

10.2.2 Atomic-Resolution TEM

Within an aberration-corrected TEM instrument, it has now become routine to obtain atomic-scale high resolution TEM (HRTEM) or STEM imaging with sub-angstrom resolution. HRTEM imaging is also known as phase-contrast imaging, in which the ray diagram is generally of the normal configuration shown in Figure 10.2(a), local atomic potentials alter the incident planar electron wavefront, and the phase change information is employed to study the atomic structures of materials. Techniques designed to measure the phase of the electron wave include, for example, exit wave reconstruction from a series of HRTEM images taken at different focuses, and negative spherical-aberration (C_S) phase-contrast HRTEM imaging. These techniques have been effectively used to resolve cation columns as well as lightweight oxygen columns (Jia et al. 2003, 2007). In the more general case of HRTEM imaging, however, due to its coherent nature, the phase contrast is not necessarily related to real atomic structures, and the interpretation of such contrasts is complex. Furthermore, only a slight variation of sample thickness or orientation, or aberrations in the objective lens, will drastically affect the final image's contrast, making the imaging condition and specimen preparation of HRTEM quite demanding.

Compared to HRTEM, STEM is more widely used to study ferroelectric materials, due to its easier operation and the fact that the image contrast can readily be related to local atomic structures. In STEM imaging, a fine electron probe is focused and raster-scanned across the specimen, and detectors are placed below the specimen to form the image (Figure 10.2(d)). The contrast of the STEM image depends on the detector's geometry. In most applications, a dark-field detector and a bright-field detector are combined to obtain dark-field and bright-field STEM images. The dark-field STEM image, especially the high-angle annular dark-field (HAADF) image, is currently the most commonly used STEM method for quantitative studies of ferroelectric microstructures, because it has several unsurpassed advantages: first, the image contrast is mainly caused by incoherent and elastically scattered electrons with large scattering angles, and is directly proportional to the atomic number Z as $Z^{1.7}$ (Pennycook 2011). Second, HAADF imaging is robust to variations in sample thickness and small deviations of the sample zone axis with respect to the incident electron beam. However, direct observation of light elements, e.g. hydrogen, nitrogen, carbon, and oxygen, in HAADF images is challenging and practically limited by their low atomic numbers (Krivanek et al. 2010).

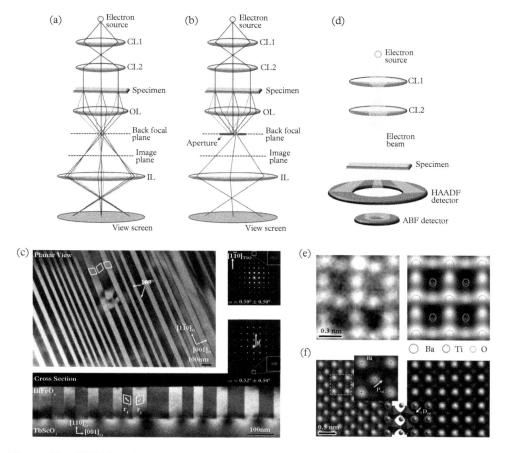

Figure 10.2 *TEM imaging techniques. (a) Ray diagram of normal TEM imaging with parallel illumination, which incorporates two condenser lenses (CL1 and CL2), an objective lens (OL), and an intermediate lens (IL) to image the specimen. Solid lines show ray traces following the specific diffraction angle from each region. (b) Ray diagram of diffraction-contrast TEM imaging whereby an aperture is inserted into the back focal plane. In this case, only one set of ray traces diffracted to the right side at the focal plane is selected to form a dark-field diffraction-contrast image. Alternatively, the centered traces can be selected to form a bright-field diffraction-contrast image. (c) Ray diagram of STEM imaging in which a convergent beam is focused by a condenser lens system and scanned sequentially on the specimen. This technique incorporates various specialized detectors placed below the specimen to measure specific specimen-electron interaction signals. Here a high-angle annular dark-field (HAADF) detector and an annular bright-field (ABF) detector are used. (d) Planar view and cross-sectional dark-field diffraction-contrast images of ordered 109° DW structures in a 100 nm BiFeO$_3$ film, and the corresponding electron diffraction patterns. (e) HRTEM images of BaTiO$_3$ involving four unit cells: experimental image (left) and the computer simulated image using the refined atomic structure which best fits the experimental one (right). The spontaneous polarization can be deduced based on the simulated structure. (f) Polarization mapping based on aberration-corrected HRTEM/STEM imaging. In BiFeO$_3$, the polarization is caused by an offset of the oxygen octahedron and the central Fe cation from the body center of Bi cation sublattice, and Fe is displaced relative to Bi along the same direction as the oxygen octahedron. This offset can be directly measured in an HRTEM image (the phase component of the exit wave determined from an HRTEM focal series) of a BiFeO$_3$ thin film (left). Since the ion charge centers are all collinear, a much more convenient determination of the polarization direction can be obtained from HAADF STEM images such as by using the relative offsets of the Bi and Fe cations alone (-D_{FB}) (right). Note: P, polarization.*

Source: (c,f) Reproduced with permission from Nelson et al. 2011. Copyright 2011 American Chemical Society. (e) Reproduced with permission from Tian et al. 2001. Copyright 2001 Materials Research Society.

To overcome this limitation, another powerful STEM technique called annular bright-field (ABF) imaging has been developed (Okunishi et al. 2009; Ishikawa et al. 2011), in which coherent and elastically scattered electrons are used for imaging. Its capability of imaging both light atomic columns (oxygen, and even hydrogen) and heavy ones simultaneously makes measurement of all ions' positions in a crystalline oxide possible.

From atomic-scale HRTEM or STEM images, quantitative information such as lattice constants and polarization-induced atomic displacements can be measured. The first demonstration of measuring polarization-induced displacements by using HRTEM imaging is reported by Tian et al. in a proceeding paper published in 2001, where they compared an experimental non-aberration-corrected HRTEM image of $BaTiO_3$ to a series of simulated images with different atomic structures and deduced the spontaneous polarization with the structure that presents the best fitting condition (Figure 10.2(e)) (Tian et al. 2001). For aberration-corrected HRTEM/STEM images, quantitative measurements of atomic positions can be much more straightforward, as each bright-contrast dot representing the atomic column in the image can be fitted with a two-dimensional Gaussian function, and the fitted center then corresponds to the atomic column's position. The precision of such measurements can reach up to several picometers (Jia et al. 2007, 2008; Kimoto et al. 2010; Nelson et al. 2011b). Based on the measured atomic positions, unit-cell-by-unit-cell maps of strains (i.e. lattice constants) could be readily extracted; and the atomic displacements between different atomic columns generally allow determination of local polarization vectors (Figure 10.2(f)). Although the absolute polarization value is not directly obtainable unless the positions of both anions (oxygens) and cations are measured, the polarization direction, for a material whose structure is already well known, can be determined unambiguously from the heavy cations only. In $BiFeO_3$, for example, an atomic displacement vector, D_{FB}, can be defined as the atomic displacement in the image plane of the Fe cation from the center of the unit cell formed by its four Bi neighbors, and can be directly measured in a HAADF STEM image (Nelson et al. 2011b). This D_{FB} vector, as the dominant manifestation of the ferroelectric polarization in $BiFeO_3$, points toward the center of the negative oxygen charges and thus is exactly opposite to the polarization component in the image plane. Therefore, $-D_{FB}$ vectors can be used to estimate the ferroelectric polarization and, in particular, to gain information about the inner structure of ferroelectric DWs and details about how the polarization re-orients across such walls.

10.2.3   In Situ TEM

As the domain and DW structures and their evolution can be readily observed by various imaging methods in standard TEM instruments, the major challenge in in situ TEM study of ferroelectric DW switching phenomena is, of course, the method by which an electric field can be applied across the TEM specimen. In early in situ TEM studies of ferroelectric switching, the electric field could only be applied parallel to the electron beam by using two parallel TEM copper grids (Figure 10.3(a)) (Yamamoto et al. 1980; Randall et al. 1987; Snoeck et al. 1994). In this configuration, the magnitude of the actual field applied to the imaging area of the specimen is difficult to estimate due to

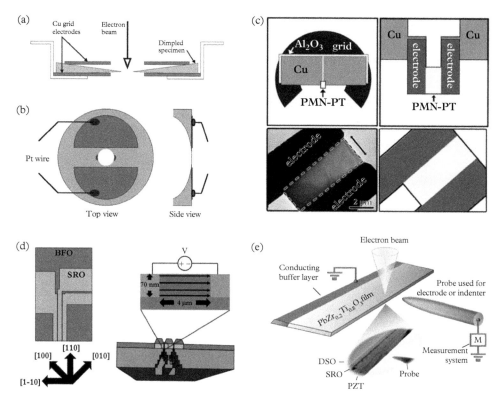

Figure 10.3 *Several methods to apply an electric field in in situ TEM. (a) Two parallel copper grid electrodes are used to apply an electric field parallel to the electron beam. (b) Two half-circle shaped gold electrodes are directly deposited on the surface of the TEM specimen to generate an electric field across the region of interest in TEM observations—the electron-transparent areas at the two small dark sites at the edge of the central perforation. (c) A specially designed in situ TEM grid. Here, a small piece of lead magnesium niobate-lead titanate (PMN-PT) specimen is cut and attached between the two parallel electrodes by using the focused ion beam (FIB) "lift-out" method. Top left and right panels show the whole view and enlargement around the specimen part, respectively. Bottom left and right panels show the low-magnification image of the TEM specimen and the corresponding schematic, respectively. In the bottom left panel, the electric field is applied between the two electrodes along the direction indicated by the arrow, and the dotted square indicates the region of interest in TEM observations. (d) A specially designed heterostructure sample for in situ experiments: left and right panels show a planar view and a side view of the BiFeO₃ (BFO)/SrRuO₃ (SRO) heterostructure, respectively. The electric field is applied between the two electrodes along the direction indicated by the arrows. (e) Integration of scanning probe into an in situ TEM specimen holder. Here, the beam is incident to the cross-sectional specimen of $PbZr_{0.2}Ti_{0.8}O_3$ (PZT) grown on DyScO₃ (DSO) substrate, an epitaxial back electrode of SrRuO₃ (SRO) is grounded, and the probe is the movable electrode.*

Source: (a,b) Reproduced with permission from Tan et al. 2005. Copyright 2005 Cambridge University Press. (c) Reproduced with permission from Sato et al. 2011. Copyright 2011 American Physical Society. (d) Reproduced with permission from Winkler et al. 2012. Copyright 2012 Elsevier. (e) Reproduced with permission from Gao et al. 2014. Copyright 2014 Nature Publishing Group.

the presence of vacuum gaps between the copper grid electrodes and the specimen. Along with recent advances in miniaturization of the electrical system for in situ TEM holders, a number of alternative approaches have been developed that allow application of an electric field perpendicular to the beam direction and quantification of the actual applied electric field. For example, Tan et al. directly deposited gold films on the TEM specimen surface to generate an electric field (Figure 10.3(b)) (He and Tan 2005; Tan et al. 2005; Qu et al. 2007). Sato et al. cut a small piece of specimen out from a single crystal of lead magnesium niobate-lead titanate and attached it between two parallel electrodes on a specially designed TEM grid, by using the focused ion beam (FIB) "lift-out" method (Figure 10.3(c)). The TEM grid was then loaded on an electrical biasing specimen holder, and an electric field could be applied through the sample (Sato et al. 2011). Alternatively, epitaxial electrodes deposited during thin film growth within a precisely controlled heterostructure have also been widely used to apply an electric field inside the in situ TEM. Winkler et al. prepared samples by growing an epitaxial electrode $SrRuO_3$ layer on $SrTiO_3$, followed by lithography and a precise ion-milling step to define the complex planar electrode structure, and a second growth of an epitaxial $BiFeO_3$ layer on top of the device structure (Figure 10.3(d)) (Winkler et al. 2012a, 2012b). However, this method only permits the application of in-plane electric fields rather than the out-of-plane electrical fields, which is more commonly used in practical applications involving ferroelectric thin film materials. The integration of a movable probe into a TEM specimen holder allows easy application of out-of-plane electrical fields to switch a local area of interest in a ferroelectric film (Chang et al. 2011; Nelson et al. 2011a; Zhang et al. 2011; Gao et al. 2014). Using such an in situ holder, an electric field along the normal of the film can be introduced by applying a voltage between the surface probe and a planar epitaxial bottom electrode, as is depicted schematically in Figure 10.3(e).

10.3 DW Structures in BiFeO₃ Thin Films

10.3.1 Boundary-Condition Engineering of DW Patterns

$BiFeO_3$ is an extremely rare case of a single-phase room-temperature multiferroic exhibiting coupled ferroelectric and antiferromagnetic ordering (Zavaliche et al. 2006; Chu et al. 2007b; Catalan and Scott 2009). Its bulk crystal has a rhombohedral structure, which can also be envisioned as two pseudocubic (PC) perovskite unit cells connected along the body diagonal, with the two oxygen octahedra in the connected perovskite units rotated clockwise and counterclockwise around this axis by 13.8° (Zavaliche et al. 2006). Structures of epitaxial $BiFeO_3$ films with tensile or moderate compressive ($<\sim$4.5%) misfit strain imposed by underlying substrates are monoclinically distorted, but closely resemble the bulk rhombohedral phase, and are therefore usually referred to as "*rhombohedral*-like (*R*-like)" structures (Christen et al. 2011). In PC unit cells of the *R*-like structure, the oxygen octahedra and the central Fe cation are displaced from their respective positions at the face and body centers, giving rise to a large spontaneous polarization ($\sim$100 μC cm^{-2}) along the <111>$_{PC}$ directions. This results

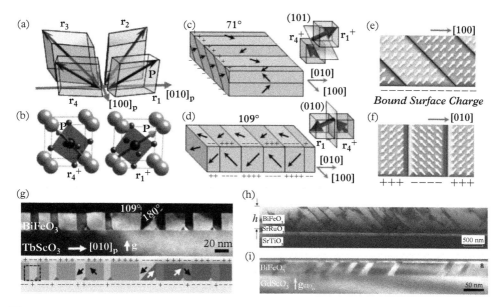

Figure 10.4 *Periodic DW structures in BiFeO₃ thin films. (a) The four ferroelastic variants of the BiFeO₃ pseudocubic (PC) unit cell, r_1-r_4. (b) Atomic models of upward polarized pseudocubic unit cells, r_1^+ and r_4^+. (c) Schematic of periodic $(101)_{PC}$-oriented $71°$ DWs, where the polarization directions are marked by the arrows. (d) Schematic of periodic $(010)_{PC}$-oriented $109°$ DWs. If uncompensated, these two patterns produce the bound surface charges shown in (e) and (f), respectively. (g) Cross-sectional dark-field TEM image of periodic $109°$ DW structures in a BiFeO₃ film grown on TbScO₃ substrate, and corresponding schematic. (h) Cross-sectional bright-field TEM image of periodic $71°$ DW structures in a BiFeO₃ film grown on SrTiO₃ substrate with the insertion of a SrRuO₃ bottom electrode. (i) Cross-sectional dark-field TEM image of periodic $180°$ DW structures in a BiFeO₃ film grown on GdScO₃ substrate. Note: P, polarization.*

Source: (a–g) Reproduced with permission from Nelson et al. 2011. Copyright 2011 American Chemical Society. (h) Reproduced with permission from Chen et al. 2007. Copyright 2007 American Physical Society. (i) Reproduced with permission from Chen et al. 2015. Copyright 2015 American Chemical Society.

in four different ferroelastic variants (r_1-r_4) (Figure 10.4(a)) and eight ferroelectric variants polarized up (+) or down (−) (Figure 10.4(b)). Rotations between polarization variants in the *R*-like BiFeO₃ can be $71°$ (ferroelastic-ferroelectric), $109°$ (ferroelastic-ferroelectric), or $180°$ (ferroelectric), yielding three types of DWs (Chen et al. 2007; Nelson et al. 2011b; Chen et al. 2015).

In ferroelectric and ferroelastic thin films, patterns of DWs depend strongly on the boundary conditions at the two surfaces or interfaces. First, the lattice mismatch between the film and substrate leads to a biaxial-strain mechanical boundary condition (Choi et al. 2004; Ederer and Spaldin 2005; Schlom et al. 2007; Schlom et al. 2014), which can be altered by choosing substrates spanning a wide range of lattice parameters (Adamo et al. 2009; Bea et al. 2009; Zeches et al. 2009; Infante et al. 2010; Nelson et al. 2011b). Second, the electrical boundary condition is critically dependent on free charge compensation at the interfaces and can be tailored by choosing substrates or epitaxial buffer layers with different conductivities (Li et al. 2002; Chisholm et al. 2010; Lichtensteiger et al. 2014). Additional restrictions on the boundary condition can be made by changing the vicinality or atomic termination of the substrates (Chu et al. 2007a; Jang et al. 2009). By modifying these boundary conditions, different types of ordered DW patterns (i.e. 71° DWs as shown in Figure 10.4(c), 109° DWs as shown in Figure 10.4(d), and 180° DWs) can be fabricated in BiFeO$_3$ thin films. The 71° DW structures have a single out-of-plane polarization vector resulting in uniform positive and negative bound charge on the opposite surfaces of the film (Figure 10.4(e)). This bound charge produces a large depolarization field when the film is very thin (<30–50 nm), which would destabilize the domain structure unless it can be screened by free charge carriers. It follows that the 71° domain structures are usually stabilized in thick (>30–50 nm) BiFeO$_3$ thin films, or with the insertion of thick (>20 nm) bottom electrodes (Chu et al. 2009; Li et al. 2013, 2016). In contrast, if the surfaces of BiFeO$_3$ thin films are uncompensated by free charges, the system favors the formation of 109° or 180° DW patterns with alternating out-of-plane polarization that can reduce the total electrostatic energy (the case for 109° DWs is shown in Figure 10.4(f), the case for 180° DWs is similar) (Nelson et al. 2011b). Diffraction-contrast TEM observation of periodically ordered patterns of the three types of DWs in different BiFeO$_3$ thin films are shown in Figures 10.4(g–i). Examination of atomic structures of DWs based on atomic-resolution imaging can be directly coupled to first-principle calculations, as demonstrated by a study of 109° and 180° DWs in BiFeO$_3$ thin films (Figure 10.5) (Wang et al. 2013). While the 109° DW is found to be about a single transitional Fe atomic column wide (Figures 10.5 (a–c)), the 180° DW spans two transitional Bi/Fe atomic columns (Figures 10.5(d–f)). For both types of DWs, the observed displacement normal to the DW (*dx*) remains roughly constant, and the change in displacement occurs almost entirely parallel to DW plane.

In addition to the three regular types of DWs, exotic DW structures and polarization structures associated with DWs have recently attracted intense research interest, especially polarization vortices and charged DWs (CDWs). In a ferroelectric vortex, the polarization continuously rotates about one point, forming a flux-closure structure that is as small as a few nanometers in diameter (Naumov et al. 2004). TEM observation of vortex-like domains in ferroelectrics can be traced back to the 1980s (Pan et al. 1985; Pan and Unruh 1990), but it was only recently demonstrated that periodic arrays of polarization vortices can be artificially created and engineered in ferroelectric heterostructures through atomic-scale control of the boundary conditions (Nelson et al. 2011b; Tang et al. 2015; Yadav et al. 2016). The first observation of arrays of polarization vortices was demonstrated in a BiFeO$_3$ thin film grown on TbScO$_3$ substrate by using the polarization mapping technique (mapping of -$\boldsymbol{D}_{FB}$ vectors) based on atomic-scale HAADF STEM imaging (Figure 10.6(a), in which only one vortex above the BiFeO$_3$/TbScO$_3$ interface is shown; this type of vortex and related DW structures are periodically present along the interface). The polarization vectors exhibit continuous rotations and form flux-closure vortex domains at the intersection of the 109° DW and the BiFeO$_3$/TbScO$_3$ interface. Phase-field simulation results demonstrate that the vortices near the interfaces result from the bound charge density waves, i.e. alternate

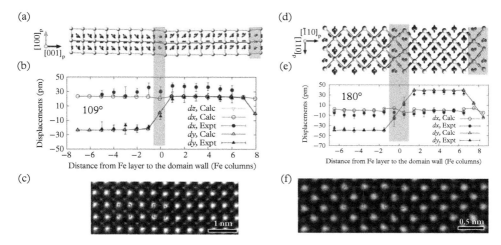

Figure 10.5 *Atomic structures of 109° and 180° DWs in BiFeO₃. (a) Atomic positions and the -D_{FB} vectors (shown by the arrows) across a 109° DW in BiFeO₃ calculated by density functional theory (DFT). (b) Comparison of x, y, z components of the -D_{FB} vectors across the 109° DW measured by DFT calculation and HAADF STEM imaging. The DW regions are highlighted by the green shadows. (c) HAADF STEM image of a 109° DW in a BiFeO₃ thin film. (d) Atomic positions and the -D_{FB} vectors across a 180° DW in BiFeO₃ calculated by DFT. (e) Comparison of x, y, z components of the -D_{FB} vectors across the 180° DW measured by DFT calculation and HAADF STEM imaging. (f) HAADF STEM image of a 180° DW in a BiFeO₃ thin film. Note: HAADF, high-angle annular dark-field. Source: Reproduced with permission from Wang et al. 2013. Copyright 2013 American Physical Society.*

positive and negative bound charges at the interfaces, and the vortices possess a net curl of the polarization vector that can be switched by an irrotational electric field (Xue et al. 2016). The feasibility of manipulating of these nanoscale vortices by electric fields make them especially attractive for the use as functional elements in memory devices (Yadav et al. 2016; Damodaran et al. 2017).

Ferroelectric CDWs carrying a net bound charge form as a result of direct "head-to-head" or "tail-to-tail" polarization configurations. They are stabilized by defect pinning or charge compensation from free charge carriers (Vul et al. 1973; Jia et al. 2008; Gureev et al. 2011; Qi et al. 2012; Sluka et al. 2013; Wang et al. 2015) (see also Chapters 1, 3, and 6). CDWs are electrically active and in some cases can be the host of a quasi-two-dimensional electron gas (q-2DEG), which leads to a steady metallic conductivity that is 10^9 times that of the bulk (Sluka et al. 2013). Besides the enhanced conductivity, the bound charge at the CDWs can also lead to an accumulation of oxygen vacancies, effectively lowering the local energy bandgap and enhancing the photocurrent (Lee et al. 2012), or produce a depolarization field that can lead to an increased electromechanical response and therefore improved piezoelectric properties (Sluka et al. 2012). Arrays of 71° CDWs have been observed in a 5-nm-thick BiFeO₃ film grown on a TbScO₃ substrate (Figure 10.6(b)). The stability of these CDWs relies on not only the formation of a high density of 109°/180° triangular DW junctions in the film but also charge

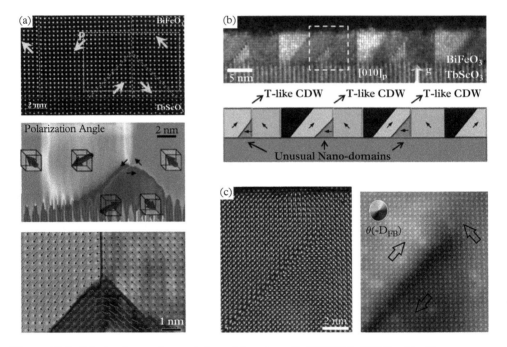

Figure 10.6 *Polarization vortices and charged domain walls (CDWs) in BiFeO₃ thin films. (a) HAADF STEM image of the intersection of two 109° DWs with the TbScO₃ surface (top panel). The arrows indicate the polarization direction, and dashed lines indicate the position of the DWs. A corresponding map of the polarization angle showing the domain structure of the BiFeO₃ (center panel). Polarization vector map of the vortex domain structure in local region of BiFeO₃ as highlight by the rectangle in the top panel (bottom panel). (b) Cross-sectional HRTEM image and corresponding schematic revealing the DW patterns containing an array of CDWs in a 5-nm-thick BiFeO₃ film grown on a TbScO₃ substrate. The smaller arrows in the schematic show the polarization directions. (c) Polarization vector (-D_{FB} vector) map overlaid on a HAADF STEM image of a CDW near the top surface of a 20-nm-thick BiFeO₃ film (left panel), and the corresponding map of polarization angle (right panel). Note: HAADF, high-angle annular dark-field; P, polarization.*

Source: (a) Reproduced with permission from Nelson et al. 2011. Copyright 2011 American Chemical Society. (b,c) Reproduced with permission from Li et al. 2013. Copyright 2013 American Chemical Society.

compensation from the adsorbed charged species at the free surface. As shown in the map of polarization vectors (-D_{FB} vectors) near a similar 71° CDW below the surface of a 20-nm-thick BiFeO₃ film (Figure 10.6(c)), the polarization rotates gradually from the <111>$_{PC}$ directions beside the CDW to the out-of-plane orientation at the CDW, indicating the formation of a tetragonal-like (*T*-like) structure at the CDW. The tip region of the triangular domain below the CDW is found to possess an unexpected ferroelectric state, as its polarization has rotated from the ⟨111⟩$_{PC}$ direction to the in-plane direction. This avoids a direct "tail-to-tail" configuration at the triangular tip and thus releases some of the electrostatic energy. Due to the unique polarization configuration in this nanosized tip region, the rotation angle of polarization across the DWs formed with the neighboring domains is no longer 180° (left side) or 109° (right side), although they do return to those angles in the region far below the *T*-like CDW.

10.3.2 Defect Engineering of DW Patterns

While boundary-condition engineering has demonstrated tremendous success in controlling DW patterns in ferroelectric thin films, one major limitation of such a method has been that once the variable of the boundary condition is set, such as the choice of substrate, further modification to control or alter DW patterns during material synthesis becomes difficult. This reduces the parameter space for creating more complex structures with ordered DW patterns and thus imposes restrictions on the functionalities of the system. On the other hand, defects in ferroelectric oxides can also have a remarkable impact on ferroelectric DW structures. It is well known that some common types of defects, such as dislocations and vacancies, can interact with ferroelectric DWs, pinning metastable polarization configurations that are electrostatically charged or elastically unfavored by the mechanical boundary conditions (Yang et al. 1999; Klaui et al. 2005). In Figures 10.7(a,b), for example, two steps are observed at a vertical 109° DW in $BiFeO_3$ as a result of defect pinning, which leads to an overall curved DW configuration (Lubk et al. 2012). At the steps, the polarization changes across a 45° inclined line with a gradual polarization rotation across more than four pseudocubic unit cells in width, which is much wider than the charge-neutral sections at the DW (Lubk et al. 2012).

Recently, it was discovered that nanoscale impurity defects with structures different from host materials can induce dramatic changes of DW structures in ferroelectric thin films (Li et al. 2017; Xie et al. 2017). In Figure 10.7(c), for example, it is shown that a nanoscale planar charged defect—Bi_2FeO_{6-x}, composed of a single layer of FeO_6 octahedra sandwiched between two Bi_2O_2 layers that are inserted in the $BiFeO_3$ matrix—can induce novel mixed-phase polarization structures with a head-to-head CDW (Li et al. 2017). The map of the lattice parameters shows the existence of large strain gradients along the out-of-plane direction, as the out-of-plane lattice parameter to the in-plane lattice parameter ratio ($a_\perp/a_{//}$) is very large (~1.20) in the lattice layers that are in contact with the defect but relaxes to the bulk value (~1.06) within only a few unit cells above and below the defect. The opposite out-of-plane strain gradients near the planar defect lead to a strong head-to-head built-in field ($\sim 3.7 \times 10^3$ MV/m) pointing to the defect through the flexoelectric effect (Li et al. 2017), which in turn results in stabilization of the head-to-head CDW at the defect, as evidenced by the map of the polarization vectors (-D_{FB} vectors) (Figure 10.7(d)). An enhancement in the magnitude of the polarization is also observed in the $BiFeO_3$ lattice in contact with the defect, which is clearly coupled to the local increase in tetragonality, i.e. the enhanced $a_\perp/a_{//}$ratio in these layers. Furthermore, the polarization map shows an interesting change in the $BiFeO_3$ lattice symmetry in the two regions separated by the defect. While the lattice above the defect possesses a *T*-like structure with polarization oriented along the $[00\bar{1}]_{PC}$ direction, the lattice below the defect is possibly in the *R*-like phase, with a mostly attenuated polarization oriented along the diagonal, pointing upward in the

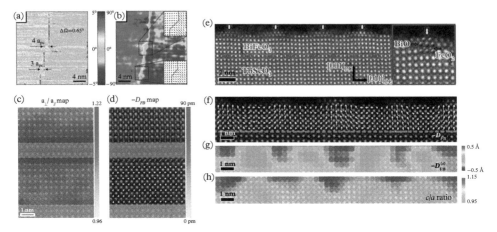

Figure 10.7 *Interaction of polarization and defects in BiFeO$_3$ thin films. (a) Lattice and (b) polarization vector (-$\mathbf{D}_{FB}$ vector) rotation map at a stepped 109° DW in BiFeO$_3$. The enlarged squares inserted in (b) show the distribution of -$\mathbf{D}_{FB}$ vectors across the steps, where diffuse polarization rotation and attenuation are observed. (c) Color map of the out-of-plane lattice parameter (a$_\perp$) over the in-plane lattice parameter (a$_{||}$) ratio (a$_\perp$/a$_{||}$) overlaid on an HAADF STEM image, in which one planar defect is located above the BiFeO$_3$/La$_{0.7}$Sr$_{0.3}$MnO$_3$ interface. The defect and the La$_{0.7}$Sr$_{0.3}$MnO$_3$ substrate are overlaid with uniform green and silver colors, respectively. (d) Spatial distribution of the -$\mathbf{D}_{FB}$ vectors overlaid on the same HAADF STEM image. The color scale indicates the magnitude of the -$\mathbf{D}_{FB}$ vector. (e–h) HAADF STEM image (e) of an ultrathin BiFeO$_3$ film grown on a TbScO$_3$ substrate, where island-like monolayer defects are observed at the film surface, and the same image overlaid with -$\mathbf{D}_{FB}$ map (f), -$\mathbf{D}_{FB}{}^{(z)}$ (out-of-plane component of -$\mathbf{D}_{FB}$) map (g), c/a (which is same as a$_\perp$/a$_{||}$) ratio map (h) of the film. The green dashed line in (e) marks the position of the BiFeO$_3$/TbScO$_3$ interface. Note: HAADF, high-angle annular dark-field.*

Source: (a,b) Reproduced with permission from Lubk et al. 2012. Copyright 2012 American Physical Society. (c,d) Reproduced with permission from Li et al. 2017. Copyright 2017 American Chemical Society. (e–h) Reproduced with permission from Xie et al. 2017. Copyright 2017 John Wiley and Sons.

image plane. Such mixed-phase structures can be associated with the defect. As a previous study has shown, Bi$_2$O$_2$ layers existing at the substrate interfaces can induce the formation of T-like structures in moderately strained (~1.5%) BiFeO$_3$ films grown on SrTiO$_3$ substrates, and the Bi$_2$O$_2$ layers within the defect observed here could also be responsible for the stabilization of the T-like phase above the local R-like structures. The upward polarization below the defect is further suppressed by the downward Schottky built-in field from the back-electrode interface, resulting in the observed attenuated polarization.

In another example, the existence of island-like monolayer bismuth-oxide defects at the surface of an ultrathin BiFeO$_3$ film can stabilize unusual polarization structures with a very large variation of strain states (Figures 10.7(e–h)) (Xie et al. 2017). Figure 10.7(e) shows a STEM HAADF image of a 2-nm-thick-BiFeO$_3$ film grown on insulating TbScO$_3$ substrates, where discontinuous single-atomic-layer defects are observed at the film surface. The mapping results of -$\mathbf{D}_{FB}$ vectors (Figure 10.7(f)) and dz (the out-of-plane component of -D_{FB}) (Figure 10.7(g)) and the out-of-plane lattice parameter to the in-plane lattice parameter ratios (which are called c/a ratios in Figure 10.7(h), and are identical to the $a_\perp/a_\parallel$ ratios in the previous discussion) show that unlike the conventional R-phase BiFeO$_3$, the polarization points either in the out-of-plane [001]$_{PC}$ direction or in the in-plane direction, more similar to those of the tetragonal (T) phase ferroelectrics. As a result, undulating domain structures resembling c-type (the c-axis of the unit cell lies parallel to the film growth direction) and a-type (the a-axis of the unit cell lies parallel to the film growth direction) T-like domains are observed, with lattice dilations at the c-type domains clearly seen in the maps of the c/a ratios. While the a-type domains with in-plane polarization are commonly expected in ferroelectric ultrathin films due to depolarization effects, the large out-of-plane polarization in the c-type domains is quite unusual. The fact that the surface defects appear exclusively on top of the c-type domains indicates that a strong polarization-defect interaction could be the origin of such unusual polarization structures. Further study using first-principle calculations and phase-field simulations have confirmed such an interaction, i.e. a strong built-in field pointing to the defects, which can give rise to a large polarization enhancement (Xie et al. 2017).

Although the impurity defects in the above two findings are accidentally introduced during the material synthesis, these results suggested the possibility of using engineered impurity defects in combination with suitable interface boundary conditions to control domain formation and create complex DW structures. Inspired by these findings, an intriguing work has demonstrated a route to deliberately creating an array of non-stoichiometric nanoregions as charged defects into the BiFeO$_3$ matrix (Figure 10.8(a)) (Li et al. 2018a). By initiating the film growth with an increased substrate temperature (slightly higher than the optimal temperature) and then reducing the temperature to the optimal point, an array of circle-like defects can be introduced into the BiFeO$_3$ film at a constant level that is several nanometers above the interface. Due to their non-stoichiometric nature, the defects are negatively charged and produce strong built-in fields forcing the polarization to point toward them. These defects can thus be used as nano-building-blocks to stabilize novel hedgehog/antihedgehog nanodomains with complex DW structures. Shown in Figures 10.8(b,c) are typical examples of such defect-induced polarization states. A tiny hedgehog polarization state with a diameter of four unit cells is observed within the nanoregion encircled by the defect. Antihedgehog polarization states are stabilized at the shell regions surrounding the defects. The circle-like defect thus plays as a head-to-head CDW. On larger length scales outside of these defects, the overall continuous polarization rotation patterns lead to flux-closure vortex structures.

While the previous example clearly demonstrated that nanosized defects can be used to stabilize novel polarization structures at the nanoscale, further study shows that the DW patterns in the bulk of ferroelectric films can also be effectively controlled by introducing engineered defects (Figures 10.8(d–i)). With a similar method of control-ling the substrate temperature, an array of atomically thin, negatively charged defects were introduced into a 400-nm-thick BiFeO$_3$ film at a nearly constant depth that is 110–130 nm above the bottom interface. The interaction between the charged defects and the polarization led to a reconfigured DW pattern in the BiFeO$_3$ film. As shown in

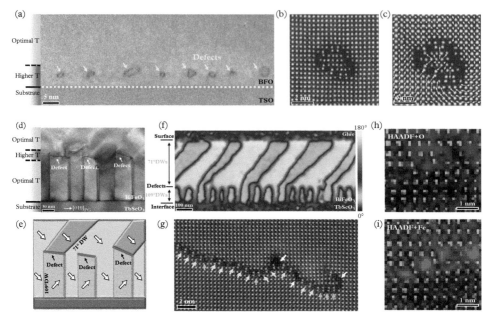

Figure 10.8 *Control of DW structures in BiFeO₃ thin films through defect engineering. (a) HAADF STEM image of an array of circle-like defects in a BiFeO₃ film grown on TbScO₃ substrate, which are at a constant level that is several nanometers above the BiFeO₃/TbScO₃ (BFO/TSO) interface. The array of defects is introduced by initiating the film growth with an increased substrate temperature (slightly higher than the optimal temperature) and then reducing the temperature to the optimal point. (b) Magnified HAADF STEM image of a circle-like defect, and (c) corresponding polarization vectors (-D$_{FB}$ vectors) overlaid on the BiFeO₃ lattice showing exotic polarization structures stabilized the defect. (d–i) Periodic DW structures in a 400 nm BiFeO₃ thin film engineered by defects. (d) Cross-sectional bright-field TEM image and (e) corresponding schematic of polarization structures showing ordered 71° and 109° DWs separated by an array of defects in the BiFeO₃ film. The polarization directions are marked by the white arrows in the schematic. (f) PFM phase image showing the same periodically ordered DW patterns. (g) Atomic-scale HAADF STEM image of a linear defect inserted in the BiFeO₃ matrix. Stepped units on the defect are indicated by the yellow arrows; planar units are indicated by green arrows; and white arrows indicate several defect segments with consecutive pairs of Fe atoms with no Bi atoms in between them. (h,i) EELS spectrum images of a segment on the defect. O: blue color, Fe: red color, HAADF signal (i.e. Bi): green color. Note: HAADF, high-angle annular dark-field.*

Source: (a–c) Reproduced with permission from Li et al. 2018a. Copyright 2018 American Physical Society. (d–i) Reproduced with permission from Li et al. 2018b. Copyright 2018 John Wiley and Sons.

Figures 10.8(d,e), a linear defect is observed on each 109° domain that is polarized upward (green color in Figure 10.8(e)), and the horizontal dimension of the defect matches exactly with the domain width. Above some of the defects, 71° domains (blue color in Figure 10.8(e)) are stabilized with the DWs located in the $(00\overline{1})_{PC}$ planes, which are different from the $(101)_{PC}$-oriented 71° domain stripes previously reported in BiFeO₃ thin films grown on orthorhombic substrates. In contrast, the downward-polarized domains (yellow color in Figure 10.8(e)) in the lower portion of the film are not blocked by the defects and can extend their volume to the upper portion. As a result, a transformation from 109° to 71° DW patterns across the defects is observed, as well as an array of "head-to-head" positively CDWs located exactly at the negatively charged defects. Periodic ordering of these DW patterns is further confirmed by a piezoresponse force microscopy (PFM) image of the same heterostructure in the cross section, in which the defects are not resolvable (Figure 10.8(f)). Atomic structures of a typical linear defect inserted in the $BiFeO_3$ matrix is shown in a HAADF STEM image (Figure 10.8(g)), and chemical compositions across a segment of the defect are revealed by EELS spectrum mapping (Figures 10.8(h,i)).

10.4   DW Dynamics in $BiFeO_3$ Thin Films

10.4.1   Effects of Defects on the Switching of DWs

Ferroelectric switching generally occurs in three steps: first, the nucleation of a reversed polarization domain, which usually takes place at the interfaces or defects due to the local high free energy; second, fast forward propagation of domains in the direction of the electric field, as a result of the high energy of the forward CDWs; and last, lateral domain growth in the direction perpendicular to the electric field, enabled by a sidewise motion of the nominal charge-neutral DW (Chapter 13). Some common types of defects, such as dislocations and vacancies, usually act as pinning centers hindering the DW motion and enhancing the stability of DWs. The in situ TEM study on ferroelectric domains and DW switching in $BiFeO_3$ thin films was first done by Nelson et al. (Figure 10.9) (Nelson et al. 2011a), where simultaneously high spatial (~1 nm) and temporal (30 ms) resolution are achieved. Multiple metastable 71° nuclei formed at the bottom interface at an applied voltage of 0.9 V and the nuclei did not continue their growth before surpassing a critical voltage of 2.2 V (Figure 10.9(a)). This is contrary to conventionally held kinetic growth models (Dawber et al. 2005) and points to a thermodynamic restriction on domain growth (Nelson et al. 2011a). A local energy well favoring the "down" polarized domains in the vicinity of the electrode could be induced by the downward Schottky built-in field at the interface. Further increasing the voltage to 4 V resulted in growth of the switched 71° domain. However, this created domain did not reach the top surface but was pinned midway through the film, forming a horizontal "tail-to-tail" CDW. In another switching event, a thin layer of $BiFeO_3$ film along the bottom electrode layer undergoes an independent 180° switching process from the primary 71° switching (Figure 10.9(b)) (Nelson et al. 2011a). These interfacial 180° domains form several hundred nanometers away from the tip, and they have a low nucleation bias but require much larger negative voltages to erase (Nelson et al. 2011a). Such large delocalization of the interfacial switching from the tip and the large negative voltage offset of the switching threshold suggest substantial contributions by the downward built-in electric fields of the bottom interface Schottky junction. The lateral edges of the interfacial domains are charge-neutral DWs.

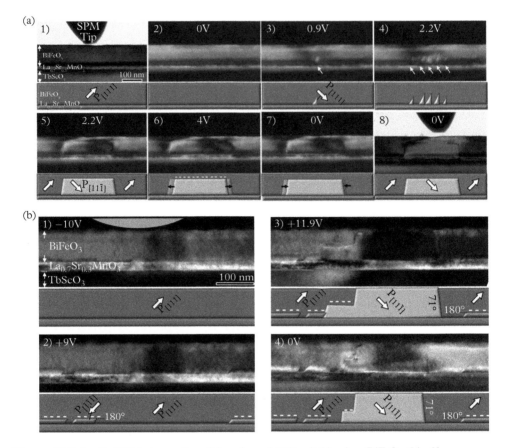

Figure 10.9 *In situ TEM observation of domain and DW switching in a BiFeO₃ thin film.*
(a) Interface-assisted nucleation and growth of a switched domain in a BiFeO₃ thin film grown on TbScO₃ substrate with an epitaxial La₀.₇Sr₀.₃MnO₃ electrode: a chronological dark-field diffraction-contrast TEM image series depicts the evolution of a "down" polarized domain from an "up" polarized single-domain film. Below each TEM image, a schematic of the corresponding domain configuration of the system is shown. Nucleation occurred at the electrode interface at 0.9 V (image 3), producing a metastable stationary domain. Additional bias increased the size and number of these nucleated domains (white arrows in the micrographs) up to a bias of 2.2 V (image 4). At 2.2 V, the domains merged and propagated forward until they stabilized just short of the surface (image 5), where the top horizontal DW became pinned with the accumulation of negative bound charge. At 4.0 V, the lateral extent of the domain increased slightly, but the top horizontal DW remained pinned (image 6). The lateral extent of the domain is reduced slightly with the removal of the bias (image 7) and remains unchanged after the tip is removed (image 8). (b) Independent switching of the interfacial layer in a BiFeO₃ thin film grown on a TbScO₃ substrate with an epitaxial La₀.₇Sr₀.₃MnO₃ electrode: a chronological dark-field diffraction-contrast TEM image series depicts the domain and DW evolution as the voltage increases from −10V to +12 V and falls to 0 V. Below each TEM image, a schematic of the corresponding domain configuration of the system is shown. At +9 V (image 2), 180° interfacial domains appear first. At +11.9 V (image 3), a 71° primary domain appear subsequently. After the voltage is removed (image 4), both the 71° primary domain and the 180° interfacial domains remain stable. Note: SPM, scanning probe microscopy; P, polarization.

Source: Reproduced with permission from Nelson et al. 2011. Copyright 2011 The American Association for the Advancement of Science.

However, the horizontal ones are electrostatically unstable CDWs carrying a net negative bound charge due to their "tail-to-tail" polarization configuration. Despite the energy cost of these CDWs, the interfacial domains exhibit long-term stability (>6 months) (Nelson et al. 2011a). Such stability indicates strong pinning effects by structural defects on the DWs (Scott and Dawber 2000; Yuan and Uedono 2009; Zhang et al. 2009; Folkman et al. 2010).

Recent in situ TEM studies have also demonstrated a profound effect exerted by nanoscale impurity defects with structures different from the host materials. The impurity defects can be atomically thin in dimension and are usually buried in the bulk material (MacLaren et al. 2013; Deniz et al. 2014; MacLaren et al. 2014). They can induce strong local perturbations, such as charge and strain, resulting in a dramatic change of the ferroelectric properties. An example can be found in the switching of the same $BiFeO_3$ film with the planar charged Bi_2FeO_{6-x} defects (the switching phenomena are shown in Figure 10.10, and the atomic structures can be found in Figures 10.7(c,d)). As discussed previously, strong flexoelectric fields $(\sim 3.7 \times 10^3 MV/m)$ are induced by the strain gradients above and below the defects. Such fields are two orders of magnitude higher than the typical value of the Schottky built-in field at the $BiFeO_3/La_{0.7}Sr_{0.3}MnO_3$ interface (Nelson et al. 2011a), and can flip the polarization adjacent to the defect. Since the planar impurity defects are located just above the $BiFeO_3/La_{0.7}Sr_{0.3}MnO_3$ interface, the strong built-in fields induced by the defects should be predominantly pointing downward through the $BiFeO_3$ film, destabilizing any upward polarized domains that may be created by applying a bias (Li et al. 2017). A local switching event in the presence of an array of such planar impurity defects shows that although a large domain with upward polarization can be created in the downward-polarized matrix with an applied bias, the created domain shrank back to a smaller metastable state within 15 s after the voltage was removed, with a concurrent back-switching initiated at the interfacial sublayer adjacent to the planar defects (Figure 10.10). Consequently, the metastable domain only penetrated from the top surface to about 35 nm in depth, resulting the formation of a horizontal tail-to-tail CDW at the middle of the film. Then this metastable state further relaxed to smaller sizes through slow thermodynamic processes, and eventually annihilated in 20 min, leading to a complete retention failure.

10.4.2 Switching of Charged DWs

Since in most cases DWs are more conductive than the bulk domains, switching of DW patterns can in principle lead to a change in the conductivity of ferroelectric materials, opening the possibilities of designing novel resistance-switching memory devices. Most observed DWs, however, are nearly charge-neutral due to minimization of the electrostatic energy, and only slightly enhanced conductivity at such uncharged DWs has been shown (Seidel et al. 2009, 2010; Farokhipoor and Noheda 2011; Guyonnet et al. 2011; Maksymovych et al. 2011). Electronically active CDWs can gather compensating free charges and allow much higher conductivity, holding more promise

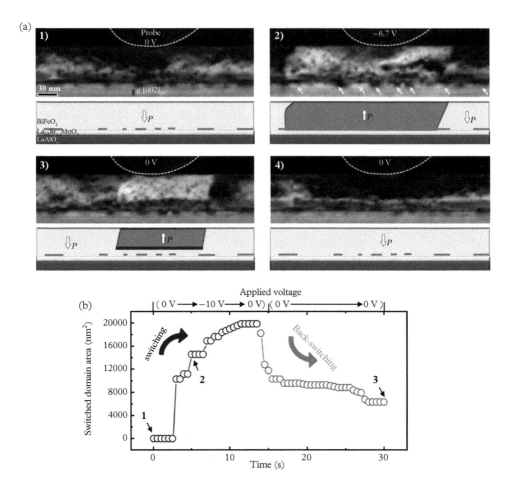

Figure 10.10 *Defect-induced ferroelectric retention failure observed in a BiFeO₃ thin film.*
(a) Chronological TEM dark-field image series and corresponding schematics depicting the evolution of an upward polarized domain from a downward polarized matrix in the presence of defects in BiFeO₃ thin film grown on an LaAlO₃ substrate with an epitaxial La₀.₇Sr₀.₃MnO₃ electrode: (1) the original state, (2) switched domain under an applied voltage, (3) metastable domain state 15 s after the voltage was removed, and (4) final stable state 20 min after the voltage was removed. Planar defects just above the BiFeO₃/La₀.₇Sr₀.₃MnO₃ interface are indicated by the small yellow arrows in (2) and are shown by the short discontinuous lines in the schematics. In (3), a horizontal tail-to-tail CDW is marked by the longer red solid line in the middle of the film in the schematic. (b) Plot of the measured area of the switched domain as a function of time (bottom axis) and applied voltage (top axis). The locations of (1), (2), and (3) are marked. Note: P, polarization.
Source: Reproduced with permission from Li et al. 2017. Copyright 2017 American Chemical Society.

for nanodevice applications. The exploration of CDWs has been largely enabled by conventional scanning probe microscopy, which only allows access to the CDW structure and conductivity from the surface (Kalinin et al. 2010). In fact, due to the requirement of free charge compensation, some strongly charged domain walls (sCDWs) present at the top surface of ferroelectric films do not penetrate to the bottom interface. For example, in one of the previous examples of $BiFeO_3$ thin films (Figures 10.6(b,c)), the cross-sectional TEM image shows that the CDWs form only at local areas below the film surface, and beneath the CDWs there are triangular charge-neutral 109°/180° DW junctions (Li et al. 2013). These triangular DW junctions have a small size approaching the resolution limit of surface probe microscopy but can significantly affect the film conductivity at the local region.

The results of an in situ TEM study on the dynamic behaviors of an sCDW in a 20-nm-thick $BiFeO_3$ thin film are shown in Figure 10.11, where the sCDW can be created from a triangular 109°/180° DW junction with an applied positive voltage.

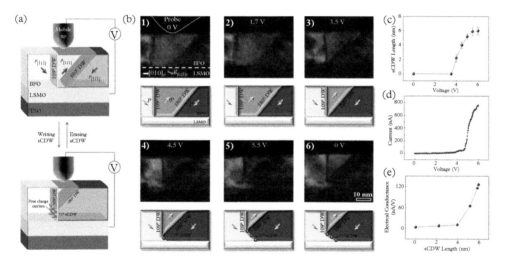

Figure 10.11 *Creation of a strongly charged domain wall (sCDW) caused by applying a positive ramp bias using in situ TEM. (a) Schematic of experimental setup: a $BiFeO_3$ (BFO) thin film was grown on $La_{0.7}Sr_{0.3}MnO_3/TbScO_3$ (LSMO/TSO), and a movable tungsten tip acted as one electrode for electrical switching with the LSMO layer being grounded. Through the application of an electrical bias, a sCDW can be written and erased in the BFO film. (b) A chronological series of dark-field diffraction-contrast TEM images depicting switching of the sCDW. (1) The original stable state without a sCDW in the BFO film. (2) At the critical bias, the DW started to move. (3) The 109° and 180° DWs intersect at the substrate interface as a result of DW motion. (4,5) Formation and growth of an sCDW as a result of upward motion of the tip of the triangular 180°/109° DW junction. (6) After removal of the bias, the sCDW relaxed to be shorter than that observed in (5). (c) The measured length of the sCDW as a function of the applied voltage. (d) In situ measured current during the creation of the sCDW. (e) The measured local electrical conductance of the film as function of the sCDW length.*

Source: Reproduced with permission from Li et al. 2016. Copyright 2016 John Wiley and Sons.

With a combination of in situ TEM imaging and simultaneous electrical measurement, the DW configuration can be directly correlated to the film conductivity in real time (Li et al. 2016). During the switching, the shrinkage of the triangular domain led to an upward motion of the triangular domain tip and resulted in the formation and elongation of an sCDW (Figure 10.11(b)). The simultaneous in situ electrical measurements show a strongly enhanced conductivity with currents of 640 to 760 nA occurred at voltages higher than 5.5 V as the sCDW penetrated more than a critical length (~4 nm in this case) (Figures 10.11(c–e)). After the bias was removed, the system relaxed to a stable state with high conductivity, where the length of the sCDW remained ~5 nm. This sCDW can also be erased, with the local film reverting to a high-resistance state. Using this effect, ultra-high-density information storage based on sCDWs may be possible, as shown by the observation of arrays of sCDWs with a spacing of 10 nm in films produced by molecular-beam epitaxy (Figure 10.6(b)) (Li et al. 2013).

10.5   Summary and Outlook

Recent advances in fabrication of ferroelectric oxides, with improvements in thin film growth on oxide substrates, facilitate the rapid development of ferroelectric-DW-based devices, including high-density non-volatile memories and a myriad of multifunctional nanodevices. Development in new ferroelectric-DW-based materials, such as multiferroic and magnetoelectric materials, ferroelectric nanostructures, and organic/perovskite hybrid systems, continues apace. Advanced TEM imaging techniques, with spatial and time resolution unavailable by any other methods, has been a quantitative and powerful method for probing DW structures and its switching dynamics. Atomic-scale imaging based on aberration-corrected TEM/STEM has offered opportunities to explore the complex interplay of spin, charge, orbital, and lattice degrees of freedom in ferroelectric systems. Further study will lead to a more comprehensive understanding of the interaction between DWs and interfaces, defects, or other related microscopic structures in nanoscale ferroelectrics. This will allow the strategy for fabricating complex, multilayer DW structures to become more flexible through boundary-condition engineering and defect engineering. In situ TEM has opened paths for directly accessing the microscopic details of highly inhomogeneous dynamic processes of domain switching and DW motion, which form the basis for designing next-generation ferroelectric memory devices. Concurrently, much progress has been made to improve the performance of TEM-based imaging techniques and analytical tools such as EDS and EELS. Refinements of in situ methods, including new holders for application of electrical bias, force, and light, and for control of liquid and atmosphere environment, open the door for the examination of the behavior of ferroelectric DWs under many different conditions, all within the TEM. These developments provide wide opportunities for studying the properties and functionalities of ferroelectric DWs and facilitate the development of ferroelectric-DW-based devices in information and energy technologies.

Acknowledgments

The work was supported by the US Department of Energy (DOE), Office of Basic Energy Sciences, Division of Materials Sciences and Engineering under Grant DE-SC0014430. The authors would like to acknowledge the financial support provided by the National Science Foundation (NSF) under grant numbers DMR-1506535 and DMR-1629270.

···

REFERENCES

Adamo C., Ke X., Wang H. Q., Xin H. L., Heeg T., Hawley M. E., Zander W., Schubert J., Schiffer P., Muller D. A., Maritato L., Schlom D. G., "Effect of biaxial strain on the electrical and magnetic properties of (001) $La_{0.7}Sr_{0.3}MnO_3$ thin films," *Appl. Phys. Lett.* 95(11), 112504 (2009).

Bea H., Dupe B., Fusil S., Mattana R., Jacquet E., Warot-Fonrose B., Wilhelm F., Rogalev A., Petit S., Cros V., Anane A., Petroff F., Bouzehouane K., Geneste G., Dkhil B., Lisenkov S., Ponomareva I., Bellaiche L., Bibes M., Barthélémy A., "Evidence for room-temperature multiferroicity in a compound with a giant axial ratio," *Phys. Rev. Lett.* 102(21), 217603 (2009).

Catalan G., Scott J. F., "Physics and applications of bismuth ferrite," *Adv. Mater.* 21(24), 2463–2485 (2009).

Chang H. J., Kalinin S. V., Yang S., Yu P., Bhattacharya S., Wu P. P., Balke N., Jesse S., Chen L. Q., Ramesh R., Pennycook S. J., Borisevich A. Y., "Watching domains grow: in-situ studies of polarization switching by combined scanning probe and scanning transmission electron microscopy," *J. Appl. Phys.* 110(5), 052014 (2011).

Chen Y. B., Katz M. B., Chen Y. B., Katz M. B., Pan X. Q., Das R. R., Kim D. M., Baek S. H., Eom C. B., "Ferroelectric domain structures of epitaxial (001) $BiFeO_3$ thin films," *Appl. Phys. Lett.* 90(7), 072907 (2007).

Chen Z. H., Liu J., Qi Y., Chen D., Hsu S. L., Damodaran A. R., He X., N'Diaye A. T., Rockett A., Martin L. W., "180 degrees ferroelectric stripe nanodomains in $BiFeO_3$ thin films," *Nano Lett.* 15(10), 6506–6513 (2015).

Chisholm M. F., Luo W. D., Oxley M. P., Pantelides S. T., Lee H. N., "Atomic-scale compensation phenomena at polar interfaces," *Phys. Rev. Lett.* 105(19), 197602 (2010).

Choi K. J., Biegalski M., Li Y. L., Sharan A., Schubert J., Uecker R., Reiche P., Chen Y. B., Pan X. Q., Gopalan V., Chen L. Q., Schlom D. G., Eom C. B., "Enhancement of ferroelectricity in strained $BaTiO_3$ thin films," *Science* 306(5698): 1005–1009 (2004).

Christen H. M., Nam J. H., Kim H. S., Hatt A. J., Spaldin N. A., "Stress-induced R-M-A-M-C-T symmetry changes in $BiFeO_3$ films," *Phys. Rev. B* 83(14), 144107 (2011).

Chu Y. H., Cruz M. P., Yang C.-H., Martin L. W., Yang P.-L., Zhang J.-X., Lee K., Yu P., Chen L.-Q., Ramesh R., "Domain control in mulfiferroic $BiFeO_3$ through substrate vicinality," *Adv. Mater.* 19(18): 2662–2666 (2007a).

Chu Y. H., He Q., Yang C. H., Yu P., Martin L. W., Shafer P., Ramesh R., "Nanoscale control of domain architectures in $BiFeO_3$ thin films," *Nano Lett.* 9(4), 1726–1730 (2009).

Chu Y. H., Martin L. W., Holcomb M. B., Ramesh R., "Controlling magnetism with multiferroics," *Mater. Today* 10(10), 16–23 (2007b).

Damodaran A. R., Clarkson J. D., Hong Z., Liu H., Yadav A. K., Nelson C. T., Hsu S.-L., McCarter M. R., Park K.-D., Kravtsov V., Farhan A., Dong Y., Cai Z., Zhou H., Aguado-Puente P., García-Fernández P., Íñiguez J., Junquera J., Scholl A., Raschke M. B., Chen L.-Q., Fong D. D., Ramesh R., Martin L. W., "Phase coexistence and electric-field control of toroidal order in oxide superlattices," *Nat. Mater.* 16(10), 1003 (2017).

Dawber M., Rabe K. M., Scott J. F., "Physics of thin-film ferroelectric oxides," *Rev. Mod. Phys.* 77(4), 1083–1130 (2005).

Deniz H., Bhatnagar A., Pippel E., Hillebrand R., Hähnel A., Alexe M., Hesse D., "Nanoscale Bi_2FeO_{6-x} precipitates in $BiFeO_3$ thin films: a metastable Aurivillius phase," *J. Mater. Sci.* 49(20), 6952–6960 (2014).

Ederer C., Spaldin N. A., "Effect of epitaxial strain on the spontaneous polarization of thin film ferroelectrics," *Phys. Rev. Lett.* 95(25), 257601 (2005).

Farokhipoor S., Noheda B., "Conduction through 71 degrees domain walls in $BiFeO_3$ thin films," *Phys. Rev. Lett.* 107(12), 127601 (2011).

Folkman C. M., Baek S. H., Nelson C. T., Jang H. W., Tybell1 T., Pan X. Q., Eom C. B., "Study of defect-dipoles in an epitaxial ferroelectric thin film," *Appl. Phys. Lett.* 96(5) (2010).

Gao P., Britson J., Nelson C. T., Jokisaari J. R., Duan C., Trassin M., Baek S.-H., Guo H., Li L., Wang Y., Chu Y.-H., Minor A. M., Eom C.-B., Ramesh R., Chen L.-Q., Pan X., "Ferroelastic domain switching dynamics under electrical and mechanical excitations," *Nat. Commun.* 5, 3801 (2014).

Gureev M. Y., Tagantsev A. K., Setter N., "Head-to-head and tail-to-tail 180 degrees domain walls in an isolated ferroelectric," *Phys. Rev. B* 83(18), 184104 (2011).

Guyonnet J., Gaponenko I., Gariglio S., Paruch P., "Conduction at domain walls in insulating $Pb(Zr_{0.2}Ti_{0.8})O_3$ thin films," *Adv. Mater.* 23(45), 5377 (2011).

He H., Tan X., "Electric-field-induced transformation of incommensurate modulations in antiferroelectric $Pb_{0.99}Nb_{0.02}[(Zr_{1-x}Sn_x)_{(1-y)}Ti_y]_{(0.98)}O_3$," *Phys. Rev. B* 72(2), 024102 (2005).

He Q., Yeh C. H., Yang J. C., Singh-Bhalla G., Liang C. W., Chiu P. W., Catalan G., Martin L. W., Chu Y. H., Scott J. F., Ramesh R., "Magnetotransport at domain walls in $BiFeO_3$," *Phys. Rev. Lett.* 108(6) (2012).

Infante I. C., Lisenkov S., Dupé B., Bibes M., Fusil S., Jacquet E., Geneste G., Petit S., Courtial A., Juraszek J., Bellaiche L., Barthélémy A., Dkhil B., "Bridging multiferroic phase transitions by epitaxial strain in $BiFeO_3$," *Phys. Rev. Lett.* 105(5), 057601 (2010).

Ishikawa R., Okunishi E., Sawada H., Kondo Y., Hosokawa F., Abe E., "Direct imaging of hydrogen-atom columns in a crystal by annular bright-field electron microscopy," *Nat. Mater.* 10(4), 278–281 (2011).

Jang H. W., Ortiz D., Baek S.-H., Folkman C. M., Das R. R., Shafer P., Chen Y., Nelson C. T., Pan X., Ramesh R., Eom C.-B., "Domain engineering for enhanced ferroelectric properties of epitaxial (001) $BiFeO_3$ thin films," *Adv. Mater.* 21(7), 817–823 (2009).

Jia C. L., Lentzen M., Urban K., "Atomic-resolution imaging of oxygen in perovskite ceramics," *Science* 299(5608), 870–873 (2003).

Jia C. L., Mi S. B., Urban K., Vrejoiu I., Alexe M., Hesse D., "Atomic-scale study of electric dipoles near charged and uncharged domain walls in ferroelectric films," *Nat. Mater.* 7(1), 57–61 (2008).

Jia C. L., Nagarajan V., He J. Q., Houben L., Zhao T., Ramesh R., Urban K., Waser R., "Unit-cell scale mapping of ferroelectricity and tetragonality in epitaxial ultrathin ferroelectric films," *Nat. Mater.* 6(1), 64–69 (2007).

Kalinin S. V., Morozovska A. N., Chen L. Q., Rodriguez B. J., "Local polarization dynamics in ferroelectric materials," *Rep. Progr. Phys.* 73(5), 056502 (2010).

Kimoto K., Asaka T., Yu X., Nagai T., Matsui Y., Ishizuka K., "Local crystal structure analysis with several picometer precision using scanning transmission electron microscopy," *Ultramicroscopy* 110(7), 778–782 (2010).

Klaui M., Ehrke H., Rüdiger U., "Direct observation of domain-wall pinning at nanoscale constrictions," *Appl. Phys. Lett.* 87(10), 102509 (2005).

Krivanek O. L., Chisholm M. F., Nicolosi V., Pennycook T. J., Corbin G. J., Dellby N., Murfitt M. F., Own C. S., Szilagyi Z. S., Oxley M. P., Pantelides S. T., Pennycook S. J., "Atom-by-atom structural and chemical analysis by annular dark-field electron microscopy," *Nature* 464(7288), 571–574 (2010).

Lee W. M., Sung J. H., Chu K., Moya X., Lee D., Kim C.-J., Mathur N. D., Cheong S.-W., Yang C.-H., Jo M.-H., "Spatially resolved photodetection in leaky ferroelectric BiFeO$_3$," *Adv. Mater.* 24(10), Op49–Op53 (2012).

Li L. Z., Britson J., Jokisaari J. R., Zhang Y., Adamo C., Melville A., Schlom D. G., Chen L.-Q., Pan X., "Giant resistive switching via control of ferroelectric charged domain walls," *Adv. Mater.* 28(31), 6574 (2016).

Li L. Z., Cheng X. X., Jokisaari J. R., Gao P., Britson J., Adamo C., Heikes C., Schlom D. G., Chen L. Q., Pan X., "Defect-induced hedgehog polarization states in multiferroics," *Phys. Rev. Lett.* 120(13), 137602 (2018a).

Li L. Z., Gao P., Nelson C. T., Jokisaari J. R., Zhang Y., Kim S. J., Melville A., Adamo C., Schlom D. G., Pan X., "Atomic scale structure changes induced by charged domain walls in ferroelectric materials," *Nano Lett.* 13(11), 5218–5223 (2013).

Li L. Z., Jokisaari J. R., Zhang Y., Cheng X., Yan X., Heikes C., Lin Q., Gadre C., Schlom D. G., Chen L. Q., Pan X. Q., "Control of domain structures in multiferroic thin films through defect engineering," *Adv. Mater.* 30(38), 1802737 (2018b).

Li L. Z., Xie L., Pan X., "Real-time studies of ferroelectric domain switching: a review," *Rep. Progr. Phys.* 82(12), (2019).

Li L. Z., Zhang Y., Xie L., Jokisaari J. R., Beekman C., Yang J. C., Chu Y. H., Christen H. M., Pan X., "Atomic-scale mechanisms of defect-induced retention failure in ferroelectrics," *Nano Lett.* 17(6), 3556–3562 (2017).

Li Y. L., Hu S. Y., Liu Z. K., Chen L. Q., "Effect of electrical boundary conditions on ferroelectric domain structures in thin films," *Appl. Phys. Lett.* 81(3), 427–429 (2002).

Lichtensteiger C., Fernandez-Pena S., Weymann C., Zubko P., Triscone J., "Tuning of the depolarization field and nanodomain structure in ferroelectric thin films," *Nano Lett.* 14(8), 4205–4211 (2014).

Lubk A., Rossell M. D., Seidel J., He Q., Yang S. Y., Chu Y. H., Ramesh R., Hÿtch M. J., Snoeck E., "Evidence of sharp and diffuse domain walls in bifeo$_3$ by means of unit-cell-wise strain and polarization maps obtained with high resolution scanning transmission electron microscopy," *Phys. Rev. Lett.* 109(4), 047601 (2012).

MacLaren I., Wang L. Q., Morris O., Craven A., Stamps R., Schaffer B., Ramasse Q., Miao S., Kalantari K., Sterianou I., Reaney I., "Local stabilisation of polar order at charged antiphase boundaries in antiferroelectric $(Bi_{0.85}Nd_{0.15})(Ti_{0.1}Fe_{0.9})O_3$," *APL Mater.* 1(2), 021102 (2013).

MacLaren I., Wang L. Q., Craven A., Ramasse Q., Schaffer B., Kalantari K., Reaney I., "The atomic structure and chemistry of Fe-rich steps on antiphase boundaries in Ti-doped $Bi_{0.9}Nd_{0.15}FeO_3$," *APL Mater.* 2(6), 066106 (2014).

Maksymovych P., Seidel J., Chu Y. H., Wu P., Baddorf A. P., Chen L. Q., Kalinin S. V., Ramesh R., "Dynamic conductivity of ferroelectric domain walls in BiFeO$_3$," *Nano Lett.* 11(5), 1906–1912 (2011).

Naumov I. I., Bellaiche L., Fu H., "Unusual phase transitions in ferroelectric nanodisks and nanorods," *Nature* 432(7018), 737–740 (2004).

Nelson C. T., Gao P., Jokisaari J. R., Heikes C., Adamo C., Melville A., Baek S. H., Folkman C. M., Winchester B., Gu Y., Liu Y., Zhang K., Wang E., Li J., Chen L. Q., Eom C. B., Schlom D. G., Pan X., "Domain dynamics during ferroelectric switching," *Science* 334(6058), 968–971 (2011a).

Nelson C. T., Winchester B., Zhang Y., Kim S. J., Melville A., Adamo C., Folkman C. M., Baek S. H., Eom C. B., Schlom D. G., Chen L. Q., Pan X., "Spontaneous vortex nanodomain arrays at ferroelectric heterointerfaces," *Nano Lett.* 11(2), 828–834 (2011b).

Okunishi E., Ishikawa I., Sawada H., Hosokawa F., Hori M., Kondo Y., "Visualization of light elements at ultrahigh resolution by STEM annular bright field microscopy," *Microsc. Microanal.* 15, 164–165 (2009).

Pan X. Q., Hu M. S., Yao M.-H., Feng D., "A TEM study of the incommensurate phase and related phase-transitions in barium sodium niobate," *Phys. Status Solidi A: Appl. Res.* 92(1), 57–68 (1985).

Pan X. Q., Unruh H. G., "Electron-microscopy study of discommensurations in K$_2$ZnCl$_4$," *J. Phys.:Condens. Mat.* 2(2), 323–329 (1990).

Pennycook S. J., *Scanning Transmission Electron Microscopy: Imaging and Analysis* (Springer, New York, 2011).

Privratska J., Janovec V., "Spontaneous polarization and or magnetization in non-ferroelastic domain walls: symmetry predictions," *Ferroelectrics* 222(1–4), 23–32 (1999).

Qi Y. J., Chen Z. H., Huang C., Wang L., Han X., Wang J., Yang P., Sritharan T., Chen L., "Coexistence of ferroelectric vortex domains and charged domain walls in epitaxial BiFeO$_3$ film on (110)$_O$ GdScO$_3$ substrate," *J. Appl. Phys.* 111(10), 104117 (2012).

Qu W., Zhao X., Tan X., "Evolution of nanodomains during the electric-field-induced relaxor to normal ferroelectric phase transition in a Sc-doped Pb(Mg$_{1/3}$Nb$_{2/3}$)O$_3$ ceramic," *J. Appl. Phys.* 102(8), 084101 (2007).

Randall C. A., Barber D. J., Whatmore R. W., "Insitu Tem experiments on Perovskite-structured ferroelectric relaxor materials," *J. Microsc.: Oxford* 145, 275–291 (1987).

Sato Y., Hirayama T., Ikuhara Y., "Real-time direct observations of polarization reversal in a piezo-electric crystal: Pb(Mg$_{1/3}$Nb$_{2/3}$)O$_3$-PbTiO$_3$ studied via in situ electrical biasing transmission electron microscopy," *Phys. Rev. Lett.* 107(18), 187601 (2011).

Schlom D. G., Chen L. Q., Eom C.-B., Rabe K. M., Streiffer S. K., Triscone J.-M., "Strain tuning of ferroelectric thin films," *Annu. Rev. Mater. Res.* 37, 589–626 (2007).

Schlom D. G., Chen L.-Q., Fennie C. J., Gopalan V., Muller D. A., Pan X., Ramesh R., Uecker R., "Elastic strain engineering of ferroic oxides," *MRS Bull.* 39(02), 118–130 (2014).

Scott J. F., Dawber M., "Oxygen-vacancy ordering as a fatigue mechanism in perovskite ferro-electrics," *Appl. Phys. Lett.* 76(25), 3801 (2000).

Seidel J., Maksymovych P., Batra Y., Katan A., Yang S.-Y., He Q., Baddorf A. P., Kalinin S. V., Yang C.-H., Yang J.-C., Chu Y.-H., Salje E. K. H., Wormeester H., Salmeron M., Ramesh R., "Domain wall conductivity in La-doped BiFeO$_3$," *Phys. Rev. Lett.* 105(19), 197603 (2010).

Seidel J., Martin L. W., He Q., Zhan Q., Chu Y.-H., Rother A., Hawkridge M. E., Maksymovych P., Yu P., Gajek M., Balke N., Kalinin S. V., Gemming S., Wang F., Catalan G., Scott J. F., Spaldin N. A., Orenstein J., Ramesh R., "Conduction at domain walls in oxide multiferroics," *Nat. Mater.* 8(3), 229–234 (2009).

Sluka T., Tagantsev A. K., Damjanovic D., Gureev M., Setter N., "Enhanced electromechanical response of ferroelectrics due to charged domain walls," *Nat. Commun.* 3, 748 (2012).

Sluka T., Tagantsev A. K., Bednyakov P., Setter N., "Free-electron gas at charged domain walls in insulating BaTiO$_3$," *Nat. Commun.* 4, 1808 (2013).

Snoeck E., Normand L., Thorel A., Roucau C., "Electron-microscopy study of ferroelastic and ferroelectric domain-wall motions induced by the in-situ application of an electric-field in BaTiO$_3$," *Phase Transit.* 46(2), 77–88 (1994).

Tan X. L., He H., Shang J.-K., "In situ transmission electron microscopy studies of electric-field-induced phenomena in ferroelectrics," *J. Mater. Res.* 20(7), 1641–1653 (2005).

Tang Y. L., Zhu Y. L., Ma X. L., Borisevich A. Y., Morozovska A. N., Eliseev E. A., Wang W. Y., Wang Y. J., Xu Y. B., Zhang Z. D., Pennycook S. J., "Observation of a periodic array of flux-closure quadrants in strained ferroelectric PbTiO$_3$ films," *Science* 348(6234), 547–551 (2015).

Tian W., Haeni J. H., Schlom D. G., Pan X. Q., "Strain-induced elevation of the spontaneous polarization in BaTiO$_3$ thin films," *Mat. Res. Soc. Symp. Proc.* 655, CC7.8.1 (2001).

Vul B. M., Guro G. M., Ivanchik I. I., "Encountering domains in ferroelectrics," *Ferroelectrics* 6(1–2), 29–31 (1973).

Wang W. Y., Tang Y. L., Zhu Y.-L., Xu Y.-B., Liu Y., Wang Y.-J., Jagadeesh S., Ma X.-L., "Atomic level 1D structural modulations at the negatively charged domain walls in BiFeO$_3$ films," *Adv. Mater. Interf.* 2(9), 1500024 (2015).

Wang Y., Nelson C., Melville A., Winchester B., Shang S., Liu Z. K., Schlom D. G., Pan X., Chen L. Q., "BiFeO$_3$ domain wall energies and structures: a combined experimental and density functional theory plus U study," *Phys. Rev. Lett.* 110(26), 267601 (2013).

Wei X. K., Tagantsev A. K., Kvasov A., Roleder K., Jia C.-L., Setter N., "Ferroelectric translational antiphase boundaries in nonpolar materials," *Nat. Commun.* 5, 3031 (2014).

Winkler C. R., Damodaran A. R., Karthik J., Martin L. W., Taheri M. L., "Direct observation of ferroelectric domain switching in varying electric field regimes using in situ TEM," *Micron* 43(11), 1121–1126 (2012a).

Winkler C. R., Jablonski M. L., Damodaran A. R., Jambunathan K., Martin L. W., Taheri M. L., "Accessing intermediate ferroelectric switching regimes with time-resolved transmission electron microscopy," *J. Appl. Phys.* 112(5), 052013 (2012b).

Wu J. G., Fan Z., Xiao D., Zhu J., Wang J., "Multiferroic bismuth ferrite-based materials for multifunctional applications: ceramic bulks, thin films and nanostructures," *Progr. Mater. Sci.* 84, 335–402 (2016).

Xie L., Li L. Z., Heikes C. A., Zhang Y., Hong Z., Gao P., Nelson C. T., Xue F., Kioupakis E., Chen L., Schlom D. G., Wang P., Pan X., "Giant ferroelectric polarization in ultrathin ferroelectrics via boundary-condition engineering," *Adv. Mater.* 29(30), 1701475 (2017).

Xue F., Li L., Britson J., Hong Z., Heikes C. A., Adamo C., Schlom D. G., Pan X., Chen L. Q., "Switching the curl of polarization vectors by an irrotational electric field," *Phys. Rev. B* 94(10), 100103 (2016).

Yadav A. K., Nelson C. T., Hsu S. L., Hong Z., Clarkson J. D., Schlepütz C. M., Damodaran A. R., Shafer P., Arenholz E., Dedon L. R., Chen D., Vishwanath A., Minor A. M., Chen L. Q., Scott J. F., Martin L. W., Ramesh R., "Observation of polar vortices in oxide superlattices," *Nature* 530(7589), 198 (2016).

Yamamoto N., Yagi K., Honjo G., "Electron-microscopic studies of ferroelectric and ferroelastic Gd$_2$(MoO$_4$)$_3$. IV. Polarization reversal and field-induced phase-transformation," *Phys. Status Solidi A: Appl. Res.* 62(2), 657–664 (1980).

Yang S. Y., Seidel J., Byrnes S. J., Shafer P., Yang C.-H., Rossell M. D., Yu P., Chu Y.-H., Scott J. F., Ager J. W., Martin L. W., Ramesh R., "Above-bandgap voltages from ferroelectric photovoltaic devices," *Nat. Nanotechnol.* 5(2), 143–147 (2010).

Yang T. J., Gopalan V., Swart P. J., Mohideen U., "Direct observation of pinning and bowing of a single ferroelectric domain wall," *Phys. Rev. Lett.* 82(20), 4106–4109 (1999).

Yuan G. L., Uedono A., "Behavior of oxygen vacancies in $BiFeO_3/SrRuO_3/SrTiO_3(100)$ and $DyScO_3(100)$ heterostructures," *Appl. Phys. Lett.* 94(13), 132905 (2009).

Zavaliche F., Yang S. Y., Zhao T., Chu Y. H., Cruz M. P., Eom C. B., Ramesh R., "Multiferroic $BiFeO_3$ films: domain structure and polarization dynamics," *Phase Transit.* 79(12), 991–1017 (2006).

Zeches R. J., Rossell M. D., Zhang J. X., Hatt A. J., He Q., Yang C.-H., Kumar A., Wang C. H., Melville A., Adamo C., Sheng G., Chu Y.-H., Ihlefeld J. F., Erni R., Ederer C., Gopalan V., Chen L. Q., Schlom D. G., Spaldin N. A., Martin L. W., Ramesh R., "A strain-driven morphotropic phase boundary in $BiFeO_3$," *Science* 326(5955), 977–980 (2009).

Zhang J., Rutkowski M., Martin W., Conry T., Ramesh R., Ihlefeld J. F., Melville A., Schlom D. G., Brillson L. J., "Surface, bulk, and interface electronic states of epitaxial $BiFeO_3$ films," *J. Vac. Sci. Technol. B* 27(4), 2012–2014 (2009).

Zhang J. X., Xiang B., He Q., Seidel J., Zeches R. J., Yu P., Yang S. Y., Wang C. H., Chu Y. H., Martin L. W., Minor A. M., Ramesh R., "Large field-induced strains in a lead-free piezoelectric material," *Nat. Nanotechnol.* 6(2), 97–101 (2011).

Zhang Y., Lu H. D., Xie L., Yan X., Paudel T. R., Kim J., Cheng X., Wang H., Heikes C., Li L., Xu M., Schlom D. G., Chen L. Q., Wu R., Tsymbal E. Y., Gruverman A., Pan X., "Anisotropic polarization-induced conductance at a ferroelectric-insulator interface," *Nat. Nanotechnol.* 13(12), 1132 (2018).

11

Nanoscale Ferroelectric Switching: A Method to Inject and Study Non-equilibrium Domain Walls

A. V. Ievlev[1], A. Tselev[2], R. Vasudevan[1],
S. V. Kalinin[1], A. Morozovska[3], and P. Maksymovych[1]

[1]Center for Nanophase Materials Sciences, Oak Ridge National Laboratory, Oak Ridge, Tennessee, United States
[2]University of Aveiro, Aveiro, Portugal
[3]Institute of Physics, National Academy of Sciences of Ukraine, Kyiv, Ukraine

11.1 Introduction

In ferroelectric materials, cooling below a critical (Curie) temperature induces a structural phase transition into a non-centrosymmetric state with finite spontaneous polarization (Lines and Glass 1979). In an ideal single domain crystal, the polarization is uniformly oriented. However, in real crystals, multi-domain structures form due to thermodynamic (depolarization) and kinetic (finite cooling rate across the phase transition) processes, as well as disorder. Domains are separated by domain walls, which can be viewed as topological defects in the parent crystal structure that manifest as local symmetry and strain variations (see Chapters 1, 3, and 4 for details). Owing to strong polarization-strain coupling, the domain walls in ferroelectrics are typically very thin, several nanometers across (Tagantsev et al. 2010).

The emergence of domain walls in a ferroelectric crystal had originally been considered as a mechanism to cancel the macroscopic depolarizing field, and therefore, to lower the total energy of the system despite the added domain wall energy (Kittel 1972; Lines and Glass 1979). The emergence, motion, and annihilation of domain walls underpins ferroelectric switching by electric fields below the Curie temperature, and, therefore, it has been considered one of the key components of the ferroelectric phenomenology since the earliest days of ferroelectricity (Tagantsev et al. 2010).

More recently, advances in lithography and microscopy jointly enabled probing the properties of individual domains and domain walls. These studies were stimulated by the discovery as well as predictions of unique piezoelectric, electronic, and magnetic

A. V. Ievlev, A. Tselev, R. Vasudevan, S. V. Kalinin, A. Morozovska, and P. Maksymovych, *Nanoscale Ferroelectric Switching: A Method to Inject and Study Non-equilibrium Domain Walls* In: *Domain Walls: From Fundamental Properties to Nanotechnology Concepts.* Edited by: Dennis Meier, Jan Seidel, Marty Gregg, and Ramamoorthy Ramesh, Oxford University Press (2020).
© A. V. Ievlev, A. Tselev, R. Vasudevan, S. V. Kalinin, A. Morozovska, and P. Maksymovych.
DOI: 10.1093/oso/9780198862499.003.0011

properties for domain walls (Kleemann 1993; Damjanovic 1998; Mostovoy 2006; Seidel et al. 2009; Catalan et al. 2012; Chanthbouala et al. 2012; Maksymovych et al. 2009, 2012). Expansion of the domain wall phenomenology accounting for these properties has developed into the field of domain wall electronics, aiming to tune the properties of domain walls and to tailor them for prospective applications in information technology (Catalan et al. 2012). The marked rise of experimental interest in domain walls has further stimulated the rapid evolution of theoretical methodologies that can capture and predict the properties of nano- and mesoscale volumes incorporating domain walls (Catalan et al. 2012; Meier 2015; Sharma et al. 2017; Luk'yanchuk et al. 2019).

In light of these advances, the ability to create and manipulate ferroelectric domain walls on demand is a necessary logical step in the future evolution of ferroelectric materials. Yet even the term "domain wall injection" was not common in ferroelectrics until the recent series of works by the group of M. Gregg, where approaches to inject domain walls controllably were proposed and demonstrated (e.g. a domain wall diode) (Whyte and Gregg 2015). By contrast, control over the injection, motion, and annihilation of domain walls in magnetic materials and the related field of spintronics is well established, with mature concepts, such as race-track memory, heading toward practical implementation in next-generation devices for information storage (Prudkovskii et al. 2006; Parkin and Yang 2015).

Domain walls can be created in ferroelectrics by electric or stress fields. In this regard, ferroelectrics may offer several advantages over magnetic materials: electric fields can be concentrated in a nanoscale volume while the domain walls themselves are nanoscale entities. Although the respective levels of power dissipation and domain wall densities need to be rigorously compared for the two material classes, it can be argued that ferroelectrics will have significant advantages. As a supporting example, one may recollect the aJ/bit effort in a multiferroic material to control magnetic information with ferroelectric polarization for very similar reasons (Manipatruni et al. 2015; Meisenheimer et al. 2018).

In this chapter, we focus on the electric-field-induced formation of a single nanoscale polarization domain, emphasizing that the nucleation, growth, and annihilation of such a domain enables the injection of non-equilibrium domain walls. The focus on nanoscale domains is motivated in large part by advances in scanning probe microscopy (SPM), where the ever-increasing drive to improve spatial resolution was met by a natural reduction in the size of one of the electrodes in a ferroelectric capacitor down to nanoscale dimensions. In this case, the process of ferroelectric switching becomes dominated by the nucleation of a single domain of the opposite polarization, as schematically shown in Figure 11.1. Once allowed to equilibrate, the end result of the single domain switching is injection of a new domain wall(s) into the original ferroelectric volume. Likewise, reversing the polarization under the probe will in general eliminate the domain wall(s). The nanoscale ferroelectric switching becomes, therefore, an efficient approach to inject and erase domain walls on demand. However, its primary value may be to facilitate a detailed understanding of the domain wall injection mechanisms as well as investigation of the range of possible domain wall configurations that can be injected up to the highest fields achievable in a given material.

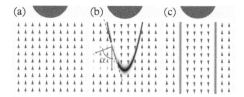

Figure 11.1 *Schematic cross section of domain wall injection in local ferroelectric switching: above-threshold electric field of the local probe (dark gray) triggers switching of initial polarization orientation (a). The injected domain wall (orange outline in (b)) separates the reversed and pristine polarization orientation. At equilibrium (c), achieved after some time, the domain wall is parallel to the polar axis, so that there is no bound charge on the wall. However, during the early stages, the injected domain wall is inclined so that $\alpha \leq 90^{\circ}$, and the domain wall carries a large density of bound polarization charge due to polarization discontinuity across the wall.*

Indeed, the central aspect of local domain wall injection is a continuum of non-equilibrium domain wall morphologies that connects the initial polarization configuration free of domain walls and a final configuration with a single, idealized, cylindrical equilibrium domain (Figure 11.1). The inclination angle of the wall relative to the polar axis (α in Figure 11.1) will span from 0° to 90° through the growth and expansion of the nanodomain. The simplest progression involves an initial half-prolate nucleus, gradually expanding along the polar axis (Figure 11.1(b)). For a thin film, the apex of the half-prolate nucleus will eventually intersect with the bottom electrode, making a truncated cone with tilted and charged domain wall segments, and eventually a cylinder.

The wall inclination relative to the polar axis is energetically very costly owing to a large polarization charge at the domain wall (Sluka et al. 2013; Bednyakov et al. 2015). The initial energy for the domain wall injection and expansion of the nucleus is supplied by the electrical circuit. However, removing the potential difference across the ferroelectric film at some intermediate stage of the domain growth will leave the charged domain walls in a metastable state. In more complicated cases, such as multi-axial $BiFeO_3$, the polarization structure of the nucleus is more intricate, giving many more possible metastable configurations. At the same time, dissipation of the energy stored in charged domain walls may include not only reversal of domain wall motion, but also generation of lattice defects (Vul et al. 1973; Gopalan et al. 2007; Gao et al. 2011) and other processes that result in screening of the polarization charge in the film. Thus, local polarization switching can be used to inject and interrogate non-equilibrium domain walls.

Understanding the processes of domain wall injection and relaxation is necessary not only for fundamental completeness. Any nanoscale ferroelectric device employing polarization switching will require these processes to be controlled to operate based on moving domain walls and avoid excessive defect creation. Moreover, as we will show below, even when measuring properties of a preexisting domain wall an electric field, non-equilibrium domain wall configuration needs to be considered as a possible source of the observable effects.

The structure of the chapter is as follows. First, we briefly summarize the experimental and analytical methods and conclusions emerging from analysis of polarization nucleation based on the experimental data obtained predominantly with SPM. This part discusses the mechanism of "domain wall breakdown" by which a macroscopic ferroelectric film can be switched with nanoscale electrodes, injecting non-equilibrium domain walls into the ferroelectric. Subsequently, we discuss the experiments probing interaction of domain wall with defects and domain wall motion in applied fields. In both cases, the domain wall morphology deviates from its idealized equilibrium geometry. We then touch upon domain wall conductance, which is a manifestation of the new property emerging from non-equilibrium domain walls. Finally, we discuss perspectives for the future development of theoretical and experimental techniques that can tackle the challenges of the non-equilibrium domain walls and domain wall injection processes, emphasizing the need for scalable data-analytic methodologies for interpretations of heterogeneous nanoscale data, and briefly survey the emerging methods for larger-scale simulations either connecting to or explicitly involving an atomistic description of ferroelectric polarization.

11.2   Generating Non-equilibrium Domain Morphologies in Ferroelectric Single Crystals

Investigating the domain growth of the locally switched polarization under an atomic force microscopy (AFM) probe provides direct evidence of non-equilibrium domain wall injection and evolution. The electric field from a voltage-biased AFM probe enables nucleation of a new polarization domain in the ferroelectric volume and its forward growth. The nonuniformity of the electric field almost guarantees a non-equilibrium domain shape, as can be verified by phase-field modeling (Figure 11. 2).

Domain formation under the charged SPM tip was originally explored by Molotskii (2003), addressing the balance of the electrostatic energy provided by the tip (interaction energy), depolarization energy of the domain, and the domain wall energy as $\Phi(r,l) = \Phi_S(r,l) + \Phi_D(r,l) - U\Phi_{UL}(r,l)$. Analytical expressions for the voltage dependence of the domain size and parameters of the tip-induced nucleation process were obtained for a prolate ellipsoid with radius r and length l. This model was further explored by Kalinin and Morozovska, who explicitly considered the dependence of the depolarization field and hence the nucleation process on the polarization screening phenomena in the rigid polarization model. It was shown that surface screening behavior can affect nucleation biases and control domain sizes by many orders of magnitude (Morozovska et al. 2006, 2007). This approach was further extended to allow in situ simulation of polarization under the action of an electric field, giving rise to the Ginzburg–Landau models of tip-induced switching. For uniaxial ferroelectric materials, it was explored analytically by Morozovska and numerically via the phase-field model by L.Q. Chen and co-authors (Morozovska et al. 2009). It was found that on clean ferroelectric surfaces, the experimentally measured nucleation biases were within 20–30% of the theoretical values predicted by the phase field with the experimentally determined tip parameters.

This illustrates that for (weakly) conductive material the depolarization field is effectively screened by the tip and polarization switching is nearly intrinsic (i.e. corresponding to the thermodynamic field). This is in contrast to polarization switching in the macroscopic systems, where the switching fields are orders of magnitude below the thermodynamic ones (Landauer paradox) due to the presence of defects and imperfections (Kalinin et al. 2007).

Further developments in understanding polarization dynamics in ferroelectric materials were greatly stimulated by the development of switching spectroscopy piezoresponse force microscopy (SS-PFM). In this technique, the hysteresis loop of the local piezoresponse is measured over a grid of pixels. SS-PFM therefore reveals the variability of switching behavior across the sample surface, providing information on the effect of local structural elements such as defects, preexisting domain walls, etc. on polarization switching (Jesse et al. 2006a, 2006b). The SS-PFM approach was used to explore polarization switching behavior in ferroelectric/non-ferroelectric composites and within individual ferroelectric nanodots (Rodriguez et al. 2007, 2008b).

SS-PFM was applied to map the energy and spatial distributions of individual nucleation centers in ferroelectrics, and to explore polarization switching down to a single defect (Jesse et al. 2008; Kalinin et al. 2008, 2010b). Piezoresponse hysteresis loops were shown to possess rich internal structure, emerging due to the interaction between the ferroelectric domain below the tip and preexisting defects. Mapping this internal fine structure has allowed the effect of a single defect on polarization switching to be mapped. Similarly, careful analysis of the hysteresis loop allowed reconstruction of the domain size and its comparison with point-by-point observations (Bdikin et al. 2008; Kalinin et al. 2008). For numerical models, the effects of the preexisting domain walls, grain boundaries, and other defects were explored (Kalinin et al. 2010a). These studies are particularly of interest for ferroelastic materials, where interactions between the dissimilar components of polarization, domain formation below the tip, and existing topological defects gives rise to a rich spectrum of polarization responses (Balke et al. 2009, 2012; Vasudevan et al. 2011; Winchester et al. 2015).

Charged domain walls and decorated nanoscale domains are other key distinctions between nanoscale ferroelectric switching and macroscopic switching in a nearly uniform electric field. For a thin film, the non-equilibrium shape of the nanodomain and its domain walls is a transient phenomenon. However, for a thick film or a single crystal, whose size is much larger than that of the AFM probe tip radius, the rapid drop-off of the electric field beyond the surface region promotes the existence of a quasi-equilibrium domain with charged domain walls, as well as nontrivial domain configurations, e.g. fractal dendrite structures, after switching by a uniform electric field (V. Shur et al. 2012) and even chaotic domain dynamics in sequential switching by an AFM probe (Ievlev et al. 2014a).

One of the remarkable consequences of the charged domain walls injected into a ferroelectric is the phenomenon of domain breakdown (Molotskii et al. 2003) wherein the growth of the conical domain is promoted by the charged configuration of the domain apex even when the external bias is removed. Model experiments carried out on nonpolar cuts of lithium niobate single crystals allowed direct study of phenomenon using lateral

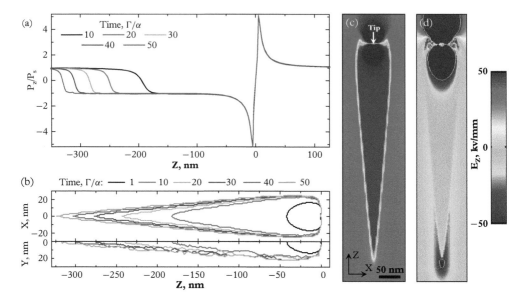

Figure 11.2 *LGD simulation of domain growth on a nonpolar cut of lithium niobate. (a) Profiles of polarization distribution for different effective times Γ/α, where Γ is the Khalatnikov coefficient, and α is the generalized dielectric stiffness; (b) time evolution of the domain shape; (c) surface distribution of the switched polarization; (d) surface distribution of the electric field strength.*

Source: Reprinted with permission from Ievlev et al. 2015a. Copyright 2015 American Chemical Society.

piezoresponse force microscopy (PFM) and numerical simulations (Alikin et al. 2015; Ievlev et al. 2015a; Morozovska et al. 2016). In particular, experiments showed forward growth of a conical domain up to 40 μm away from an AFM probe, where the electric field induced by the tip is negligible (Alikin et al. 2015).

A notable example of these studies is polarization switching on the nonpolar surfaces of the uniaxial ferroelectric LiNbO₃. In this case, polarization responds to the in-plane component of the tip electric field, and the ferroelectric domains grows in-plane, making it particularly convenient for observation. This process was numerically simulated using finite element analysis in Landau–Ginzburg–Devonshire (LGD) approximation. In this approach, the value of the normalized polarization $\tilde{P}_z = P_z/P_S$ is defined by the equation

$$\frac{\partial \tilde{P}_z}{\partial \tau} - \tilde{P}_z + \tilde{P}_z^3 - R_c^2 \left(\frac{\partial^2 \tilde{P}_z}{\partial z^2} + \frac{\partial^2 \tilde{P}_z}{\partial y^2} + \frac{\partial^2 \tilde{P}_z}{\partial x^2} \right) = \tilde{E}_z \qquad (11.1)$$

where P_S is the spontaneous polarization, τ is the normalized time, R_c is a gradient coefficient, and $\tilde{E}_z = E_z/E_c$ is the electric field normalized to coercive field of the material E_c (Ievlev et al. 2015a; Morozovska et al. 2016).

Simulation results clearly show the formation and growth of a conical domain far beyond the area of tip electric field localization (Figure 11.2(a–d)). At the same time, the distribution of the electric field demonstrates an area of a self-sustainable growth near the apex of the domain (Figure 11.2(d)), where charges from the charged domain walls induce an electric field, which promotes further domain propagation.

The presence of charges on the domain wall can also cause other types of phenomena and switching behavior. The most common effect is related to backswitching underneath a grounded tip. In this case, charged domain walls induce a charge in the grounded tip (Figures 11.3(a,b)) that in turn produces an electric field of the opposite polarity (Ievlev et al. 2015b). This leads to backswitching of the polarization inside the fresh domain with formation of the characteristic doughnut-like shape (Figure 11.3(c)) (Abplanalp et al. 2001; Kholkin et al. 2007; Lilienblum and Soergel 2011). A similar phenomenon on the nonpolar cuts of lithium niobate shows backswitching inside the primary domain and the formation of a secondary domain in the opposite direction (Figure 11.3(d)) (Alikin et al. 2015). Moreover, this "self-generated" electric field can lead to polarization switching by the grounded tip in the vicinity of a freshly switched or biased area with the formation of individual nanometer-sized domains and chains of domains along the trajectory of the tip (Ievlev et al. 2015b; Turygin et al. 2018).

The existence of charged domain walls in the ferroelectric medium leads to interaction between individual fresh domains and preexisting domain structures. As a result, a variety of types of switching dynamics during sequential switching may occur. In particular, studies show formation of uniform, quasi-periodic, and even chaotic domain structures (Figure 11.4(a)) after switching in the chain (Ievlev et al. 2014a; Turygin et al. 2018) as well as formation of symmetric structures after point switching by a sequence of triangular pulses (Figure 11.4(b)) (Ievlev et al. 2014b).

Overall, while the injection of domain walls with an AFM probe appears to be a simple intuitive idea, in reality, long-range electrostatic interactions introduce a variety of domain growth phenomena, causing breakdown, annihilation, reverse nucleation, and chaotic sequences of new domains. The processes are even more complicated in the presence of defects. Depending on the specific context, this variety can be beneficial or detrimental for the future prospects of domain wall electronics.

11.3 Non-equilibrium Domain Walls Evidenced from Electronic Conduction

The electrostatic fields of the non-equilibrium domain walls interact not only with the polarization itself but also with the electronic structure of the ferroelectric. Beginning with the original work by Vul et al. (1973), it was proposed that in real ferroelectric materials, which are mostly wide-gap semiconductors, electrostatically charged walls will gate the surrounding material, causing emission and local accumulation of compensating charges. Assuming that the charge density is sufficient to exceed the mobility edge, charged domain walls should eventually become conduits for electron transport through the bulk of the ferroelectric film. Chapters 1, 6, and 12, for example, address the rich experimental and theoretical phenomenology of the charge carrier transport along the domain walls.

More recently, field-induced processes in ferroelectric domains and domain walls were revisited, with the addition of comprehensive measurements of in-field piezoresponse and contact resonances, complementing the measured conductance (Vasudevan et al. 2017). For the case of stand-alone domain walls in BiFeO$_3$, the earlier hypothesis for enhanced domain wall conductance due to reduction of the Schottky barrier height by locally charged domain walls was put on a more robust footing via tighter integration of the observed conductance of domains and domain walls and combination with the appropriate phase-field model. One of the conclusions of that work was that even domain walls not conducting at zero bias will likely become conducting in an applied field due to the generality of field-induced deformation of the domain wall and Schottky barrier lowering (Vasudevan et al. 2017).

One of the strongest evidence for injection of non-equilibrium walls, even in very thin films, was observed in the high field switching processes of BiFeO$_3$ thin films (Figure 11.5). The key idea is that in a high enough electric field, not only can non-equilibrium domain walls be effectively injected, but even a field-driven phase transition can be

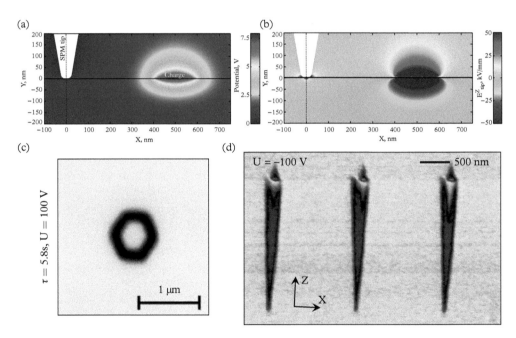

Figure 11.3 *(a,b) Finite element simulation of the electric field induced by a grounded AFM tip, distribution of (a) potential and (b) z component of the electric field. (c) Doughnut-shaped domain on the polar cut of lithium niobate. (d) Domains appearing as a result of backswitching on the nonpolar cut of lithium niobate.*

Source: (a,b) Reprinted figure with permission from Ievlev et al. 2015b. Copyright 2015 by the American Physical Society. (c) Reprinted from Lilienblum and Soergel 2011, with the permission of AIP Publishing. (d) Reprinted from Alikin et al. 2015, with the permission of AIP Publishing.

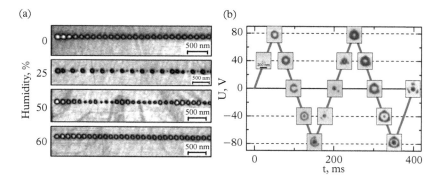

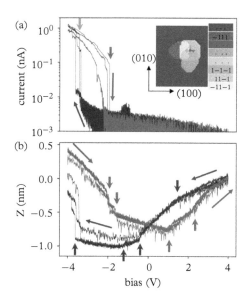

Figure 11.4 *(a) Transient, quasi-periodic, and chaotic regimes of sequential switching by an AFM tip. (b) Evolution of the domain shape produced by a sequence of triangular switching pulses applied to an AFM tip.*

Source: Reprinted from Ievlev et al. 2014a, with the permission of Springer Nature.

Figure 11.5 *Volatile resistive switching of 12 nm BiFeO$_3$ (100) film mediated by a localized rhombohedral-to-tetragonal (R-T) transition that creates locally charged phase boundaries under the SPM probe. (a) Hysteretic current–voltage curve showing abrupt increase of local conduction at −4 V and reversal back to the insulating state at −2 V. The corresponding hysteresis in the piezoelectric strain (b) clearly reveals discontinuous changes in film morphology corresponding to the phase transition. Phase-field model of the R-T nanodomain is shown in the inset of (a).*

Source: Reprinted figure with permission from Aravind et al. 2010. Copyright 2010 by the American Physical Society.

observed. Field-induced transitions between rhombohedral and tetragonal states were hypothesized by the work of Stengel and Iniguez (2015). Under these circumstances, it becomes energetically feasible to rotate the otherwise (110)-oriented polarization vector along the (100) direction (inset of Figure 11.5(a)). Obviously, this configuration is associated with large stresses, which would translate into strains and changes in elastic properties. Experimentally, the field-induced transition was convincingly detected by observing the changes in the contact resonance and very strong enhancement of piezoelectric signal in thin $BiFeO_3$ films under an applied electric bias (Figures 11.5(a,b)) (Maksymovych et al. 2011). Similar results have been shown in a proto-typical relaxor-ferroelectric, $0.72PbMg_{1/3}Nb_{2/3}O_3$-$0.28PbTiO_3$ (PMN-PT), for a field-induced transition between rhombohedral to tetragonal phases (Vasudevan et al. 2016).

Phase-field modeling indicates that when such a transition occurs in a localized electric field, a strong polarization discontinuity develops in the sub-surface volume, pointing to the possibility of increased conductance of the junction. Indeed, increased conductance, yielding local currents up to 10 nA, has been recently observed in $BiFeO_3$ films down to 12 nm under bias, and its origin was traced to the signature of the local field-induced R-T phase transition (Cao et al. 2018). Notably, because this effect is observed in an applied electric field, the conductive state is volatile, creating a function necessary for a neuristor: a volatile resistive switching element, that can be used to simulate an axon. Rewardingly, the field-induced states have been employed to create practically feasible memory devices operating on domain walls (Jiang et al. 2017; Sharma et al. 2017).

11.4   Domain Wall Contributions to Electromechanical Response Studied via Local Methods

Domain walls are prone to motion under applied fields, well below the critical field for nucleation of a new nanodomain. In fact, polarization switching in general can occur at any field strength, but kinetics prevents this from occurring on realizable time scales. High-resolution atomic-scale imaging of domain walls confirms that even extended domain walls in a material are rough, with often charged segments and deviations from ideal crystal symmetry due to the presence of pinning sites (Jia et al. 2008). Domain walls can be considered to be classical objects that possess momentum, and their behavior when subjected to fields has been described as such via canonical descriptions of an elastic medium migrating through a disordered material with distributed pinning defects (Figure 11.6). Since wall motion occurs, this can result in an enhanced piezoelectric response, when the moving walls are ferroelastic. Similar behavior has been known in magnetic systems since the work of Lord Rayleigh (1887). These studies motivated the exploration of the interplay between polarization switching and preexisting domain walls both in model single domain wall systems, and in multiple domain wall systems where collective phenomena and responses may emerge.

A detailed study of domain wall dynamics in uniaxial ferroelectric material was reported by Aravind et al. (2010). In this case, the authors performed a set of switching studies on the preexisting 180° domain wall in $LiNbO_3$ with progressively increasing bias windows in hysteresis loop measurements. For sufficiently low biases, the applied

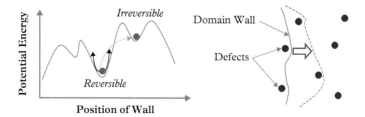

Figure 11.6 *Potential energy profile of domain wall (left) and schematic of moving domain wall (right). Domain walls can vibrate reversibly within deep minima, and if given enough energy, shift position irreversibly. This is shown by the solid and dashed lines (right), respectively.*

Source: Reprinted figure with permission from Aravind et al. 2010. Copyright 2010 by the American Physical Society.

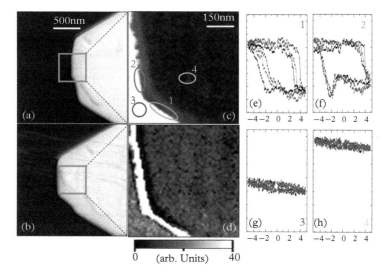

Figure 11.7 *(a) Mixed piezoresponse force microscopy images of the domain structure before and the SS-PFM scan with ±5 V voltage window in LiNbO$_3$ thin single crystal. Panels (b), (c), and (d) represent the area under the central square in (a). Note that although the domain wall moved during the experiment, the high veracity of the SS-PFM map indicates that no significant wall rearrangement occurred during single pixel or scan line acquisition. (b) The piezoresponse, (c) imprint, and (d) work of switching the SS-PFM map in the domain wall region. (e)–(h) Hysteresis loops from selected locations, where different colors correspond to different switching cycles.*

Source: Reprinted figure with permission from Vasudevan et al. 2013. Copyright 2013 by John Wiley and Sons.

the deformation of the wall was local and largely reversible. This behavior is shown on the set of the SS-PFM maps in Figure 11.7. Here, a SS-PFM map was acquired over the region containing a 180° wall in LiNbO$_3$. Notably, the wall displaced only insignificantly after multiple switching cycles. It is obvious that hysteresis loops acquired in the positive and negative domains are closed, and the corresponding vertical position corresponds to

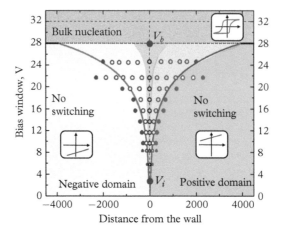

Figure 11.8 *(a) Switching phase diagram showing polarization dynamics as a function of the bias window and tip–wall separation. Shown are the regions of no switching, wall-mediated switching with asymmetric loops and symmetric loops (region near x = 0), and bulk nucleation. Solid lines correspond to first-order phase transitions across which the switching loops change discontinuously from open to closed.*

the (unswitched) polarization of the domain. At the same time, the loops are open at the domain, indicative of wall dynamics.

This behavior is further illustrated in Figure 11.8. Here, polarization switching was explored as a function of the bias window in hysteresis loop measurements and the distance from the domain wall. For low voltages (below 3 V), switching does not occur anywhere. For high voltages (>28 V), switching occurs everywhere on the surface and is bulk-like. In the interval 3–28 V switching is possible only in the vicinity of the wall, and the corresponding switching bias corresponds to the formation of the domain wall kink (Miller-Weinreich nucleus) (Morozovska et al. 2008; Aravind et al. 2010).

An insight into switching behavior at the non-180° walls was obtained in a series of studies by Jesse et al. Nucleation was shown to be enhanced in the vicinity of preexisting a-c domain walls. The mechanism for this effect was explored by Choudhury et al., who demonstrated that the effect is anisotropic and related to the interaction of the growing domain and tilted ferroelastic wall (Choudhury et al. 2008; Jesse et al. 2008).

The effect of the ferroelastic 71° and 109° domain walls on switching was explored by Balke et al. It was shown that preexisting domain walls reduce the nucleation bias for one direction of switching and leave it invariant for the opposite direction. This effect was traced to the interactions between the in-plane components of the tip field and the in-plane polarization component that may result in the characteristic wall bending. This work was further extended to realize the symmetry breaking of the thermodynamically equivalent states on the $BiFeO_3$ surface and create controllable in-plane ferroelastic domain structures, eventually patterning ferroelectric domain vortices with enhanced switching at the core (Balke et al. 2009, 2012).

Finally, the effect of individual well-defined defects on the domain nucleation and growth was studied by Rodriguez et al., particularly with polarization switching across the engineered bicrystal grain boundary. Non-monotonic dependence of switching parameters across the grain boundary was shown, with significant reduction of the nucleation bias on one side and weak enhancement of nucleation on the other side. The phase-field modeling studies have shown that this behavior emerges from the specific interaction of the nucleating domain and the grain boundary, where the lowest nucleation bias corresponds to the formation of a characteristic petal-like ferroelastic domain anchored at the grain boundary. These studies illustrated that the deterministic tip-induced domain switching mechanisms can be deciphered via phase-field modeling, and the predicted mesoscopic observables such as the nucleation bias and the area under the hysteresis loop can be directly compared with the experiment (Rodriguez et al. 2008a, 2009).

The fundamentally new behaviors can emerge in the systems with a high defect density and a large number of ferroelectric walls. The interaction between a single domain wall and multiple defects was explored in a series of works by Paruch et al. (2005a, 2005b).

Experiments on a magnetic system have shown that the susceptibility was a function of the applied field, which also holds for many ferroelectrics, leading to the (phenomenological) equation

$$d_{33} = d_{init} + \alpha_D E \qquad (11.2)$$

where d_{33} is the converse piezoelectric coefficient, d_{init} is a contribution from intrinsic mechanisms such as the lattice response and includes reversible domain wall motion, α_D is a so-called Rayleigh coefficient that arises from an extrinsic contribution(s), and E is the applied electric field. The relationship was found to hold for small fields. For magnetic systems, this leads to quadratic dependence of the magnetization on the applied magnetic field. For ferroelectrics, the equivalent is

$$s(E) = (d_{init} + \alpha_D E) E \pm \frac{\alpha_D}{2} \left(E^2 - E_{max}^2 \right) \qquad (11.3)$$

where s is the strain, and E_{max} is the maximum field in the cycle. The sign is positive when the field increases, and negative when it decreases. The question of the origins of this behavior has been explored through both experiment and theory. Notable contributions by Hilzinger and Kronmüller (1976) and later Boser (1987) reproduced the Rayleigh relationships, assuming that domain walls move independently of each other, and that the defect distribution responsible for exerting force on the domain wall is spatially random. In the latter case, Boser (1987) was able to show that the irreversible changes to polarization could be effectively approximated by a quadratic equation, and furthermore, provided a relationship between the Rayleigh constant and the physical descriptors of the domain wall and defect density. Moreover, it was shown that the Rayleigh relationship would hold for fields up to about half of the coercive field. This relationship was found to hold in some cases, although not all (Damjanovic 1997).

The statistical theories posit mechanisms that need to be confirmed via local studies, not just of the macroscopic observables. Moreover, it is already known that some of the assumptions in the theory, such as the independence of the motion of domain walls, are not appropriate in all cases (for example, see cascade, avalanching of domains (Harrison and Salje 2010; Salje et al. 2017), etc.). Recently, local studies have provided fresh insights into the nature of domain wall evolution under applied fields.

A natural first question is whether the nonlinearity caused by irreversible domain wall contributions can be observed on the level of a single domain wall. Li et al. (2016) probed this question through a combination of local band and excitation spectroscopic experiments, and found a strong correlation between measurements of the local piezoelectric nonlinearity and that of the loop area in a $(K,Na)NbO_3$ single crystal.

At the same time, if the distribution of pinning sites is not random (i.e. may be correlated due to defect–defect interactions), then it is expected that the nonlinearity may not obey the same relationship. Experiments and analysis by Vasudevan et al. (2013) on PZT thin films showed that as opposed to quadratic strain-field behavior, a dual

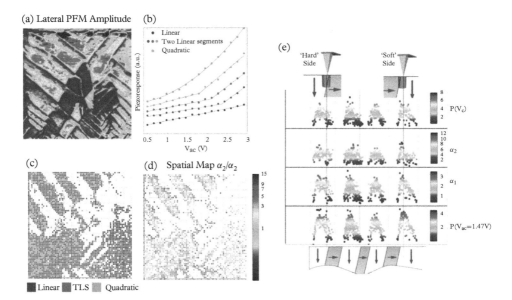

Figure 11.9 *Piezoresponse nonlinearity measurements of epitaxial PZT thin film. (a) Lateral PFM amplitude of a thick (~2 um) PZT thin film showing complex a/c domain patterns. Nonlinearity measurements on a spatial grid were performed in this area. (b) Selected representative piezoresponse curves as a function of AC excitation, captured with the band-excitation method. (c) Fitting-type map showing regions of the dual linear ("two linear segment (TLS)") behavior. (d) Spatial map of the slope ratio α_2/α_1, where α refers to the slope of the second and first segments of the two linear segments.*

Source: Reprinted figure with permission from Ye et al. 2018. Copyright 2018 by the American Physical Society.

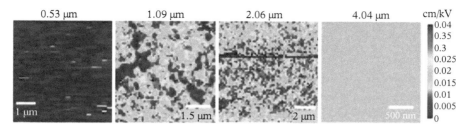

Figure 11.10 *Band-excitation piezoelectric nonlinearity study on ferroelectric capacitors of polycrystalline PZT of different thickness. The maps show the degree of nonlinear behavior in the recorded strain. At small thickness, no nonlinearity is observed. At micron thickness, the collective motion of walls in volumes larger than the grain size appears to be at the heart of the observed nonlinearity, which is almost uniform past 4 um in thickness.*
Source: From Bintachitt et al. 2010.

linear behavior was observed, consisting of a critical field after which the piezoelectric contribution was still linear, but with a pronounced higher slope. The magnitude of the ratio of slopes of these two linear regions increased with decreasing distance to the ferroelastic walls, as would be expected if their activated motion was the underlying cause. Moreover, distributions of the activation voltage, presumably due to distributions of defects, were shown to recover "on average" Rayleigh-like behavior. This is summarized in Figure 11.9.

At the same time, the majority of commercially used ferroelectrics are polycrystalline ceramics. In situ X-ray diffraction has proved to be an indispensable tool to determine the origins of Rayleigh behavior in such systems, and a good review is available on the subject (Pramanick et al. 2011). The local behavior of domain walls in polycrystalline ferroelectrics was explored by the Trolier-McKinstry group (Bintachitt et al. 2010). In that case, it was found that micron-size clusters of domain walls were responsible for nonlinearity in polycrystalline PZT ceramics, as shown in Figure 11.10. In other words, the microscopic mechanisms at play governing the motion of domain walls are complex, and dependent on the type of sample (epitaxial/crystal or ceramic), distribution and concentration of defects, and the type of domain walls, and more studies will be needed to ascertain the microscopic origins of Rayleigh behavior in a wider array of ferroics (for example, in improper or relaxor ferroelectrics).

11.5 Chemical Aspects of Ferroelectric Switching

In most studies, ferroelectric switching is considered to not disrupt the chemical composition of the material. Yet stabilization of the single domain state requires compensation of the charge discontinuity at surfaces, interfaces, and non-equilibrium domain walls. In the simple case of a metal–ferroelectric contact, screening of polarization charges is caused by the charge redistribution in the metal conduction band. Screening at bare ferroelectric surface is more complex, and can be supported by semiconductor-like band bending (Fridkin 1980), or adsorption of ionic species (Wang et al. 2009;

Highland et al. 2011). The crucial role of the state of the surface layer in the switching behavior was shown in numerous investigations carried out with the biased AFM tip (Kalinin et al. 2011, 2012; Kalinin and Spaldin 2013). In particularly, it was shown that the size of the local domains (Dahan et al. 2006; Brugère et al. 2011; Shur et al. 2011) and the behavior of the local hysteresis loops (Ievlev et al. 2014c; Blaser and Paruch 2015) depend on the value of the relative humidity inside the AFM chamber.

Chemically controlled switching in ferroelectrics has been recently using synchrotron X-ray, which provided direct evidence of ionic screening on near-equilibrium surfaces (Fong et al. 2006; Wang et al. 2009; Highland et al. 2010, 2011; Stephenson and Highland 2011). Furthermore, ample evidence of the presence of ionic screening charges was obtained from variable temperature Kelvin probe force microscopy studies of oxide surfaces including ferroelectrics and grain boundaries. In grain boundary systems, it was shown that the size of the grain boundary space charge region is much larger than expected from the estimates based on capacitance and current voltage measurements. Furthermore, the sign of the grain boundary potential feature is opposite that expected from the electric measurements: i.e. the observed grain boundary potential is positive, although the resistive grain boundary in n-doped material should be negative to be depleted. Using lateral biasing experiments and variable temperature measurements, this behavior was shown to arise from the screening on the grain boundary–surface junction which inverts the sign of the grain boundary potential and broadens the corresponding feature (Kalinin and Bonnell 2000b).

Even more drastic are the effects of the ionic screening on ferroelectric surfaces. It was shown that ionic screening leads to inversion of the observed domain potential; i.e. domains with a positive polarization charge are negative on surface potential images, and vice versa. This is because the microscope detects the polarity of the dipole formed by the polarization charge and screening charge above it. The ionic screening charge dynamics is rather slow and can be directly observed in variable temperature experiments. While the polarization charge responds instantly to temperature as evidenced by changes in the surface topography (e.g. the angles of the surface corrugations at the ferroelastic domain walls), the screening charge response is on the time scale of tens of minutes, giving rise to phenomena such as potential retention above the Curie temperature on heating and temperature-induced potential inversion on cooling. Interestingly, careful analysis of this phenomenon allows the thermodynamic parameters of the screening process to be extracted (Kalinin and Bonnell 2000a; Kalinin et al. 2002).

The coupling between the ionic screening and ferroelectric instability in ferroelectric has been proposed to give rise to "ferroionic" states. These states produce a broad spectrum of polarization values, and characteristic relaxational and pressure dynamics, as explored by Morozovska et al. (2017a, 2017b), Cao et al. (2017), and Yang et al. (2017).

One downside of traditional AFM is the inability to directly image and identify the chemical species. Alternatives can be found in mass spectrometry imaging techniques (Belianinov et al. 2018), such as time-of-flight secondary ion mass spectrometry (ToF-SIMS). In combination with AFM, ToF-SIMS is able to provide information about local chemical phenomena associated with ferroelectric switching. In particular, investigation with this multimodal approach demonstrated local cation intermixing in the 3-nm-thick

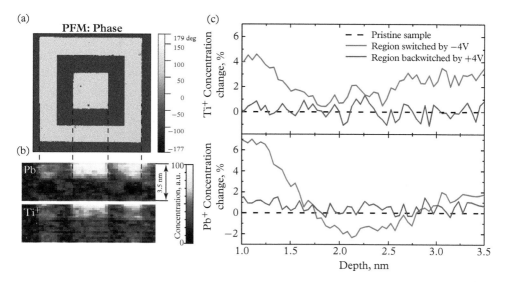

Figure 11.11 *Local chemical changes in the bulk of PZT film, induced by local polarization switching. (a) PFM phase of polled regions; (b) corresponding ToF-SIMS X–Z map of Pb+ and Ti+; (c) depth profiles of local concentration changes with respect to the pristine sample of Ti+ (top plot) and Pb+ (bottom plot) within the outer square switched by negative bias (red line) and the middle square backswitched by a positive bias (blue line).*

Source: Reprinted with permission from Ievlev et al. 2015a. Copyright 2015 American Chemical Society.

surface of $PbZr_{0.3}Ti_{0.7}O_3$ thin film (Figure 11.11) with formation of the concentration Pb^+ double layer (Ievlev et al. 2018b). In the more extreme regimes, with the electric field produced by the AFM tip significantly exceeding the coercive voltage of the film, this approach demonstrated the possibility of alteration in the structure of the studied sample, by intermixture of the ions of the adsorption layer, film, and even substrate (Ievlev et al. 2016). Similarly, chemical changes in the structure of the ferroelectric film were found while investigating ferroelectric fatigue in $PbZr_{0.3}Ti_{0.7}O_3$ thin films. It was concluded that Cu ions were injected from the top electrode into structure of the film with polarization cycling. A corresponding decrease in switchable polarization was attributed to the formation of pinning centers within the modified regions (Ievlev et al. 2019).

The extension of these studies to charged domain walls is necessarily quite challenging, because the latter represent an embedded interface, concealing possibly rich chemical phenomena accompanying or perhaps controlling ferroelectric switching. The role of oxygen vacancies in the compensation of polarization charge and electronic conductance at the domain walls is discussed in Chapter 3. However, it can be argued that chemical modifications of the materials with abundant charged domain walls are an intriguing opportunity for future studies.

11.6 Perspective: Understanding Domain Walls with Modeling, and Statistical and Machine Learning

By definition, interpretation of the measured dynamics and properties of domain walls and interference with the underlying physical mechanisms constitutes a challenging inverse problem. The nanoscale measurements by SPM, in particular, are inherently noisy and inhomogeneous. Most of the progress to date has relied on simulations, such as phase-field (Li et al. 2001) and other thermodynamic modeling that attempts to link the local polarization structure with the observed response, be it electronic, elastic, electromechanical, optical, etc. There also exist other models that are statistical in nature, for example, the Preisach models or more complex hysteresis models that consider kinetic effects, and which can fit macroscopic observables very well. However, such models may not always have a sound physical basis. One of the problems has to do with the length scale: how can we bridge experiments at multiple length scales, with theory, also at multiple length scales, in a consistent and integrated fashion? And how do we make sense of the experimental data that is inherently noisy, within this framework? Some perspectives are offered below.

One approach is to consider continuum models based on force fields (FFs), from which all properties can, at least in principle, be derived. A ReaxFF model is an example, and recently it has been applied to $BaTiO_3$ (Akbarian et al. 2018). These models can incorporate the adsorbates on the surface, which has been shown to greatly influence the phase diagram of thin-film $BaTiO_3$ (below 30 nm) (Morozovska et al. 2017b). A combination of X-ray photoemission spectroscopy, and atomic-scale imaging with scanning tunneling microscopy could be sufficient to determine the nature of the surface reconstruction (if any) and the type and distribution of the surface adsorbates. This could further be confirmed via measurements of the surface potential through the transition temperature, with Kelvin probe force microscopy. These could be directly supplied to the ReaxFF model to predict the full phase diagram, which could be confirmed by experiment, e.g. for a range of different thicknesses. Such studies are currently underway, attempting to bridge mesoscopic and atomic length scales and to provide greater validation of the ReaxFF modeling approach. If successful, inclusion of the applied fields may be possible in a more comprehensive version of the models, so that the dynamics of polarization switching and its interaction with the lattice potential and defects could be investigated.

The relative success of such approaches also relies on the reliability of the experimental data. In most SPM measurements, for example, the noise is typically high, and spatial averaging is used to attempt to reduce the noise floor to acceptable levels. This can be avoided to some degree by implementing statistical methods such as principal component analysis. Still, instrumental limitations will be present, leading to uncertainty when the data is analyzed and fitted to different models. Therefore, we suggest using a Bayesian approach when noisy data is fit to models. The Bayesian method, in contrast to a frequentist statistics approach, can incorporate prior information and provide uncertainty quantification in a (relatively) straightforward manner. For example, suppose

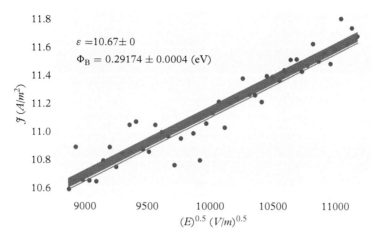

Figure 11.12 *Bayesian estimation on I–V data on BiFeO₃. The raw I–V data (dots) was taken near a domain wall in BiFeO₃. The fit was performed using a Bayesian method that includes a prior on the noise (assumed to follow a normal distribution). The probability density for the dielectric constant ε (i.e. derived from the slope of the graph) is essentially a delta function with no variance. A small variance exists in the barrier height ϕ_B, as indicated. The solid lines indicate samples from the posterior.*

one is fitting the curve in Figure 11.12 to one of several conduction models. The correct approach from a Bayesian perspective is to consider the probability of the different models given the data, y, i.e. $p(M_i|y)$, where M_i is the model and y is the observable (in this case, an I–V curve). Correspondingly, each model will be parameterized by some vector of parameters θ, over which probabilities should be computed (i.e. it is our aim to determine $p(\theta|y)$. Estimating these distributions ("posteriors") is often computationally difficult due to high dimensionality, and generally relies on sampling methods such as Markov-chain Monte-Carlo.

Finally, when dynamic models exist, there is a question of the comparison between experiment and theory. If imaging is utilized, then the general method to compare the theory and experiment is to compute some macroscopic statistics of the process, and then proceed by fitting the model to the observed statistics. This, of course, is highly dependent on the choice of statistics used for the model matching. An alternative approach can be via learning the manifolds. Since the data comes from a low-dimensional manifold (given that it arises from a relatively simple physical processes), the manifold learning methods allow visualization of the form of the manifold in low-dimensional space. A similar manifold can be learned for the theoretical model, and the parameters of the model can be tuned such that the shapes of the manifolds are most similar. This ensures that the model is not overfit to particular descriptors, and predicts well across the thermodynamic range, rather than to the most heavily sampled points. Such manifold learning methods (Wang and Suter 2007; Li et al. 2019) are just now becoming utilized within the materials imaging community.

11.7 Summary

Domain walls in ferroelectrics are unique objects of interest due to their structural distinction compared with the domains they separate, and therefore can host unique functional properties. Recent advances in experimental microscopy methods are enabling injection of individual domain walls and manipulation of existing domain walls with increasing precision (see Chapters 10 and 13). In these scenarios, domain wall formation requires a traversal through many states, many of which are non-equilibrium. Studying the dynamics of this process at the mesoscopic and atomic level, how they are affected by elastic and electric fields, and how these walls interact with surface and bulk defects remains a tantalizing problem both in terms of a fundamental physical understanding, as well as for future device miniaturization based on domain walls. At the same time, manipulation of domain walls has revealed emergent properties, such as domain wall conductance, chaotic domain switching, and domain wall breakdown, which are likewise rooted in domain walls dynamics and non-equilibrium states. Continued understanding of the complex physics that underpins domain wall dynamics will require integration of modeling and experiment across length scales, leveraging advances in scalable theoretical methods, data science, and the latest instrumental techniques. This understanding will drive continued progress in the design and implementation of single domain wall-based electronics for speed, reliability, and efficiency, and, at the same time enable new materials functionalities arising from dense and designer domain wall arrays.

REFERENCES

Abplanalp M., Fousek J., Günter P., "Higher order ferroic switching induced by scanning force microscopy," *Phys. Rev. Lett.* 86, 5799–5802 (2001).

Akbarian D., Yilmaz D., Van Duin A. C., *Evaluation and Extension of ReaxFF Reactive Force Field Method for Applications to Dielectric Oxides and their Multi-Material Interfaces* (Penn State University, University Park, United States, 2018).

Alikin D. O., Ievlev A. V., Turygin A. P., Lobov A. I., Kalinin S. V., Shur V. Y., "Tip-induced domain growth on the non-polar cuts of lithium niobate single-crystals," *Appl. Phys. Lett.* 106 (2015).

Aravind V. R., Morozovska A. N., Bhattacharyya S., Lee D., Jesse S., Grinberg I., Li Y. L., Choudhury S., Wu P., Seal K., Rappe A. M., Svechnikov S. V., Eliseev E. A., Phillpot S. R., Chen L. Q., Gopalan V., Kalinin S. V., "Correlated polarization switching in the proximity of a 180 degrees domain wall," *Phys. Rev. B* 82 (2010).

Balke N., Choudhury S., Jesse S., Huijben M., Chu Y. H., Baddorf A. P., Chen L. Q., Ramesh R., Kalinin S. V., "Deterministic control of ferroelastic switching in multiferroic materials," *Nat. Nanotechnol.* 4, 868–875 (2009).

Balke N., Winchester B., Ren W., Chu Y. H., Morozovska A. N., Eliseev E. A., Huijben M., Vasudevan R. K., Maksymovych P., Britson J., Jesse S., Kornev I., Ramesh R., Bellaiche L., Chen L. Q., Kalinin S. V., "Enhanced electric conductivity at ferroelectric vortex cores in BiFeO$_3$," *Nat. Phys.* 8, 81–88 (2012).

Bdikin I. K., Kholkin A. L., Morozovska A. N., Svechnikov S. V., Kim S. H., Kalinin S. V., "Domain dynamics in piezoresponse force spectroscopy: quantitative deconvolution and hysteresis loop fine structure," *Appl. Phys. Lett.* 92 (2008).

Bednyakov P. S., Sluka T., Tagantsev A. K., Damjanovic D., Setter N., "Formation of charged ferroelectric domain walls with controlled periodicity," *Sci. Rep.* 5 (2015).

Belianinov A., Ievlev A. V., Lorenz M., Borodinov N., Doughty B., Kalinin S. V., Fernandez F. M., Ovchinnikova O. S., "Correlated materials characterization via multimodal chemical and functional imaging," *ACS Nano* 12, 11798–11818 (2018).

Bintachitt P., Jesse S., Damjanovic D., Han Y., Reaney I. M., Trolier-Mckinstry S., Kalinin S. V., "Collective dynamics underpins Rayleigh behavior in disordered polycrystalline ferroelectrics," *Proc. Natl. Acad. Sci. U S A* 107, 7219–7224 (2010).

Blaser C., Paruch P., "Subcritical switching dynamics and humidity effects in nanoscale studies of domain growth in ferroelectric thin films," *New J. Phys.* 17 (2015).

Boser O., "Statistical-theory of hysteresis in ferroelectric materials," *J. Appl. Phys.* 62, 1344–1348 (1987).

Brugère A., Gidon S., Gautier B., "Finite element method simulation of the domain growth kinetics in single-crystal $LiTaO_3$: role of surface conductivity," *J. Appl. Phys.* 110, 052016 (2011).

Cao Y., Li Q., Huijben M., Vasudevan R. K., Kalinin S. V., Maksymovych P., "Electronic switching by metastable polarization states in $BiFeO_3$ thin films," *Phys. Rev. Mater.* 2, 094401 (2018).

Cao Y., Morozovska A., Kalinin S. V., "Pressure-induced switching in ferroelectrics: phase-field modeling, electrochemistry, flexoelectric effect, and bulk vacancy dynamics," *Phys. Rev. B* 96 (2017).

Catalan G., Seidel J., Ramesh R., Scott J. F., "Domain wall nanoelectronics," *Rev. Mod. Phys.* 84, 119–156 (2012).

Chanthbouala A., Garcia V., Cherifi R. O., Bouzehouane K., Fusil S., Moya X., Xavier S., Yamada H., Deranlot C., Mathur N. D., Bibes M., Barthelemy A., Grollier J., "A ferroelectric memristor," *Nat. Mater.* 11, 860–864 (2012).

Choudhury S., Zhang J. X., Li Y. L., Chen L. Q., Jia Q. X., Kalinin S. V., "Effect of ferroelastic twin walls on local polarization switching: phase-field modeling," *Appl. Phys. Lett.* 93 (2008).

Dahan D., Molotskii M., Rosenman G., Rosenwaks Y., "Ferroelectric domain inversion: the role of humidity," *Appl. Phys. Lett.* 89, 152902 (2006).

Damjanovic D., "Stress and frequency dependence of the direct piezoelectric effect in ferroelectric ceramics," *J. Appl. Phys.* 82, 1788–1797 (1997).

Damjanovic D., "Ferroelectric, dielectric and piezoelectric properties of ferroelectric thin films and ceramics," *Rep. Progr. Phys.* 61, 1267–1324 (1998).

Fong D. D., Kolpak A. M., Eastman J. A., Streiffer S. K., Fuoss P. H., Stephenson G. B., Thompson C., Kim D. M., Choi K. J., Eom C. B., Grinberg I., Rappe A. M., "Stabilization of monodomain polarization in ultrathin $PbTiO_3$ films," *Phys. Rev. Lett.* 96, 127601 (2006).

Fridkin V. M., *Ferroelectric Semiconductors* (Consultant Bureau, New York, 1980).

Gao P., Nelson C. T., Jokisaari J. R., Baek S. H., Bark C. W., Zhang Y., Wang E. G., Schlom D. G., Eom C. B., Pan X. Q., "Revealing the role of defects in ferroelectric switching with atomic resolution," *Nat. Commun.* 2 (2011).

Gopalan V., Dierolf V., Scrymgeour D. A., "Defect-domain wall interactions in trigonal ferroelectrics," *Annu. Rev. Mater. Res.* 37, 449–489 (2007).

Harrison R. J., Salje E. K., "The noise of the needle: Avalanches of a single progressing needle domain in $LaAlO_3$," *Appl. Phys. Lett.* 97, 021907 (2010).

Highland M. J., Fister T. T., Fong D. D., Fuoss P. H., Thompson C., Eastman J. A., Streiffer S. K., Stephenson G. B., "Equilibrium polarization of ultrathin PbTiO$_3$ with surface compensation controlled by oxygen partial pressure," *Phys. Rev. Lett.* 107, 5 (2011).

Highland M. J., Fister T. T., Richard M. I., Fong D. D., Fuoss P. H., Thompson C., Eastman J. A., Streiffer S. K., Stephenson G. B., "Polarization switching without domain formation at the intrinsic coercive field in ultrathin ferroelectric PbTiO$_3$," *Phys. Rev. Lett.* 105, 4 (2010).

Hilzinger H., Kronmüller H., "Statistical theory of the pinning of Bloch walls by randomly distributed defects," *J. Magnetism Magnetic Mater.* 2, 11–17 (1976).

Ievlev A. V., Alikin D. O., Morozovska A. N., Varenyk O. V., Eliseev E. A., Kholkin A. L., Shur V. Y., Kalinin S. V., "Symmetry breaking and electrical frustration during tip-induced polarization switching in the nonpolar cut of lithium niobate single crystals," *ACS Nano* 9, 769–777 (2015a).

Ievlev A. V., Brown C., Agar J. C., Velarde G., Martin L. W., Belianinov A., Maksymovych P., Kalinin S. V., Ovchinnikova O. S., "Nanoscale electrochemical phenomena of polarization switching in ferroelectrics," *ACS Appl. Mater. Interf.* 10(44), 38217–38222 (2018a).

Ievlev A. V., Brown C. C., Agar J. C., Velarde G. A., Martin L. W., Belianinov A., Maksymovych P., Kalinin S. V., Ovchinnikova O. S., "Nanoscale Electrochemical Phenomena of Polarization Switching in Ferroelectrics," *ACS Appl.Mater. Interf.* 10, 38217–38222 (2018b).

Ievlev A. V., Jesse S., Morozovska A. N., Strelcov E., Eliseev E. A., Pershin Y. V., Kumar A., Shur V. Y., Kalinin S. V., "Intermittency, quasiperiodicity and chaos in probe-induced ferroelectric domain switching," *Nat. Phys.* 10, 59–66 (2014a).

Ievlev A. V., Kc S., Vasudevan R. K., Kim Y., Lu X., Alexe M., Cooper V. R., Kalinin S. V., Ovchinnikova O. S., "Non-conventional mechanism of ferroelectric fatigue via cation migration," *Nat. Commun.* 10, 3064 (2019).

Ievlev A. V., Maksymovych P., Trassin M., Seidel J., Ramesh R., Kalinin S. V., Ovchinnikova O. S., "Chemical state evolution in ferroelectric films during tip-induced polarization and electroresistive switching," *ACS Appl. Mater. Interf.* 8, 29588–29593 (2016).

Ievlev A. V., Morozovska A. N., Eliseev E. A., Shur V. Y., Kalinin S. V., "Ionic field effect and memristive phenomena in single-point ferroelectric domain switching," *Nat. Commun.* 5 (2014b).

Ievlev A. V., Morozovska A. N., Shur V. Y., Kalinin S. V., "Humidity effects on tip-induced polarization switching in lithium niobate," *Appl. Phys. Lett.* 104, 092908 (2014c).

Ievlev A. V., Morozovska A. N., Shur V. Y., Kalinin S. V., "Ferroelectric switching by the grounded scanning probe microscopy tip," *Phys. Rev. B* 91 (2015b).

Jesse S., Baddorf A. P., Kalinin S. V., "Switching spectroscopy piezoresponse force microscopy of ferroelectric materials," *Appl. Phys. Lett.* 88, 062908 (2006a).

Jesse S., Lee H. N., Kalinin S. V., "Quantitative mapping of switching behavior in piezoresponse force microscopy," *Rev. Sci. Instrum.* 77 (2006b).

Jesse S., Rodriguez B. J., Choudhury S., Baddorf A. P., Vrejoiu I., Hesse D., Alexe M., Eliseev E. A., Morozovska A. N., Zhang J., Chen L.-Q., Kalinin S. V., "Direct imaging of the spatial and energy distribution of nucleation centres in ferroelectric materials," *Nat. Mater.* 7, 209–215 (2008).

Jia C.-L., Mi S.-B., Urban K., Vrejoiu I., Alexe M., Hesse D., "Atomic-scale study of electric dipoles near charged and uncharged domain walls in ferroelectric films," *Nat. Mater.* 7, 57 (2008).

Jiang J., Bai Z. L., Chen Z. H., He L., Zhang D. W., Zhang Q. H., Shi J. A., Park M. H., Scott J. F., Hwang C. S., Jiang A. Q., "Temporary formation of highly conducting domain walls for non-destructive read-out of ferroelectric domain-wall resistance switching memories," *Nat. Mater.* 17, 49 (2017).

Kalinin S. V., Bonnell D. A., "Effect of phase transition on the surface potential of the $BaTiO_3$ (100) surface by variable temperature scanning surface potential microscopy," *J. Appl. Phys.* 87, 3950–3957 (2000a).

Kalinin S. V., Bonnell D. A., "Surface potential at surface-interface junctions in $SrTiO_3$ bicrystals," *Phys. Rev. B* 62, 10419–10430 (2000b).

Kalinin S. V., Borisevich A., Fong D., "Beyond condensed matter physics on the nanoscale: the role of ionic and electrochemical phenomena in the physical functionalities of oxide materials," *ACS Nano* 6, 10423–10437 (2012).

Kalinin S. V., Jesse S., Rodriguez B. J., Chu Y. H., Ramesh R., Eliseev E. A., Morozovska A. N., "Probing the role of single defects on the thermodynamics of electric-field induced phase transitions," *Phys. Rev. Lett.* 100 (2008).

Kalinin S. V., Jesse S., Tselev A., Baddorf A. P., Balke N., "The role of electrochemical phenomena in scanning probe microscopy of ferroelectric thin films," *ACS Nano* 5, 5683–5691 (2011).

Kalinin S. V., Johnson C. Y., Bonnell D. A., "Domain polarity and temperature induced potential inversion on the $BaTiO_3$ (100) surface," *J. Appl. Phys.* 91, 3816–3823 (2002).

Kalinin S. V., Morozovska A. N., Chen L. Q., Rodriguez B. J., "Local polarization dynamics in ferroelectric materials," *Rep. Progr. Phys.* 73, 056502 (2010a).

Kalinin S. V., Rodriguez B. J., Borisevich A. Y., Baddorf A. P., Balke N., Chang H. J., Chen L. Q., Choudhury S., Jesse S., Maksymovych P., Nikiforov M. P., Pennycook S. J., "Defect-mediated polarization switching in ferroelectrics and related materials: from mesoscopic mechanisms to atomistic control," *Adv. Mater.* 22, 314–322 (2010b).

Kalinin S. V., Rodriguez B. J., Jesse S., Chu Y. H., Zhao T., Ramesh R., Choudhury S., Chen L. Q., Eliseev E. A., Morozovska A. N., "Intrinsic single-domain switching in ferroelectric materials on a nearly ideal surface," *Proc. Natl. Acad. Sci. U S A* 104, 20204–20209 (2007).

Kalinin S. V., Spaldin N. A., "Functional ion defects in transition metal oxides," *Science* 341, 858–859 (2013).

Kholkin A. L., Bdikin I. K., Shvartsman V. V., Pertsev N. A., "Anomalous polarization inversion in ferroelectrics via scanning force microscopy," *Nanotechnology* 18, 095502 (2007).

Kittel C., "Thickness of domain walls in ferroelectric and ferroelastic crystals," *Solid State Commun.* 10, 119–121 (1972).

Kleemann W., "Random-field induced antiferromagnetic, ferroelectric and structural domain states," *Int. J. Mod. Phys. B* 7, 2469–2507 (1993).

Li L., Yang Y., Liu Z., Jesse S., Kalinin S. V., Vasudevan R. K., "Correlation between piezoresponse nonlinearity and hysteresis in ferroelectric crystals at the nanoscale," *Appl. Phys. Lett.* 108, 172905 (2016).

Li X., Dyc O. E., Oxle M. P., Lupini A. R., Mcinnes L., Healy J., Jesse S., Kalinin S. V., "Manifold learning of four-dimensional scanning transmission electron microscopy," *Npj Comput. Mater.* 5, 5 (2019).

Li Y. L., Hu S. Y., Liu Z. K., Chen L. Q., "Phase-field model of domain structures in ferroelectric thin films," *Appl. Phys. Lett.* 78, 3878–3880 (2001).

Lilienblum M., Soergel E., "Anomalous domain inversion in $LiNbO_3$ single crystals investigated by scanning probe microscopy," *J. Appl. Phys.* 110, 052018 (2011).

Lines M., Glass A., *Principles and Applications of Ferroelectrics and Related Materials* (Clarendon Press, Oxford, 1979).

Luk'yanchuk I., Tikhonov Y., Sene A., Razumnaya A., Vinokur V. M., "Harnessing ferroelectric domains for negative capacitance," *Commun. Phys.* 2 (2019).

Maksymovych P., Huijben M., Pan M. H., Jesse S., Balke N., Chu Y. H., Chang H. J., Borisevich A. Y., Baddorf A. P., Rijnders G., Blank D. H. A., Ramesh R., Kalinin S. V., Ultrathin limit and dead-layer effects in local polarization switching of $BiFeO_3$. *Phys. Rev. B* 85, 8 (2012).

Maksymovych P., Jesse S., Yu P., Ramesh R., Baddorf A. P., Kalinin S. V., "Polarization control of electron tunneling into ferroelectric surfaces," *Science* 324, 1421–1425 (2009).

Maksymovych P., Seidel J., Chu Y. H., Wu P. P., Baddorf A. P., Chen L. Q., Kalinin S. V., Ramesh R., "Dynamic conductivity of ferroelectric domain walls in $BiFeO_3$," *Nano Lett.* 11, 1906–1912 (2011).

Manipatruni S., Nikonov D. E., Ramesh R., Li H., Young I. A., "Spin-orbit logic with magneto-electric nodes: a scalable charge mediated nonvolatile spintronic logic," *arXiv e-prints* [Online]. Available: https://ui.adsabs.harvard.edu/abs/2015arXiv151205428M [Accessed December 01, 2015] (2015).

Meier D., "Functional domain walls in multiferroics," *J. Phys.: Condens. Mat.* 27, 463003 (2015).

Meisenheimer P. B., Novakov S., Vu N. M., Heron J. T., "Perspective: magnetoelectric switching in thin film multiferroic heterostructures," *J. Appl. Phys.* 123, 240901 (2018).

Molotskii M., "Generation of ferroelectric domains in atomic force microscope," *J. Appl. Phys.* 93, 6234–6237 (2003).

Molotskii M., Agronin A., Urenski P., Shvebelman M., Rosenman G., Rosenwaks Y., "Ferroelectric domain breakdown," *Phys. Rev. Lett.* 90, 107601 (2003).

Morozovska A. N., Eliseev E. A., Kalinin S. V., "Domain nucleation and hysteresis loop shape in piezoresponse force spectroscopy," *Appl. Phys. Lett.* 89, 192901 (2006).

Morozovska A. N., Eliseev E. A., Kurchak A. I., Morozovsky N. V., Vasudevan R. K., Strikha M. V., Kalinin S. V., "Effect of surface ionic screening on the polarization reversal scenario in ferroelectric thin films: crossover from ferroionic to antiferroionic states," *Phys. Rev. B* 96 (2017a).

Morozovska A. N., Eliseev E. A., Li Y., Svechnikov S. V., Maksymovych P., Shur V. Y., Gopalan V., Chen L.-Q., Kalinin S. V., "Thermodynamics of nanodomain formation and breakdown in scanning probe microscopy: Landau-Ginzburg-Devonshire approach," *Physical Review B* 80, 214110 (2009).

Morozovska A. N., Eliseev E. A., Morozovsky N. V., Kalinin S. V., "Ferroionic states in ferroelectric thin films," *Phys. Rev. B* 95, 195413 (2017b).

Morozovska A. N., Ievlev A. V., Obukhovskii V. V., Fomichov Y., Varenyk O. V., Shur V. Y., Kalinin S. V., Eliseev E. A., "Self-consistent theory of nanodomain formation on nonpolar surfaces of ferroelectrics," *Phys. Rev. B* 93, 165439 (2016).

Morozovska A. N., Kalinin S. V., Eliseev E. A., Gopalan V., Svechnikov S. V., "Interaction of a 180 degrees ferroelectric domain wall with a biased scanning probe microscopy tip: Effective wall geometry and thermodynamics in Ginzburg-Landau-Devonshire theory," *Phys. Rev. B* 78 (2008).

Morozovska A. N., Svechnikov S. V., Eliseev E. A., Jesse S., Rodriguez B. J., Kalinin S. V., "Piezoresponse force spectroscopy of ferroelectric-semiconductor materials," *J. Appl. Phys.* 102 (2007).

Mostovoy M., "Ferroelectricity in spiral magnets," *Phys. Rev. Lett.* 96, 067601 (2006).

Parkin S., Yang S. H., "Memory on the racetrack," *Nat. Nanotechnol.* 10, 195–198 (2015).

Paruch P., Giamarchi T., Triscone J. M., "Domain wall roughness in epitaxial ferroelectric PbZr$_{0.2}$Ti$_{0.8}$O$_3$ thin films," *Phys. Rev. Lett.* 94 (2005a).

Paruch P., Tybell T., Giamarchi T., Triscone J. M., "Nanoscopic studies of disorder-controlled static and dynamic properties of domain walls in epitaxial ferroelectric films," *Abstr. Pap. Am. Chem. Soc.* 230, U2812–U2812 (2005b).

Pramanick A., Damjanovic D., Daniels J. E., Nino J. C., Jones J. L., "Origins of electro-mechanical coupling in polycrystalline ferroelectrics during subcoercive electrical loading," *J. Am. Ceram. Soc.* 94, 293–309 (2011).

Prudkovskii P. A., Rubtsov A. N., Katsnelson M. I., "Topological defects, pattern evolution, and hysteresis in thin magnetic films," *Europhys. Lett.* 73, 104–109 (2006).

Rayleigh L., "XXV. Notes on electricity and magnetism.—III. On the behaviour of iron and steel under the operation of feeble magnetic forces," *The London, Edinburgh, and Dublin Philosophical Magazine and Journal of Science* 23, 225–245 (1887).

Rodriguez B. J., Choudhury S., Chu Y. H., Bhattacharyya A., Jesse S., Seal K., Baddorf A. P., Ramesh R., Chen L. Q., Kalinin S. V., "Unraveling deterministic mesoscopic polarization switching mechanisms: spatially resolved studies of a tilt grain boundary in bismuth ferrite," *Adv. Funct. Mater.* 19, 2053–2063 (2009).

Rodriguez B. J., Chu Y. H., RAmesh R., Kalinin S. V., "Ferroelectric domain wall pinning at a bicrystal grain boundary in bismuth ferrite," *Appl. Phys. Lett.* 93 (2008a).

Rodriguez B. J., Jesse S., Alexe M., Kalinin S. V., "Spatially resolved mapping of polarization switching behavior in nanoscale ferroelectrics," *Adv. Mater.* 20, 109–114 (2008b).

Rodriguez B. J., Jesse S., Baddorf A. P., Zhao T., Chu Y. H., Ramesh R., Eliseev E. A., Morozovska A. N., Kalinin S. V., "Spatially resolved mapping of ferroelectric switching behavior in self-assembled multiferroic nanostructures: strain, size, and interface effects," *Nanotechnology* 18, 405701 (2007).

Salje E. K., Wang X., Ding X., Scott J. F., "Ultrafast switching in Avalanche-driven ferroelectrics by supersonic Kink movements," *Adv. Funct. Mater.* 27, 1700367 (2017).

Seidel J., Martin L. W., He Q., Zhan Q., Chu Y. H., Rother A., Hawkridge M. E., Maksymovych P., Yu P., Gajek M., Balke N., Kalinin S. V., Gemming, S., Wang F., Catalan G., Scott J. F., Spaldin N. A., Orenstein J., Ramesh R., "Conduction at domain walls in oxide multiferroics," *Nat. Mater.* 8, 229–234 (2009).

Sharma P., Zhang Q., Sando D., Lei C. H., Liu Y. Y., Li J. Y., Nagarajan V., Seidel J., "Nonvolatile ferroelectric domain wall memory," *Sci. Adv.* 3, e1700512 (2017).

Shur V. Y., Ievlev A. V., Nikolaeva E. V., Shishkin E. I., Neradovskiy M. M., "Influence of adsorbed surface layer on domain growth in the field produced by conductive tip of scanning probe microscope in lithium niobate," *J. Appl. Phys.* 110, 052017 (2011).

Sluka T., Tagantsev A. K., Bednyakov P., Setter N., "Free-electron gas at charged domain walls in insulating BaTiO$_3$," *Nat. Commun.* 4, 1808 (2013).

Stengel M., Iniguez J., "Electrical phase diagram of bulk BiFeO$_3$," *Phys. Rev. B* 92, 14 (2015).

Stephenson G. B., Highland M. J., "Equilibrium and stability of polarization in ultrathin ferroelectric films with ionic surface compensation," *Phys. Rev. B* 84, 064107 (2011).

Tagantsev A. K., Cross L. E., Fousek J., *Domains in Ferroic Crystals and Thin Films* (Springer, New York, 2010).

Turygin A. P., Alikin D. O., Kosobokov M. S., Ievlev A. V., Shur V. Y., "Self-organized formation of quasi-regular ferroelectric nanodomain structure on the nonpolar cuts by grounded SPM tip," *ACS Appl. Mater. Interf.* 10, 36211–36217 (2018).

Vasudevan R. K., Cao Y., Laanait N., Ievlev A., Li L., Yang J.-C., Chu Y.-H., Chen L.-Q., Kalinin S. V., Maksymovych P., "Field enhancement of electronic conductance at ferroelectric domain walls," *Nat. Commun.* 8, 1318 (2017).

Vasudevan R. K., Chen Y. C., Tai H. H., Balke N., Wu P. P., Bhattacharya S., Chen L. Q., Chu Y. H., Lin I. N., Kalinin S. V., Nagarajan V., "Exploring topological defects in epitaxial BiFeO$_3$ thin films," *ACS Nano* 5, 879–887 (2011).

Vasudevan R. K., Khassaf H., Cao Y., Zhang S. J., Tselev A., CArmichael B., Okatan M. B., Jesse S., Chen L. Q., Alpay S. P., Kalinin S. V., Bassiri-Gharb N., "Acoustic detection of phase transitions at the nanoscale," *Adv. Funct. Mater.* 26, 478–486 (2016).

Vasudevan R. K., Okatan M. B., Duan C., Ehara Y., Funakubo H., Kumar A., Jesse S., Chen L. Q., Kalinin S. V., Nagarajan V., "Nanoscale origins of nonlinear behavior in ferroic thin films," *Adv. Funct. Mater.* 23, 81–90 (2013).

Vul B. M., Guro G. M., Ivanchik I. I., "Encountering domains in ferroelectrics," *Ferroelectrics* 6, 29–31 (1973).

Wang L., Suter D., "Learning and matching of dynamic shape manifolds for human action recognition," *IEEE Trans. Image Processing* 16, 1646–1661 (2007).

Wang R. V., Fong D. D., Jiang F., Highland M. J., Fuoss P. H., Thompson C., Kolpak A. M., Eastman J. A., Streiffer S. K., Rappe A. M., Stephenson G. B., "Reversible chemical switching of a ferroelectric film," *Phys. Rev. Lett.* 102, 047601 (2009).

Whyte J. R., Gregg J. M., "A diode for ferroelectric domain-wall motion," *Nat. Commun.* 6, 7361 (2015).

Winchester B., Balke N., Cheng X. X., Morozovska A. N., Kalinin S., Chen L. Q., "Electroelastic fields in artificially created vortex cores in epitaxial BiFeO$_3$ thin films," *Appl. Phys. Lett.* 107 (2015).

Ya Shur V.Chezganov D. S., Nebogatikov M. S., Baturin I. S., Neradovskiy M. M., "Formation of dendrite domain structures in stoichiometric lithium niobate at elevated temperatures," *J. Appl. Phys.* 112, 104113 (2012).

Yang S. M., Morozovska A. N., Kumar R., Eliseev E. A., Cao Y., Mazet L., Balke N., Jesse S., Vasudevan R. K., Dubourdieu C., Kalinin S. V., "Mixed electrochemical-ferroelectric states in nanoscale ferroelectrics," *Nat. Phys.* 13, 812–818 (2017).

12

Landau–Ginzburg–Devonshire Theory for Domain Wall Conduction and Observation of Microwave Conduction of Domain Walls

A. Tselev[1], A. V. Ievlev[2], R. Vasudevan[2], S. V. Kalinin[2], P. Maksymovych[2], and A. Morozovska[3]

[1] *University of Aveiro, Aveiro, Portugal*
[2] *Center for Nanophase Materials Sciences, Oak Ridge National Laboratory, Oak Ridge, Tennessee, United States*
[3] *Institute of Physics, National Academy of Sciences of Ukraine, Kyiv, Ukraine*

12.1 Introduction

Most ferroelectric materials are wide-gap semiconductors. Beginning with the original work by Vul et al. (1973), it was proposed that electrostatically charged walls where polarizations of the adjacent domains meet head-to-head or tail-to-tail will gate the surrounding material, causing emission and local accumulation of compensating charges. Assuming that the charge density is sufficient to exceed the mobility edge, charged domain walls should eventually become conduits for electron transport through the bulk of the ferroelectric film. Due to the energy of the associated electric field, charged domain walls are generally unstable, and, in fact, different mechanisms for the reduction of the electrostatic field energy of the charged walls can become active, such as accumulation of free charge carriers for compensation of the bound charge of the wall, as proposed by Vul et al., as well as attraction of charged defects or reconstruction in order to reduce the built-in charge in strongly charged walls (Morozovska et al. 2018). Nominally uncharged domain walls were also found to be more conducting that the domain bulk, for example, in epitaxial films of $Pb(Zr_{0.2}Ti_{0.8})O_3$ (Guyonnet et al. 2011). This type of wall in proper ferroelectrics can be attractive for potential applications because of their intrinsic stability in contrast to charged walls, whose stabilization requires external factors, such as an electric field. Several chapters (see, e.g. Chapters 3, 6, and 15) of this book address in detail the rich experimental and theoretical phenomenology of the charge carrier transport along the domain walls.

A. Tselev, A. V. Ievlev, R. Vasudevan, S. V. Kalinin, P. Maksymovych, and A. Morozovska, *Landau-Ginzburg-Devonshire Theory for Domain Wall Conduction and Observation of Microwave Conduction of Domain Walls* In: *Domain Walls: From Fundamental Properties to Nanotechnology Concepts*. Edited by: Dennis Meier, Jan Seidel, Marty Gregg, and Ramamoorthy Ramesh, Oxford University Press (2020).
© A. Tselev, A. V. Ievlev, R. Vasudevan, S. V. Kalinin, P. Maksymovych, and A. Morozovska.
DOI: 10.1093/oso/9780198862499.003.0012

The variety of the material and domain systems where conductive domain walls were reported led to several mechanisms being proposed to explain domain wall conductivity. Some of them include reduction of the electronic band gap due to alteration of the lattice structure at a domain wall (Seidel et al. 2009), attraction of intrinsic charge carriers by the bound charge of charged domain walls (Seidel et al. 2009), accumulation of charged defects attracted by the bound charge of a domain wall (Guyonnet et al. 2011), reduction of the Schottky barrier at the interface between an electrode and a charged domain wall (Vasudevan et al. 2017), and local band bending due to wall-orientation-dependent electrostriction and flexoelectric coupling (Maksymovych et al. 2012; Vasudevan et al. 2012).

Generally, in the discussion of domain wall electrical conduction, it is indirectly implied that the domain wall can be considered uniform at length scales much greater that the lattice parameter. In this case, it can be described by its orientation with respect to the polar axis, and a large-scale curvature can be used to account for a gradual change of the local orientation along the wall. However, high-resolution atomic-scale imaging of stable domain walls revealed atomic-scale variations of wall orientations at a length scale of a few lattice parameters due to the presence of pinning sites and lattice disorder (Paruch et al. 2005; Jia et al. 2008, 2011; Gonnissen et al. 2016). In other words, the walls can be rough and therefore consist of few-interatomic-distances-short charged segments.

In the first part of this chapter, we address the phenomenology of charged domain walls in the context of a Landau–Ginzburg–Devonshire (LGD) model for the ferroelectric semiconductor with analysis of the domain wall conductivity associated with the accumulation of charge carriers near domain walls. It is revealed that there exists an interplay between the wall type—head-to-head or tail-to-tail—and conduction type of the semiconductor ferroelectric with a strong dependence of the domain wall conductivity on the wall orientation. For example, with n-type doping, the head-to-head walls are surrounded by space charge layers with accumulated electrons, which leads to a more than tenfold increase in the static conductivity of an inclined head-to-head wall. In turn, tail-to-tail walls are depleted of the majority carriers—electrons—and are at least an order of magnitude less conducting due to the low mobility of holes. Furthermore, the electronic band structure at the domain wall can be modified by flexoelectric coupling between the strain field and polarization with modulation of the charge carrier densities and domain wall conductivity. The analysis also shows that small nanodomains appeared conducting across the entire domain cross section, in contrast to thick domain stripes, in which carrier accumulation takes place only at the walls. This implies that in an experiment, preexisting and uncharged domain walls can become curved and, consequently, locally conducting under a non-uniform electric field. This process assumes that the field is high enough for a substantial domain wall deformation, which is close to the switching threshold.

In view of the results of the LGD analysis when applied to rough, nominally uncharged domain walls, the rough walls consist of highly and poorly conducting segments alternating at the atomic scale. The DC conductance measurements described in, e.g. Chapters 6 and 15 require carrier transport along the whole domain wall in a ferroelectric. Apparently, conduction along the nominally uncharged rough domain walls will be strongly hindered by the poorly conducting segments. In particular, to explain the DC conduction of nominally uncharged walls in $Pb(Zr_{0.2}Ti_{0.8})O_3$ thin films, Guyonnet et al. (2011) proposed that the conduction is facilitated by a larger density of charge traps (oxygen vacancies) attracted by the electric field of the local bound charge due to the wall roughness. Besides, detection of the conductivity of domain walls, both charged and nominally uncharged, inevitably involves the problem of poor electrical contact between a probing electrode and the domain wall. Indeed, most if not all the DC conduction measurements of domain walls are rectifying, indicating the strong influence of the contacts. In the right context, this perceived limitation may offer an opportunity, because the effect of polarization dynamics on the electronic processes in the contact region will be exponentially amplified by the contact-limited transport. However, both a detailed understanding of these processes and analysis of the domain wall conductance are complicated in the contact-limited transport regime. Indeed, observation of the conduction along a path with conduction barriers required application of large enough voltages, which may be close to or exceed the switching threshold of the ferroelectric. Such a measurement becomes invasive, significantly disturbing the wall geometry and polarization configuration in the wall vicinity.

In the second part of the chapter, we review observations of high-frequency—in the gigahertz frequency range—AC conductivity along the nominally uncharged 180° domain walls in a uniaxial $Pb(Zr_{0.2}Ti_{0.8})O_3$ epitaxial film. Measurements of the conduction at high frequencies are insensitive to presence of a Schottky barrier and the electrode–ferroelectric interface. Moreover, the conductivity is registered with AC voltage amplitudes that are more than tenfold smaller that the voltages used in DC measurements of the domain wall conduction, which largely exclude the involvement of domain wall deformations in the electric field and modification of the surface barrier transparency by the electric field of the charged walls. Still, high-frequency conductivity of domain walls was reliably detected in $Pb(Zr_{0.2}Ti_{0.8})O_3$, as well as $BiFeO_3$ films (Tselev et al. 2016). In disordered materials, conductivity at gigahertz frequencies can be larger by orders of magnitude than at DC with a relatively weak power-law temperature dependence (Mott 1978; van Staveren et al. 1991; Jonscher 1999; Lunkenheimer and Loidl 2003; Lunkenheimer et al. 2003). A large class of such materials can be viewed as systems of highly conducting clusters embedded in a dielectric matrix and separated by relatively narrow dielectric bridges (van Staveren et al. 1991). The DC conduction in such a system is a result of the diffusive motion of carriers hopping between conducting sites separated by energy barriers. A fundamental reason behind the conductivity enhancement at high frequencies is that charge carriers localized by energy barriers at DC can contribute to AC conduction by oscillating between the barriers at high frequencies, so that no motion through the whole conduction part is required (Pollak and Geballe 1961; Mott 1978; van Staveren et al. 1991; Lunkenheimer et al. 2003). Apparently, the nominally uncharged rough domain walls fit well into this class of disordered systems.

12.2   Landau–Ginzburg–Devonshire Theory for Domain Wall Conduction in Ferroelectric Semiconductors and Their Thin Films

12.2.1   General Formalism

Without loss of generality, for the analysis here, we consider an infinite two-dimensional ferroelectric film of thickness h and write the Gibbs energy per surface area of the film. We begin by writing down the Gibbs potential for a ferroelectric, accounting for the polarization energy as well as contributions from the polarization gradient, electrostriction, flexoelectric effect, energies of elastic deformations, and electrostatic depolarization field. The Cartesian reference system is chosen so that the y-axis is normal to the film surface and directed into the film toward the substrate; the origin is in the film surface. We write down the free energy potential as the Euler–Lagrange equations of state for polarization components P_i, and elastic stresses σ_{ij} can be derived from the minimization of the Gibbs free energy potential. The Gibbs potential has the form

$$G_b = \int_0^h \left(G_{polar} + G_{grad} + G_{striction} + G_{elastic} + G_{flexoelectric} + G_{dep} \right) dx_2. \qquad (12.1)$$

Here, i, j, and k assume values from 1 to 3, and the axes of the reference system are numbered so that x_1, x_2, and x_3 are equivalent to x, y, and z, $G_{polar} = a_i P_i^2 + a_{ij} P_i^2 P_j^2 + a_{ijk} P_i^2 P_j^2 P_k^2$ is the Landau energy, $G_{grad} = \frac{g_{ijkl}}{2} \frac{\partial P_i}{\partial x_j} \frac{\partial P_k}{\partial x_l}$ is the gradient or Ginsburg energy, $G_{striction} = -Q_{ijkl} \sigma_{ij} P_k P_l$ is the electrostriction energy, $G_{elastic} = -\frac{s_{ijkl}}{2} \sigma_{ij} \sigma_{kl}$ is the elastic energy, $G_{flexoelectric} = \frac{F_{ijkl}}{2} \left(\sigma_{ij} \frac{\partial P_k}{\partial x_l} - P_k \frac{\partial \sigma_{ij}}{\partial x_l} \right)$ is the flexoelectric energy, and $G_{dep} = -P_i E_i^d / 2$ is the contribution of electrostatic depolarization field, E_i^d. Hereafter, a_i, a_{ij}, and a_{ijk} are the LGD-expansion coefficients of the 2nd-, 4th-, and 6th-order dielectric stiffness tensors, respectively, g_{ijkl} is the gradient coefficient tensor, g_{ijkl} is the electrostriction tensor, F_{ijkl} is the flexoelectric effect tensor, s_{ijkl} is the tensor of the elastic compliance, and σ_{ij} is the eleastic stress tensor. The components of depolarization electric field are $E_k = -\partial \varphi / \partial x_k$, where φ is the electric field potential. Coefficients $a_i = \alpha_T \left(T - T_C^* \right)$, where T is the absolute temperature and T_C^* is the Curie temperature, allow for the possible shift by misfit strain caused by a substrate (Pertsev et al. 1998). All other tensors included in the free energy Equation (12.1) are supposed to be temperature independent.

The flexoelectric term was included in the free energy potential because several works have described the impact of the flexoelectric coupling on the domain wall structure, energy, and electronic properties. In particular, it has been shown that the flexoelectric coupling induces a Bloch-type polarization component with a structure that is qualitatively different from the classical Bloch-wall structure in the tetragonal (Yudin et al. 2012) and rhombohedral (Eliseev et al. 2013) phases of BaTiO$_3$.

When the ferroelectric domain walls are studied in the vicinity of a surface, the surface part of the Gibbs potential and ambient electric energy should be added (Tagantsev and Gerra 2006b):

$$G_S = \int_S dx_1 dx_3 \left(\frac{a_{ij}^S}{2} P_i P_j + \sigma_S \varphi \right), \quad G_{ext} = - \int_{\vec{r} \notin V_{FE}} d^3 r \frac{\varepsilon_0 \varepsilon_e}{2} E_i E_i. \quad (12.2)$$

Below, we use an isotropic approximation for the tensor coefficients $a_{ij}^S = a_{S0} \delta_{ij}$ and introduce the electric field via the electrostatic potential φ as $E_i = -\partial \varphi / \partial x_i$. In Equation (12.2), σ_S is a surface screening charge density.

The Euler–Lagrange equations for the polarization components P_x, P_y, and P_z along with boundary conditions for them are derived from minimization of Equations (12.1) and (12.2): $\delta G_b / \delta P_k = 0$. They are listed in Balke et al. (2012), Eliseev et al. (2012), and Eliseev et al. (2013). The solution for the elastic strain u_{ij} and stress σ_{ij} comes from the equation $u_{ij} = -\delta G_b / \delta \sigma_{kl}$, the mechanical equilibrium condition $\partial \sigma_{ij} / \partial x_i = 0$, and elastic boundary conditions, as considered in Balke et al. (2012), Eliseev et al. (2012), and Eliseev et al. (2013).

The Euler–Lagrange equations couple the polarization components with the depolarization electric field. The corresponding electrostatic potential φ is determined from the Poisson equation:

$$\varepsilon_0 \varepsilon_b \frac{\partial^2 \varphi}{\partial x_i^2} = \frac{\partial P_i}{\partial x_i} - \rho, \quad (12.3)$$

where ε_b is the background permittivity (Tagantsev and Gerra 2006a, 2006b), and $\varepsilon_0 = 8.85 \times 10^{-12}$ F/m is the universal dielectric constant.

The space charge density of an n-doped wide-gap semiconductor is $\rho = e \left(N_d^+ - n \right)$, where $e = 1.6 \times 10^{-19}$ C is the electron charge, n is the electron density in the conduction band, and N_d^+ is the concentrations of the ionized shallow donors. If the donor level can be regarded as infinitely thin with an activation energy E_d, the concentration of the ionized donors is determined by the Fermi–Dirac distribution function $N_d^+ (\varphi) = N_{d0} (1 - f (E_d - E_F - e\varphi))$, where N_{d0} is the concentration of the donors, E_F is the Fermi energy level, and E_d is the donor level. The concentration of free electrons in a continuous conduction band is

$$n (\varphi, u_{ij}) = \int_0^\infty d\varepsilon \cdot g_n (\varepsilon) f \left(\varepsilon + E_C (u_{ij}) - E_F (u_{ij}) - e\varphi \right), \quad (12.4)$$

where $g_n (\varepsilon) \approx \sqrt{2 m_n^3 \varepsilon} / (2\pi^2 \hbar^3)$ is the electron density of states, $f(x) = (1 + \exp (x/k_B T))^{-1}$ is the Fermi–Dirac distribution function, $k_B = 1.3807 \times 10^{-23}$ J/K is the Boltzmann constant, T is the absolute temperature, $E_{F,C} = E_{F,C0} + \tilde{\Xi}_{ij}^{F,C} \delta \tilde{u}_{ij}$ are the strain-dependent Fermi level energy and the energy of the bottom of the conduction band, respectively, and $\tilde{\Xi}_{ij}^{F,C}$ are the corresponding tensors of the deformation potential. The values $E_{F,C0} = E_{F,C}\left(\tilde{u}_{ij}^{S}\right)$ already account for the spontaneous strain $\tilde{u}_{ij}^{S}$ existing far from the domain wall. Estimations for the derivative $\partial E_g/\partial u_{ij} \sim 20$ eV and for $|\Xi_{ij}^{C,V}| \sim (5-20)$ eV are available, e.g. for BiFeO$_3$. The concentration of holes and, respectively, acceptors can be considered in a similar way.

The conductivity enhancement in the domain wall is caused by the electric potential and elastic field variation inside the wall. The potential well/hump leads to a higher/lower electron concentration in the domain wall due to the local band bending. The concentration of holes is considered to be negligible, which leads to a purely n-type conductivity in Equation (12.4).

For a confined ferroelectric (e.g. for a film), Equation (12.3) should be supplemented by the condition of the potential continuity at the top surface of the film, $\left(\varphi^{(e)} - \varphi^{(i)}\right)\big|_{x_2=0} = 0$ (the superscripts (e) and (i) stand for "external" and "internal," respectively). Fixed potentials are imposed at the electrodes (Landau and Lifshitz 1963). The difference between the electric displacement components $D_n^{(i)} - D_n^{(e)}$ is conditioned by the surface screening produced by the ambient free charges at the film surface $\left(D_n^{(i)} - D_n^{(i)} + \sigma_S\right)\big|_{x_2=0} = 0$, where the "effective" surface charge density σ_S is introduced to model realistic conditions of incomplete screening of the spontaneous polarization at the film surface. The effective surface charge density depends on the film ambient (vacuum, air, inert or chemically active gases, liquids, semiconductor, or imperfect electrode cover). Several theoretical studies (Eliseev et al. 2016, 2018; Morozovska et al. 2018) used the linear dependence of the charge density σ_S on the electric potential excess at the surface of a ferroelectric, $\delta\varphi = \varphi^{(i)}\big|_{x_2=0} - \varphi^{(e)}\big|_{x_2\to-\infty}$:

$$\sigma_S[\varphi] \approx -\varepsilon_0 \frac{\delta\varphi}{\Lambda}, \tag{12.5}$$

where Λ is an "effective" screening length varying in a wide range from, e.g. 0.01 nm (Wang et al. 2011) to 1 nm. It is noteworthy that Equation (12.5) is an approximation as it includes an "effective" screening charge, while the real space charge is distributed in the ultrathin layer near the interface (Batra et al. 1973). The first justification of Equation (12.5) was proposed by Wurfel and Batra (1976), who showed that the space charge distribution in imperfect electrodes with nonzero screening length could be represented by a model with ideal conducting electrodes separated from the ferroelectric by a vacuum gap, and all bound and free charges are localized at the interfaces. Later, Stengel et al. (2009) and Stengel and Spaldin (2006) have shown that the concept of an effective screening length can be generalized for a given ferroelectric/electrode interface, and that the interfacial capacitance per unit area is proportional to ε_0/Λ. Introduction of the interfacial capacitance $C_{IF} = \varepsilon_0\varepsilon_{IF}S/\Lambda$ (in the parallel-plate capacitor approximation) allows Equation (12.5) to be justified in a simple way, because the product $C_{IF}\varphi|_{x_2=0}$ is the total value of the interfacial space charge, $q = \sigma_S S$, and therefore, $\sigma_S = \dfrac{C_{IF}\varphi|_{x_2=0}}{S} \approx -\varepsilon_0\dfrac{\varphi|_{x_2=0}}{\Lambda}$

in the case of the film thickness $h >> \Lambda$ and the interfacial dielectric permittivity $\varepsilon_{IF} \approx 1$. However, it is more likely that $\varepsilon_{IF} >> 1$, as expected for an ultrathin paraelectric passive layer at the film surface, and thus, Λ should be redefined as $\Lambda^* \cong \Lambda/\varepsilon_{IF}$.

12.2.2   Static Conductivity of Ferroelectric Domain Walls and Nanodomains in Uniaxial Ferroelectric Semiconductors

The accumulated/depleted charge density, n, in the bulk of the ferroelectric, ignoring quantum and mobility edge effects, can be estimated from measurements of the current and use of the equation

$$\mathbf{j} = \sigma \mathbf{E} \sim e\eta_e n \mathbf{E}, \tag{12.6}$$

where j is the measured current density, σ is the materials electrical conductivity, and η_e is the charge carrier mobility.

Using the LGD approach, Equations (12.1)–(12.6), the static conductivity of the charged ferroelectric domain walls has been calculated for uniaxial ferroelectric semiconductors (Eliseev et al. 2011). A sketch of the charged domain walls in a uniaxial ferroelectric semiconductor is shown in Figures 12.1(a–d). The domain walls have different inclination angles θ with respect to the vector of spontaneous polarization and accumulate a bound charge proportional to $2P\sin\theta$. In order to screen the bound charge, free space charge regions appear in the vicinity of the charged walls. In this case, the quantitative characteristics of these regions (width and carrier distribution) are very different for the tail-to-tail and head-to-head walls in p- or n-type ferroelectric semiconductors.

In particular, with n-type doping, the head-to-head walls are surrounded by a space charge layer with accumulated electrons and depleted donors of equal thicknesses of about 40–100 correlation radii, defined as $r_c \cong \sqrt{-g/\alpha}$, where g is the polarization gradient coefficient, and α is the coefficient at the P^2 term in the LGD free energy expression (see Figure 12.1(e)). Far below the Curie temperature, r_c is on the order of the lattice constant. Due to the electron accumulation, the static conductivity increases at the inclined head-to-head wall by one order of magnitude for small inclination angles $\theta \sim \pi/40$ and further by up to three orders of magnitude for the perpendicular domain wall ($\theta = \pi/2$). Consequently, the static conductivity significantly increases.

Unlike the head-to-head walls, tail-to-tail walls in n-type ferroelectric semiconductors are surrounded by a thin space charge layer with accumulated holes of thickness ~5–10 correlation lengths and a thick layer with accumulated donors of thickness ~100–200 correlation lengths (Figure 12.1(f)), as well as by an electron-depleted layer of thickness ~100–200 correlation lengths. Thus, the conductivity across the tail-to-tail wall is at least an order of magnitude smaller than that of the head-to-head wall due to the low mobility of the holes.

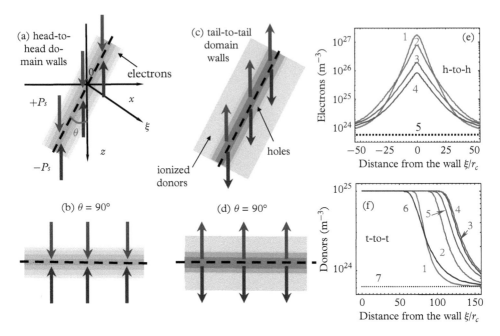

Figure 12.1 *(a–d) Sketch of charged walls in a uniaxial ferroelectric semiconductor of n-type.*
(a) Inclined head-to-head, (b) perpendicular head-to-head, (c) inclined tail-to-tail, (d) and perpendicular
tail-to-tail domain walls are shown. The gradient colors indicate the increased free carrier concentrations
in the domain wall vicinity (electrons in (a) and (b) and donors+holes in (b) and (d)). The inclination
angle of the domain wall is θ. *(e) Concentrations of electrons across the inclined "head-to-head" domain*
wall with different inclination angles: $\theta = \pi/2, \pi/6, \pi/20, \pi/40, 0$ *(curves 1–5, respectively).*
(f) Concentration of ionized donors across the inclined "tail-to-tail" domain walls with different
inclination angles: $\theta = \pi/2, \pi/4, \pi/10, \pi/20, \pi/27, \pi/40, 0$ *(curves 1–7, respectively). Material*
parameters used for the calculations correspond to LiNbO$_3$.

Source: Reprinted figures with permission from Eliseev et al. 2011. Copyright 2011 by the American Physical
Society. http://dx.doi.org/10.1103/PhysRevB.83.235313.

12.2.3 Static Conductivity and Structure of Ferroelectric Domain Walls in Multiaxial Ferroelectric Semiconductors

Experimental observations suggest that nominally uncharged, as-grown ferroelectric domain walls in ferroelectrics can be conductive, yet comprehensive theoretical models to explain this behavior have been lacking (Aravind et al. 2010; Balke et al. 2012; Eliseev et al. 2012; Maksymovych et al. 2012; Morozovska et al. 2012). The LGD approach (Equations (12.1)–(12.6)) allows studying the conductance of various domain structures in multiaxial ferroelectric semiconductors with special attention to the role of flexoelectric coupling, wall tilt, and curvature (see, e.g. Aravind et al. 2010; Balke et al.

2012; Eliseev et al. 2012; Maksymovych et al. 2012; Morozovska et al. 2012). Both tilted and straight (i.e. nominally uncharged) walls of stripe domains and cylindrical domains are of practical interest. The flexoelectric coupling, proximity, and domain size effect of the electron and donor accumulation/depletion by thin stripe domains and cylindrical nanodomains have been revealed (Balke et al. 2012; Eliseev et al. 2012; Morozovska et al. 2012). That is, small nanodomains of radius less than 5–10 correlation lengths appeared conducting across the entire cross section, in contrast to thick domain stripes, in which the carrier accumulation takes place at the walls only. Applications of such conductive nanosized channels may be promising for nanoelectronics.

It turned out that "positive" flexoelectric coupling charges the head-to-head walls of domain stripes by electrons at zero and very small tilt angles (compare the dashed, dotted, and solid curves in Figure 12.2(a)). In turn, "negative" flexoelectric coupling charges the head-to-head walls by ions at zero and very small tilt angles (compare the dashed, dotted, and solid curves in Figure 12.2(b)).

LGD theory was used to develop an analytical treatment of anisotropic carrier accumulation by domain walls in multiferroic $BiFeO_3$ that was revealed experimentally (Maksymovych et al. 2012; Vasudevan et al. 2012). In particular, ambient and ultra-high-vacuum SPM measurements demonstrated that the conductivity of the domain walls in $BiFeO_3$ can be modulated by up to 500% in the spatial dimension as a function of the domain wall curvature in the film plane. The observed strong angular dependence of the carrier accumulation by 180° domain walls originates from local band bending via angle-dependent electrostriction and flexoelectric coupling (Figures 12.3(a–c)). These theoretical results are in qualitative agreement with the experimental data and are consistent with recent first-principles calculations.

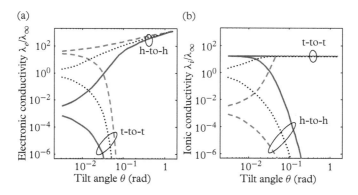

Figure 12.2 *Dependence of the relative (a) electronic and(b) ionic conductivity on the domain wall tilt (inclination) angle θ between the neighboring head-to-head (h-to-h) and tail-to-tail (t-to-t) domain stripes calculated for negative, zero, and positive flexoelectric coupling coefficients:* $F_{12} = (0.5, 0, 0.5) \times 10^{-10}$ m³/C *(solid, dotted, and dashed curves, respectively). Material parameters used for the calculations correspond to* $PbZr_{0.2}Ti_{0.8}O_3$.

Source: Reprinted figure with permission from Eliseev et al. 2012. Copyright 2012 by the American Physical Society. http://dx.doi.org/10.1103/PhysRevB.85.045312.

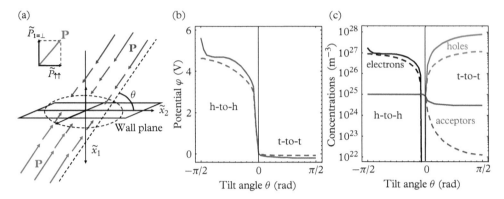

Figure 12.3 *(a) One-dimensional distribution of polarization in the vicinity of an isolated 180° domain wall in BiFeO₃. (b) Dependence of potential on the domain wall tilt angle θ. (c) Dependence of electron, hole, and acceptor concentrations (labeled near the curves) on the domain wall tilt angle θ. Solid curves correspond to the maximal values; dashed curves correspond to the values averaged over the wall width (10 nm). Material parameters used for the calculations correspond to BiFeO₃.*

Source: Reprinted with permission from Vasudevan et al. 2012. Copyright 2012 American Chemical Society.

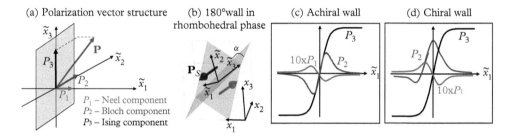

Figure 12.4 *(a) Polarization vector structure. (b) Rotated coordinate frame* $\{\tilde{x}_1, \tilde{x}_2, \tilde{x}_3\}$ *for the 180° nominally uncharged domain walls in the rhombohedral ferroelectric BTO; α is the wall rotation angle counted from crystallographic plane (101). The wall is green; the distance from the wall plane is* $\tilde{x}_1$. *(c) and (d) Schematics of the polarization component distributions inside the achiral and chiral domain walls, respectively.*

Source: Reprinted figures with permission from Eliseev et al. 2013. Copyright 2013 by the American Physical Society. http://dx.doi.org/10.1103/PhysRevB.87.054111.

The structure and electronic phenomena at the 180° domain wall in the rhombohedral phase of BaTiO₃ have been studied using the LGD approach (Yudin et al. 2012; Eliseev et al. 2013). Two types of the domain wall behavior were revealed as a function of the wall orientation. Both showed Néel components (see the polarization vector structure shown in Figures 12.4(a,b)). The low-energy "achiral" phase occurs in the vicinity of the (110) wall orientation and has an odd polarization profile invariant with respect to the inversion around the wall center (Figure 12.4(c)). The "chiral" phase appears around the (211) wall orientations and corresponds to the mixed parity domain walls (Figure 12.4(d)). The temperature-induced transformation between the "achiral" and "chiral" domain walls is abrupt and accompanied by a 20–30% change in the domain wall thickness. This process gives rise to significant changes of the wall electronic structure (Figures 12.5(a,b)). Depending on the temperature and flexoelectric coupling strength, the relative conductivity of the walls increases by at least one order of magnitude in comparison with the single-domain region (Figures 12.5(c,d)).

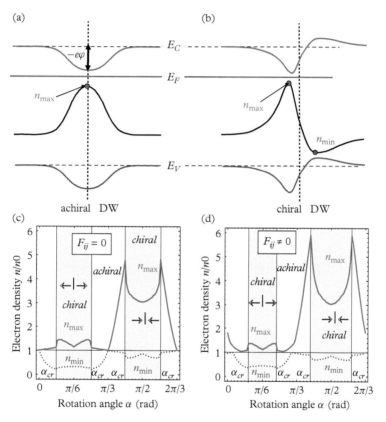

Figure 12.5 *Sketches of local band bending for (a) achiral and (b) chiral walls, where the spatial regions with maximal (n_{max}) and minimal (n_{min}) electron density are indicated. (c) Relative maximal n_{max}/n_0 and (d) minimal n_{min}/n_0 electron density versus the domain wall rotation angle α calculated in $BaTiO_3$ at 180 K (a) without flexoelectric coupling $F_{ij} = 0$ and (b) with flexoelectric coupling $F_{11} = 2.46 \times 10^{-11} m^3/C, F_{12} = 0.48 \times 10^{-11} \ m^3/C, and \ F_{44} = 0.05 \times 10^{-11} \ m^3/C.*

Source: Reprinted figures with permission from Eliseev et al. 2013. Copyright 2013 by the American Physical Society. http://dx.doi.org/10.1103/PhysRevB.87.054111.

12.3 Microwave Conductance of Ferroelectric Domain Walls

12.3.1 Microwave Impedance Microscopy of 180° Domain Walls in a Uniaxial Ferroelectric Film

As mentioned in the beginning of the chapter, high-frequency AC measurements are largely insensitive to contact resistance, thereby allowing one to probe the conductance of domain walls in the absence of applied DC electric fields, with small AC electric fields, and deep in the volume of the film, thus overcoming many of the limitations of DC current measurements. Therefore, AC current measurements arguably enable the most direct experimental verification among the theoretical methodologies outlined above of intrinsic electronic conductance of charged domain walls. However, the measurement and interpretation of AC signals, particularly on the nanoscale, requires careful attention and methodological developments, since several competing mechanism can give rise to signals similar to those expected of AC conductivity. Below we outline in detail the corresponding AC measurements and the logic behind the interpretation of the results from several representative ferroelectric thin films.

We start with the simplest equivalent electrical circuit model of the metal-semiconductor junction as shown in Figure 12.6(a) to describe its AC response. It consists of a parallel $G_b \| C_b$ pair representing the conductance and dielectric permittivity of the ferroelectric bulk and an analogous $G_s \| C_s$ pair representing the resistive Schottky barrier at the interface. Due to the small thickness of the barrier layer, $C_s >> C_b$. Since $G_b >> G_s$, the resistor for the Schottky barrier can be omitted to simplify the analysis. At frequencies f such that $2\pi f C_s >> G_b$, the capacitance of the Schottky barrier is effectively short-circuited, and the current through the sample is determined by the $G_s \| C_s$ of the bulk. The high-frequency measurements can be performed with an AC voltage of an amplitude significantly below the threshold for the domain wall displacement, thus ensuring the non-invasiveness of the probing.

Non-destructive microwave detection of ferroelectric domain walls at high frequencies >100 MHz has been demonstrated a number of times using a scanning microwave impedance microscope (sMIM) (Tselev et al. 2016; Wu et al. 2017). The microscope outputs calibrated values of the real $Re(Y)$ and imaginary $Im(Y)$ parts of the complex admittance $Y = G + I2\pi f C$ of the probe-sample system in two signal channels corresponding to the system conductance, sMIM-G, and the capacitance, sMIM-C, respectively. In our work (Tselev et al. 2016), the imaging was performed on a 100-nm-thick epitaxial thin film of $Pb(Zr_{0.2}Ti_{0.8})O_3$ (100) grown on a single-crystalline $SrTiO_3$ (001) substrate with a 50-nm-thick film of $SrRuO_3$ as a bottom electrode. Figures 12.6(b,c) show the topography and piezoresponse force microscopy (PFM) images of the film area where a box-in-box domain structure was created with the sMIM probe. Figures 12.6(d,e) respectively show sMIM-C and sMIM-G images of the same area. The PFM image in Figure 12.6(c) reveals the box-in-box domain structure as well as spontaneous domains in the film outside the structure. The contrast in the sMIM-C

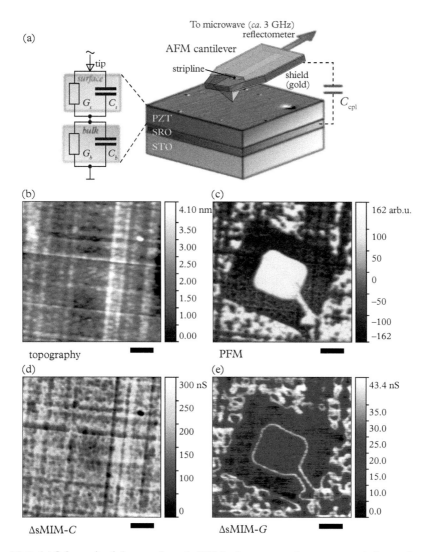

Figure 12.6 *(a) Schematic of the sample and sMIM microwave probe on an atomic force microscope (AFM) cantilever. An 100-nm-thick epitaxial thin film of Pb(Zr₀.₂Ti₀.₈)O₃ with a 50-nm-thick SrRuO₃ (SRO) bottom electrode is deposited on an SrTiO₃(001) (STO) substrate. (b)–(e) Images obtained from the same area of the Pb(Zr₀.₂Ti₀.₈)O₃ film: (b) topography, (c) combined out-of-plane piezoresponse, (d) sMIM-C, and (e) sMIM-G images of a box-in-box domain structure written by the sMIM probe. Spontaneous domains and associated domain walls visible outside the structure were erased inside the structure; all domain walls in the images are conducting. Scale bars in (b)–(e) are 1 μm. Source: Adapted from Tselev et al. 2016.*

image in Figure 12.6(d) is strongly dominated by the signal from the topographic features with a very weak signal from the big square domain and without any signal from the ferroelectric domain walls. In turn, the conductance, sMIM-G, image clearly shows the conducting domain walls both of the created box-in-box structure and of the spontaneous domains. The sMIM images in Figures 12.6(d,e) were taken with a zero DC bias at the probe and an AC voltage amplitude $\lesssim 300$ mV.

Domain structures of more complicated geometries were created as well that also showed microwave conductance of the domain walls. Furthermore, the domain wall contrast remained after a tenfold reduction of the AC voltage amplitude.

12.3.2 Microwave Signatures of Charged Domain Walls

Measurements of the sMIM response as a function of the DC bias at fixed locations on the $Pb(Zr_{0.2}Ti_{0.8})O_3$ film surface help to gain insight into the mechanism of microwave conductance of the domain walls in the film. As displayed in Figure 12.7(a), the DC bias sweeps between -8 V and $+8$ V show a response due to the dielectric tunability of $Pb(Zr_{0.2}Ti_{0.8})O_3$ in the sMIM-C channel with a curve shape typical for a ferroelectric. Kinks and jumps in the signal correspond to polarization switching under the probe. In turn, the sMIM-G channel (Figure 12.7(b)) reveals an abrupt increase in conductivity at the switching events. The signal at the peak maximum is larger than the contrast from the individual domain walls of stable large domains. The conductivity jumps can be identified as arising from the conductance of tilted and charged domain walls of ferroelectric nanodomains (Maksymovych et al. 2012). During domain growth, the AC conductance gradually decreases down to the bulk value, indicating that walls relax to the weakly charged state and eventually move out of the probed volume due to the growth of

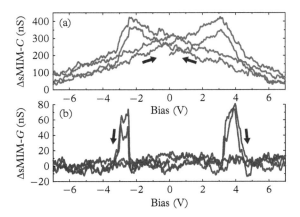

Figure 12.7 *Single-point signal from the (a) sMIM capacitance channel and (b) sMIM conductance channel as a function of the DC bias applied to the probe for the annealed film. Arrows indicate direction of the hysteresis. The signals are shown with respect to a reference signal approximately corresponding to the domain bulk. Adapted from Tselev et al. 2016.*

the domains. Note that microwave conduction displayed negligible dependence on the DC bias away from polarization switching events. This is in stark contrast to rectifying DC *I–V* curves acquired with the same film, which are strongly nonlinear and hysteretic, with current seen only at a positive bias above a threshold of ~1.3 V. This behavior clearly indicates the negligible influence of the contact effects dominating the *I–V* response in DC measurements. Furthermore, it is important that ac conduction in the $Pb(Zr_{0.2}Ti_{0.8})O_3$ film also showed negligible temperature dependence up to 115 °C (Tselev et al. 2016). This is unlike DC conduction of domain walls, which could be both linear (Crassous et al. 2015) and exponential (Maksymovych et al. 2012) versus temperature.

12.3.3 Modeling of Microwave Response

Microwave measurements, by virtue of a straightforward detection of tip-sample admittance variations, are amenable to quantitative calibration and further interpretation with numerical modeling. For example, the dielectric permittivity and conductivity of the $Pb(Zr_{0.2}Ti_{0.8})O_3$ film at 3 GHz were determined through calibrated measurements with the microscope supported by finite-element modeling of the tip-sample system. Details of the numerical model can be found in Tselev et al. (2016). With $\varepsilon_{PZT} = 70$, measurements of the film conductivity yielded values in the range 0.4–0.7 S/m. The AC conductivity of the domain walls at 3 GHz was estimated with the help of further measurements and numerical simulations of the tip-sample admittance with the probe in contact with a domain wall. Assuming a 3 nm wall thickness, the domain wall AC conductivity fell in the range 4–8 S/m, i.e. about 10 times above the 3 GHz conductivity of the domain bulk.

Figure 12.8 displays the results of numerical simulations of the tip-sample admittance change for the probe in contact with a domain wall in respect to the domain bulk (Figure 12.8(a)) as a function of the frequency for several values of the domain wall conductivity σ_{wall}. For the calculations, the conductivity of the domain bulk σ_{bulk} was set to 0.6 S/m, and the poorly conducting Schottky barrier was modeled by an insulating layer of 1.5 nm thickness with a permittivity of $\varepsilon_{PZT} = 70$. The plots represent the expected contrast in the sMIM-G and sMIM-C images depending on the probing frequency for the studied $Pb(Zr_{0.2}Ti_{0.8})O_3$ film. At the measurement frequency of 3 GHz (denoted by the dashed line), the response in the sMIM-C channel ($\delta Im(Y)$ is about 15 times smaller than in the sMIM-G channel ($\delta Re(Y)$), and hence, the expected signal from the domain wall in the sMIM-C channel is significantly below the noise floor of the microscope. This explains the absence of contrast from the domain walls in the sMIM-C image in Figure 12.6(d). With increasing σ_{wall}, the contrast grows in both the channels at a fixed imaging frequency. With increasing frequency, the response in sMIM-G saturates, and the response in sMIM-C drops to zero. The contrast due to the wall conductivity is expected in both the channels in a frequency range below 3 GHz for the considered combinations of σ_{bulk} and σ_{wall}. Experimentally, such a behavior was observed in Tselev et al. (2007) during broadband microwave microscopy of Ga-doped films of $(Ba_{0.6}Sr_{0.4})TiO_3$. A contrast in the capacitance channel from the conductivity of

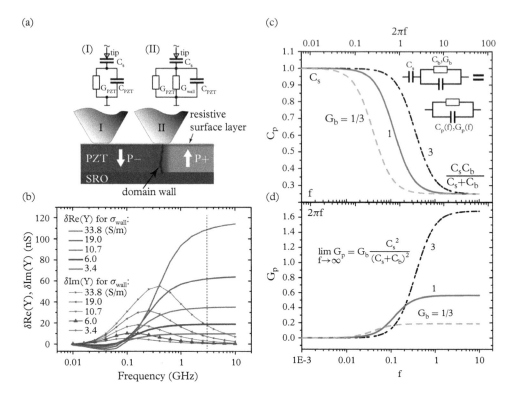

Figure 12.8 *(a) Lumped-element high-frequency equivalent circuits corresponding to the position (I) and (II) of the probe over the film (bottom). C_s is the capacitance of the resistive surface layer, C_{PZT} and G_{PZT} are the capacitance and conductance of the ferroelectric film, respectively, and G_{wall} is the conductance of the domain wall. (b) Numerically simulated variations of the real, $\delta Re(Y)$, and imaginary, $\delta Im(Y)$, parts of the tip-sample system admittance versus frequency for different values of the domain wall conductivity, σ_{wall}, shown in the legend. Curves for $\delta Im(Y)$ are with symbols. Curves for $\delta Re(Y)$ and $\delta Im(Y)$ of the same color correspond to the same value of σ_{wall}. (c) and (d) Calculated lumped element values of parallel, frequency-dependent, C_p and the G_p, respectively, as functions of frequency, f, for the circuit with frequency-independent C_s, C_b, and R_b shown in the inset. The curves are presented in a dimensionless coordinate system: The capacitance and conductance are measured in units of C_s, and the unit of the frequency f is $(2\pi)^{-1}$. Each panel contains three curves corresponding to three values of the conductance G_s: 1/3 (dashed, green), 1 (solid, red), and 3 (dot-dashed, black).*

Source: (a) Adapted from Tselev et al. 2016. (b,d) Adapted from Tselev et al. 2007. (c) Reprinted from Tselev et al. 2007, with the permission of AIP Publishing.

a leaky dielectric film was seen at lower imaging frequencies and completely disappeared at high frequencies. The saturation of the response above a certain frequency in the G channel is noteworthy for potential applications of the AC conductivity of the domain walls because it substantially alleviates the requirements governing the choice of the detection frequency in a reading device.

The nature of the frequency-dependent contrast behavior can be explained using the model equivalent circuit shown in Figures 12.6(a) and 12.8(a). The C_s-$G_b \| C_b$ circuit can be represented by a $G_p \| C_p$ circuit with frequency-dependent elements as shown in the inset of Figure 12.8(c). Figures 12.8(c,d) display generic plots of the capacitance $C_p(f)$ and conductance $G_p(f)$ of the C_s-$G_b \| C_b$ circuit for three different values of the conductance G_b. The image contrast due to a conductivity change is the difference between two of the curves in a plot. Both $C_p(f)$ and $G_p(f)$ are step-like functions of frequency, and the positions of the steps against frequency are determined by the so-called pole frequency f_{pole} of the circuit, which is defined by the equation $2\pi f_{pole}(C_s + C_b)/G_b = 1$. The height of the step in $C_p(f)$ is determined only by the capacitors of the circuit. In turn, the height of the step in $G_p(f)$, a step up from zero versus frequency, is determined by all three elements, and it is directly proportional to G_b. From the plots, it is clear that with uniform permittivity of a sample and at a frequency sufficiently above the pole frequency, the contrast from the conductivity will be observed only in the G channel of the microscope. The contrast in the C channel will appear only in the frequency range approximately between the pole frequencies corresponding to the range of the conductance variations in the imaged area of a sample.

It is of interest to compare the microwave images of the conductive wall in the $Pb(Zr_{0.2}Ti_{0.8})O_3$ film with other experiments on the AC conductivity of the ferroelectric domain walls using scanning probes. Schaab et al. (2018) reported SPM imaging of conducting neutral domain walls in bulk samples of a hexagonal improper ferroelectric $ErMnO_3$. The imaging experiments were performed in the frequency range 0.02 MHz to 20 MHz. In the employed measurement scheme, the imaging contrast appeared due to the rectification properties of the Schottky barrier with a signal registered as a rectification DC voltage generated at the tip-sample junction. The signal behavior was explained with the same equivalent circuit that was discussed above, and therefore, the results for $ErMnO_3$ can be directly compared with those for the $Pb(Zr_{0.2}Ti_{0.8})O_3$ film. In the experiments with $ErMnO_3$, the DC signal from the probe disappeared above a certain cut-off frequency that depends on the local sample conductivity. By the definition used in Schaab et al. (2018), the cut-off frequency is higher than the pole frequency of the equivalent circuit defined above, and it is close to the high-frequency side of the G_p step in Figure 12.8(d), above which the Schottky barrier is effectively short-circuited and the generation of the rectification voltage does not occur. The cut-off frequencies for the domain bulk followed the bulk sample conductivity at DC, and conducting domain walls were observed due to higher cut-off frequencies compared to those of the domains. For a sample with a bulk conductivity of 0.19 S/m, the cut-off frequency in images for domains was 3.58 MHz. Using this result, the cut-off frequencies for the domain and domain walls of the $Pb(Zr_{0.2}Ti_{0.8})O_3$ films ($\sigma_{bulk} = 0.6$ S/m, $\sigma_{wall} = 6$ S/m) can be estimated to be about 11 MHz and 110 MHz, respectively. The position of the peak in the $\delta Re(Y)$ plot for $\sigma_{wall} = 6$ S/m in Figure 12.8(b) is at about 100–200 MHz, which is in reasonable agreement with this estimate.

12.3.4   Interpretation of the Origin of Microwave Conductance

From the behavior of the AC domain wall conductance as a function of the bias voltage, temperature, and time, it was concluded that the cloud of free carriers forming around the transient charged domain walls during polarization switching incompletely dissipates after the domain stabilization and partially remains at the wall due to the rough wall morphology promoted by the disorder potential in the film. Such a configuration reduces DC electrical conduction but does not hinder ac conduction. As was pointed in the chapter introduction, in disordered materials, the conductivity at gigahertz frequencies can be larger by orders of magnitude than at DC with a relatively weak power-law temperature dependence since charge carriers localized by energy barriers at DC can contribute to AC conduction by oscillating between the barriers at high frequencies (Pollak and Geballe 1961; Mott 1978; van Staveren et al. 1991; Jonscher 1999; Lunkenheimer and Loidl 2003; Lunkenheimer et al. 2003). Supporting this interpretation, the DC conductivity of the domain bulk measured by means of c-AFM was found to be 100–200 times smaller than the 3 GHz AC conductivity. The same ratio for the domain wall yielded a DC conductivity ~50 times lower than the AC value. The undetectable temperature dependence of the AC conductivity corroborates this picture as well.

Consideration of ac conduction of the ferroelectric domain walls would be incomplete without mentioning a fundamentally different mechanism that can lead to a domain wall contrast in sMIM images. Generally, looking only at the conduction channel of the microwave microscope at a single frequency, it is not possible to distinguish between the two mechanisms. In contrast to the conduction caused by mobile charge carriers and lattice defects as in the $Pb(Zr_{0.2}Ti_{0.8})O_3$ film, the second mechanism stems from bound charge motion and is intrinsic in nature.

Observations of a wall contrast were reported in Wu et al. (2017) in sMIM experiments with neutral domain walls of relatively weakly conducting, single-crystalline hexagonal manganites. The observed contrast was explained by acoustic-like oscillatory sliding of the domain walls, which show a resonance in the gigahertz frequency range. Later, Prosandeev et al. (2018) proposed an alternative mechanism associated with low-frequency soft modes in the crystal lattice at a domain wall, although the DFT modeling by them was performed for another ferroelectric, $BiFeO_3$ (Prosandeev et al. 2018). In both the proposed mechanisms, the forced harmonic wall vibrations result in polarization oscillations at the domain wall. Generally, since the polarization, P, follows the applied field (with some phase difference), experimentally, this is effectively seen as an alteration of the local material polarization under the probe and, hence, as an altered polarizability $\chi = dP/\varepsilon_0 dE$ and a locally altered displacement current $j = dP/dt = P_0 i\omega \exp(i\omega t)$ at the domain wall with $\omega = 2\pi f$ (Prosandeev et al. 2018). This current is sensed by the microscope in the capacitance channel, resulting in the domain wall contrast in the sMIM-C images. The energy loss from the atomic vibrations into the lattice at the domain walls produces a peak in $Re(Y)$ against frequency at the frequency corresponding to the most efficient coupling of the microwave field into the vibrations, that is, at the resonant frequency of the vibrations. This energy dissipation results in the domain wall contrast in the sMIM-G channel.

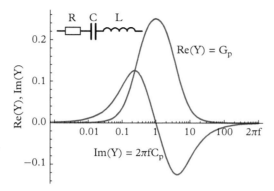

Figure 12.9 *Frequency behavior of the real, Re(Y), and imaginary, Im(Y), parts of the admittance for the in-series LCR circuit shown in the inset. The admittance, Y, and frequency, f, are dimensionless, $L = C = 1, R = 4$.*

This behavior can be captured by an in-series RCL circuit as illustrated in Figure 12.9. The changes in the polarization are represented by the charge on the capacitor C of the circuit. At frequencies well below the resonance, the polarization follows the applied voltage in phase but with a small amplitude. The polarizability change due to the vibrations is positive. With increasing frequency approaching the resonance, the oscillation amplitude grows, but the polarization increasingly lags the applied bias in phase. The change in the polarizability seen by the microscope depends both on the amplitude $P(\omega)$ and the phase lag ϕ as $\delta\chi \sim \text{Re}\,[P(\omega)\exp(i\phi)]$. At the resonance, while the oscillation amplitude is at a maximum, the vibrational distortions are in quadrature with the applied field, and the resulting apparent polarizability is equal to that of the domain bulk. Above the resonance, the phase lag is greater than 90°, which means the polarizability change is negative. Accordingly, the Re(Y) of the model RCL circuit peaks at the resonance frequency, while Im(Y) shows a maximum with increasing frequency, then drops to negative values and after a minimum, asymptotically returns to zero (Figure 12.9). It is clear that the increase in Re(Y) can be observed only in a relatively narrow frequency range near the resonance frequency of the vibrations, and except in the very vicinity of the resonance frequency, simultaneously there should be seen a nonzero response in Im(Y). In Wu et al. (2017), such behavior was observed during broadband multi-frequency imaging of the domain walls on the (001) surface of single-crystalline hexagonal rear-earth manganites $YMnO_3$ and $ErMnO_3$. The trend in the Re(Y) and Im(Y) signals versus frequency followed that depicted by the model in Figure 12.9 below the resonance frequency. While the domain bulk conductivity at DC in $YMnO_3$ was 3×10^{-4} S/m, the apparent conductivity of the domain walls deduced from the calibrated signals of the microscope reached a very large value of about 500 S/m at a frequency of 4.5 GHz, that is, nearly two orders of magnitudes above the AC conductivity of the domain walls in the $Pb(Zr_{0.2}Ti_{0.8})O_3$ film. Importantly, in $YMnO_3$, the charge-neutral domain walls are known to be more resistive than the domain bulk at DC (Choi et al. 2010).

In principle, both the extrinsic and intrinsic mechanisms of the domain wall response can coexist. However, apparently, the conditions for the operation of the two mechanisms are mutually exclusive, and conditions favoring one mechanism weaken the response due to the other. For example, lattice defects reduce the intensity of the bound change response to the high-frequency electric field, while fewer mobile change carriers exist in a perfect lattice. Therefore, the observation of the two mechanisms operating simultaneously in one domain wall with the current equipment is rather improbable.

Overall, conducting domain walls have progressed from theoretical concepts to numerous experimental demonstrations and even practical device embodiments in less than a decade. Future progress requires a better understanding of the phenomenology of charged domain walls, down to atomic scales, and particularly interaction with defects. Scalable and more accurate theoretical methodologies are urgently required in this regard. At the same time, new approaches to inject and sustain charged domain walls, possibly involving disorder fields, should be pursued. Also, high-frequency conductance measurements provide a new window into domain wall conductance, together with low-frequency currents, thereby allowing new device implementations and deeper insight into the properties of domain walls in ferroelectric and related materials.

....

REFERENCES

Aravind V. R., Morozovska A. N., Bhattacharyya S., Lee D., Jesse S., Grinberg I., Li Y. L., Choudhury S., Wu P., Seal K., Rappe A. M., Svechnikov S. V., Eliseev E. A., Phillpot S. R., Chen L. Q., Gopalan V., Kalinin S. V., "Correlated polarization switching in the proximity of a 180° domain wall," *Phys. Rev. B* 82, 024111 (2010).

Balke N., Winchester B., Ren W., Chu Y. H., Morozovska A. N., Eliseev E. A., Huijben M., Vasudevan R. K., Maksymovych P., Britson J., Jesse S., Kornev I., Ramesh R., Bellaiche L., Chen L. Q., Kalinin S. V., "Enhanced electric conductivity at ferroelectric vortex cores in BiFeO$_3$," *Nat. Phys.* 8, 81–88 (2012).

Batra I. P., Wurfel P., Silverma B. D., "New type of first-order phase-transition in ferroelectric thin-films," *Phys. Rev. Lett.* 30, 384–387 (1973).

Choi T., Horibe Y., Yi H. T., Choi Y. J., Wu W., Cheong S. W., "Insulating interlocked ferroelectric and structural antiphase domain walls in multiferroic YMnO$_3$," *Nat. Mater.* 9, 253 (2010).

Crassous A., Sluka T., Tagantsev A. K., Setter N., "Polarization charge as a reconfigurable quasi-dopant in ferroelectric thin films," *Nat. Nanotechnol.* 10, 614–618 (2015).

Eliseev E. A., Fomichov Y. M., Kalinin S. V., Vysochanskii Y. M., Maksymovich P., Morozovska A. N., "Labyrinthine domains in ferroelectric nanoparticles: manifestation of a gradient-induced morphological transition," *Phys. Rev. B* 98, 054101 (2018).

Eliseev E. A., Morozovska A. N., Svechnikov G. S., Gopalan V., Shur V. Y., "Static conductivity of charged domain walls in uniaxial ferroelectric semiconductors," *Phys. Rev. B* 83, 235313 (2011).

Eliseev E. A., Morozovska A. N., Svechnikov G. S., Maksymovych P., Kalinin S. V., "Domain wall conduction in multiaxial ferroelectrics," *Phys. Rev. B* 85, 045312 (2012).

Eliseev E. A., Semchenko A. V., Fomichov Y. M., Glinchuk M. D., Sidsky V. V., Kolos V. V., Pleskachevsky Y. M., Silibin M. V., Morozovsky N. V., Morozovska A. N., "Surface and finite size effects impact on the phase diagrams, polar, and dielectric properties of $(Sr,Bi)Ta_2O_9$ ferroelectric nanoparticles," *J. Appl. Phys.* 119, 204104 (2016).

Eliseev E. A., Yudin P. V., Kalinin S. V., Setter N., Tagantsev A. K., Morozovska A. N., "Structural phase transitions and electronic phenomena at 180-degree domain walls in rhombohedral $BaTiO_3$," *Phys. Rev. B* 87, 054111 (2013).

Gonnissen J., Batuk D., Nataf G. F., Jones L., Abakumov A. M., Van Aert S., Schryvers D., Salje E. K. H., "Direct observation of ferroelectric domain walls in $LiNbO_3$: wall-meanders, kinks, and local electric charges," *Adv. Funct. Mater.* 26, 7599–7604 (2016).

Guyonnet J., Gaponenko I., Gariglio S., Paruch P., "Conduction at domain walls in insulating $Pb(Zr_{0.2}Ti_{0.8})O_3$ thin films," *Adv. Mater.* 23, 5377–5382 (2011).

Jia C.-L., Mi S.-B., Urban K., Vrejoiu I., Alexe M., Hesse D., "Atomic-scale study of electric dipoles near charged and uncharged domain walls in ferroelectric films," *Nat. Mater.* 7, 57 (2008).

Jia C.-L., Urban K. W., Alexe M., Hesse D., Vrejoiu I., "Direct observation of continuous electric dipole rotation in flux-closure domains in ferroelectric $Pb(Zr,Ti)O_3$," *Science* 331, 1420–1423 (2011).

Jonscher A. K., "Dielectric relaxation in solids," *J. Phys. D: Appl. Phys.* 32, R57–R70 (1999).

Landau L. D., Lifshitz E. M., *Theoretical Physics, Electrodynamics of Continuous Media*, Volume VIII (Pergamon, Oxford, 1963).

Lunkenheimer P., Loidl A., "Response of disordered matter to electromagnetic fields," *Phys. Rev. Lett.* 91, 207601 (2003).

Lunkenheimer P., Rudolf T., Hemberger J., Pimenov A., Tachos S., Lichtenberg F., Loidl A., "Dielectric properties and dynamical conductivity of $LaTiO_3$: from dc to optical frequencies," *Phys. Rev. B* 68, 245108 (2003).

Maksymovych P., Morozovska A. N., Yu P., Eliseev E. A., Ch, Y. H., Rames, R., Baddor, A. P., Kalinin S. V., "Tunable metallic conductance in ferroelectric nanodomains," *Nano Lett.* 12, 209–213 (2012).

Morozovska A. N., Fomichov Y. M., Maksymovych P., Vysochanskii Y. M., Eliseev E. A., "Analytical description of domain morphology and phase diagrams of ferroelectric nanoparticles," *Acta Mater.* 160, 109–120 (2018).

Morozovska A. N., Vasudevan R. K., Maksymovych P., Kalinin S. V., Eliseev E. A., "Anisotropic conductivity of uncharged domain walls in $BiFeO_3$," *Phys. Rev. B* 86 (2012).

Mott N., "Electrons in glass," *Rev. Mod. Phys.* 50, 203–208 (1978).

Paruch P., Giamarchi T., Triscone J. M., "Domain wall roughness in epitaxial ferroelectric $PbZr_{0.2}Ti_{0.8}O_3$ thin films," *Phys. Rev. Lett.* 94, 197601 (2005).

Pertsev N. A., Zembilgotov A. G., Tagantsev A. K., "Effect of mechanical boundary conditions on phase diagrams of epitaxial ferroelectric thin films," *Phys. Rev. Lett.* 80, 1988–1991 (1998).

Pollak M., Geballe T. H., "Low-frequency conductivity due to hopping processes in silicon," *Phys. Rev.* 122, 1742–1753 (1961).

Prosandeev S., Yang Y., Paillard C., Bellaiche L., "Displacement current in domain walls of bismuth ferrite," *Npj Comput. Mater.* 4, 8 (2018).

Schaab J., Skjærvø S. H., Krohns S., Dai X., Holtz M. E., Cano A., Lilienblum M., Yan Z., Bourret E., Muller D. A., Fiebig M., Selbach, S. M., Meier D., "Electrical half-wave rectification at ferroelectric domain walls," *Nat. Nanotechnol.* 13, 1028–1034 (2018).

Seidel J., Martin L. W., He Q., Zhan Q., Chu Y. H., Rother A., Hawkridge M. E., Maksymovych P., Yu P., Gajek M., Balke, N., Kalinin S. V., Gemming S., Wang F., Catalan G., Scott J. F., Spaldin N. A., Orenstein J., Ramesh R., "Conduction at domain walls in oxide multiferroics," *Nat. Mater.* 8, 229 (2009).

Stengel M., Spaldin N. A., "Origin of the dielectric dead layer in nanoscale capacitors," *Nature* 443, 679–682 (2006).

Stengel M., Vanderbilt D., Spaldin N. A., "First-principles modeling of ferroelectric capacitors via constrained displacement field calculations," *Phys. Rev. B* 80 (2009).

Tagantsev A. K., Gerra G., "Interface-induced phenomena in polarization response of ferroelectric thin films," *J. Appl. Phys.* 100, 051607 (2006a).

Tagantsev A. K., Gerra G., "Interface-induced phenomena in polarization response of ferroelectric thin films," *J. Appl. Phys.* 100 (2006b).

Tselev A., Anlage S. M., Ma Z., Melngailis J., "Broadband dielectric microwave microscopy on micron length scales," *Rev. Sci. Instrum.* 78, 044701 (2007).

Tselev A., Yu P., Cao Y., Dedon L. R., Martin L. W., Kalinin S. V., Maksymovych P., "Microwave A.C. conductivity of domain walls in ferroelectric thin films," *Nat. Commun.* 7 (2016).

Van Staveren M. P. J., Brom H. B., De Jongh L. J., "Metal-cluster compounds and universal features of the hopping conductivity of solids," *Phys. Rep.: Rev. Sect. Phys. Lett.* 208, 1–96 (1991).

Vasudevan R. K., Cao Y., Laanait N., Ievlev A., Li L., Yang J.-C., Chu Y.-H., Chen L.-Q., Kalinin S. V., Maksymovych P., "Field enhancement of electronic conductance at ferroelectric domain walls," *Nat. Commun.* 8, 1318 (2017).

Vasudevan R. K., Morozovska A. N., Eliseev E. A., Britson J., Yang J. C., Chu Y. H., Maksymovych P., Chen L. Q., Nagarajan V., Kalinin S. V., "Domain wall geometry controls conduction in ferroelectrics," *Nano Lett.* 12, 5524–5531 (2012).

Vul B. M., Guro, G. M., Ivanchik I. I., "Encountering domains in ferroelectrics," *Ferroelectrics* 6, 29–31 (1973).

Wang J., Tagantsev A. K., Setter N., "Size effect in ferroelectrics: competition between geometrical and crystalline symmetries," *Phys. Rev. B* 83, 014104 (2011).

Wu X. Y., Petralanda U., Zheng L., Ren Y., Hu R. W., Cheong S. W., Artyukhin S., Lai K. J., "Low-energy structural dynamics of ferroelectric domain walls in hexagonal rare-earth manganites," *Sci. Adv.* 3, e1602371 (2017).

Wurfel P., Batra I. P., "Depolarization effects in thin ferroelectric-films," *Ferroelectrics* 12, 55–61 (1976).

Yudin P. V., Tagantsev A. K., Eliseev E. A., Morozovska A. N., Setter N., "Bichiral structure of ferroelectric domain walls driven by flexoelectricity," *Phys. Rev. B* 86, 134102 (2012).

13

Control of Ferroelectric Domain Wall Motion using Electrodes with Limited Conductivity

P. V. Yudin[1,2] and L. J. McGilly[3]

[1] Institute of Physics, Academy of Sciences of the Czech Republic, Na Slovance 2, 18221 Praha 8, Czech Republic
[2] Kutateladze institute of Thermophysics, Siberian Branch of Russian Academy of Science, Lavrent'eva av. 1, Novosibirsk, Russia
[3] Department of Physics, Columbia University, New York, New York 10027, United States

13.1 Introduction

The problem of a moving domain boundary, addressed in this chapter, belongs to the general class of free-boundary problems, or Stefan problems, after Josef Stefan, who mathematically described ice formation at the end of the nineteenth century (Stefan 1875) and then demonstrated the generality of his approach by applying the same technique to describe diffusion (Stefan 1889). In the framework of this approach, the position of the boundary is determined from the transport of a physical quantity (e.g. heat) flowing through and partially consumed at the boundary (latent heat) (Gupta 2003). Currently mathematical modeling of Stefan problems has developed into a rich field of knowledge where both analytical and numerical methods are applied to solve various important applied tasks. Areas where the Stefan model was useful range from classical applications, like laser materials processing (Kasuya and Shimoda 1997), to finance (Peskir 2005) and biology (Lin 2007). Here, in contrast to Chapters 1 to 4, where the internal structure of a domain wall (DW) was of importance, we will be interested in the DW position and its evolution in time. Correspondingly a DW will be treated as an infinitely thin object in the framework of a Stefan problem for the transport of electric charge.

From an applications point of view, DWs in ferroelectric materials have been touted as possible nanoscale elements in future electronic devices. There are numerous reasons why this is a promising idea; DWs are intrinsically small, typically only a few nanometers wide, they exhibit distinct functional properties, such as enhanced conduction and

P. V. Yudin and L. J. McGilly, *Control of Ferroelectric Domain Wall Motion using Electrodes with Limited Conductivity* In: *Domain Walls: From Fundamental Properties to Nanotechnology Concepts*. Edited by: Dennis Meier, Jan Seidel, Marty Gregg, and Ramamoorthy Ramesh, Oxford University Press (2020). © P. V. Yudin and L. J. McGilly.
DOI: 10.1093/oso/9780198862499.003.0013

perhaps most importantly, they are not fixed features: they can be created, displaced, and removed.

Major milestones on the way toward realization of nanoelectronic devices are the abilities to (1) control nucleation of DWs: it is important that DWs appear in a predefined position in a reproducible manner; (2) displace DWs to a given position, which can be used, e.g. for ferroelectric memory; and (3) control the speed of the DW, as the speed of operation of a device is determined largely by the velocity of the propagating DW. Therefore, knowledge of DW motion is essential to progress further toward realization of any domain-wall-based device. Useful insight here can be gained from inspection of the ferromagnetic counterpart to ferroelectric DWs. Such ferromagnetic DWs are generally larger; commonly their width is 10 times that of 180° ferroelectric DWs, and they do not offer the enhanced conduction relative to the surrounding domains that has caused such a stir in ferroelectrics. However, they have been studied for much longer and as such are well developed and understood. They also have the benefit of being able to propagate through several means, either by direct application of external magnetic fields or through spin-polarized currents and can be observed with optical methods such as the magneto-optical Kerr Effect. The latter approach has led to the development of the ferromagnetic "racetrack" and similar concepts wherein a number of walls can be propagated along narrow channels. In such devices, high wall speeds of 10^2 to 10^3 ms^{-1} are routinely achieved (Lewis et al. 2010; Yang et al. 2010), which allows for fast nanosecond operation. However, ferroelectric DWs are generally considerably slower (Meng et al. 2015) and rarely, if ever, reach comparable speeds (Li et al. 2004; Grigoriev et al. 2006).

In this chapter, we address the experimental control of ferroelectric DW motion in thin films using electron-beam-induced deposition (EBID) electrodes with limited conductivity which governs the supply of charges required for DW nucleation and propagation. The process will be described by analogy with the classical Stefan model (Gupta 2003) historically applied to the motion of phase boundaries under propagation of heat but is here applied to precisely describe DW motion under linear electrodes and the 2D growth of a circular domain.

13.2 Control of DW Nucleation Sites

In the standard picture of switching described in Figure 13.1, the process begins with "nucleation," where a small reverse domain is created at the ferroelectric–electrode interface that then propagates in a thin high-aspect-ratio "needle" through the ferroelectric along the electric field direction in what is called the "forward growth" phase. Upon reaching the opposite electrode, the next phase called "lateral motion" begins where the needle starts to radially expand through sideward motion of the DWs and completes switching through the coalescence of domains. Switching can be said to be "nucleation limited" if it is restricted by the rate of domain creation rather than by the speed of the lateral propagation of the walls.

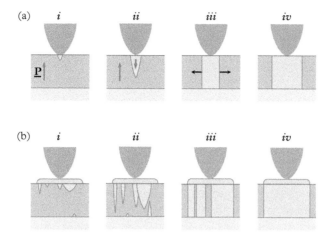

Figure 13.1 *Schematic illustration of different switching mechanisms. (a) Switching progresses underneath a biased probe tip. In the first step (i) a small reverse domain forms under the tip apex at the high electric field region, which grows forward into a high-aspect needle and (ii) then expands laterally (iii) to completely switch an area through domain wall (DW) motion (iv). When a probe is used to connect to an electrode as in (b) multiple nuclei can form at the interfaces (i) which progress in the forward growth mode (ii) coalescing during lateral growth (iii) to complete switching under the entire electroded area (iv).*

The control of domain nucleation and subsequent DW propagation is an essential step in order to harness their full potential for realization in novel devices such as reconfigurable electronics where the DWs are the rewritable current-carrying interconnects in circuits. To date, however, it has been surprisingly difficult to control, in a precise manner, the nucleation sites of domains and thus the associated walls in a solid-state-type device.

Scanning probe techniques such as piezoresponse force microscopy (PFM) are uniquely placed to define nucleation sites; however, they are limited by the highly inhomogeneous and local electric field such that DW propagation is usually over short length scales (Tybell et al. 2001; Rodriguez et al. 2005) as illustrated in Figure 13.1(a). To propagate DWs over any significant distance would require movement of the probe itself, which is generally impractical for nanoelectronic devices. Large-scale microscopic top electrodes allow for propagation of DWs over relatively large distances but at the expense of nucleation control (Gruverman et al. 2005, 2008) as is shown schematically in Figure 13.1(b). Usually, multiple DWs will be nucleated with a single voltage pulse, and their positions are dictated by structural inhomogeneities such as defects which are usually randomly located in the sample.

One example of an attempt to address this issue comes from Whyte et al. (2014a), wherein controlled nucleation sites resulted from predefined structural defects fabricated through focused ion beam milling of thin single-crystal sheets of $KTiOPO_4$, which essentially led to an engineered electric field hotspot that promoted local polarization reversal. The local electric field hotspots can also be engineered through triangular "notches" in the electrodes that successfully determine nucleation points for DWs (Whyte et al. 2014b).

13.3   DW Motion Control using Electrodes with Limited Conductivity

To overcome the difficulties faced in controlling domain nucleation, we were aiming to obtain defect-free thin films where DW motion will be fully governed by external factors, and not related to the structure of the film itself. Thin films of tetragonal (001) $PbZr_{0.1}Ti_{0.9}O_3$ (PZT) and $SrRuO_3$ (SRO) were epitaxially grown by pulsed laser deposition on $SrTiO_3$ (STO) substrates; the details can be found in McGilly et al. 2015a. This ensured a uniform out-of-plane polarization in the as-grown state. Subsequently, EBID Pt top "line" electrodes were deposited by a focused electron beam. In this technique, a precursor gas $[(CH_3)_3CH_3C_5H_4Pt]$ which decomposes under focused e-beams is introduced into the vacuum chamber, and Pt is deposited in a predefined pattern through raster scanning of the beam. The thickness is determined by the number of passes of the e-beam over the pattern. The geometry of the fabricated structure is shown in Figures 13.2(a,c). For imaging of domains (Figure 13.2(b)) an atomic force microscope was operated in PFM mode. The excellent ferroelectric properties of the PZT thin film are evidenced through the PFM hysteresis loop (Figure 13.2(e)), which shows that the switching is sharp and occurs over a small voltage range.

A series of experiments were conducted in which the same tip was used for application of voltage pulses to achieve ferroelectric switching under the electrode (the applied bias magnitude exceeds 5 V) and imaging (an applied AC bias of 1 V). In the part of the experiments that we hereafter refer to as "positive switching," the film is initially uniformly polarized in the "up" direction (away from the substrate), which we denote "domain 1," and a new reverse domain called "domain 2" was created and grown using a positively biased PFM tip in contact with the electrode at one edge. Voltage pulses of different duration were applied to nucleate and propagate DWs under the electrode, and subsequent PFM imaging was performed to determine the DW position, after which the system was reset into the initial state by application of a voltage pulse of the opposite polarity, as schematically shown in Figure 13.2(d). In this stroboscopic fashion, individual snapshots of the DW position can be combined to build up a data set of DW displacement as a function of time (Figures 13.3(a,b)).

Analysis of the PFM images revealed that DW nucleation occurred reproducibly under the tip, while the distance at which the DW was displaced was across a wide range of parameters proportional to the square root of the pulse duration (Figure 13.3(c)) and voltage magnitude, Figure 13.3(d).

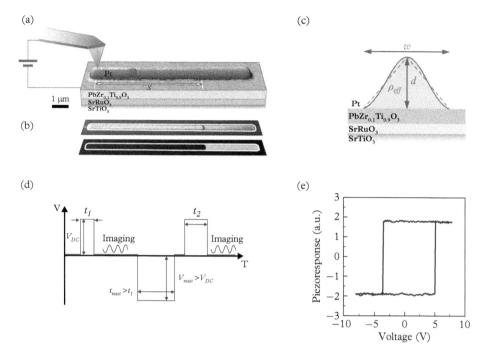

Figure 13.2 *Experimental setup for investigation of individual domain wall (DW) motion. (a) A 3D topography image and schematic shows the EBID Pt line electrode deposited on the ferroelectric thin film and contacted by a conducting probe which is biased relative to the SRO lower electrode. (b) Corresponding PFM amplitude (top) and phase (bottom) images of the domain state after the application of a voltage pulse showing a DW propagating a distance X from the nucleation point. (c) A cross-sectional schematic of the setup showing an example of the electrode shape with width w and height d, and an approximation to a triangle (dotted lines). (d) The stroboscopic approach used to investigate switching involves an initial pulse of magnitude V_{DC} and duration t_1 followed by imaging to assess the domain wall motion. Subsequently a reset pulse of opposite polarity returns the system to its original state. The cycle is repeated with incrementally increased switching pulses to build up a set of images showing the DW progression. (e) An example hysteresis loop measured by piezoresponse force microscopy (PFM) shows extremely sharp switching.*
Source: After McGilly et al. 2016.

13.4 The Stefan Model for DW Propagation

To gain insight into the observed phenomena, we performed modeling of the process in the framework of continuum theory. The central idea was that because the conductivity of the electrode is limited, the voltage u is a function of both time t and the coordinate x along the electrode (Figure 13.4(a)). In view of the DW motion, in its simplest one-dimensional (1D) statement, the problem reduces to finding a pair of functions $u(x, t)$ and $X(t)$, where X is the DW position. The governing equations and boundary conditions for the problem were obtained from the basic laws: charge conservation,

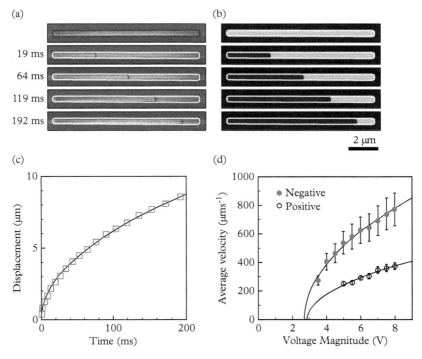

Figure 13.3 *Characterizing propagating domain walls (DWs) in line electrodes. A series of piezoresponse force microscopy (PFM) amplitude (a) and phase (b) images from the initial switched state (top row) and the DW position after pulses of duration as shown on the left-hand side. The DW displacement as a function of switching time (c) for the system shown in (a) and (b). The effect of voltage magnitude on DW velocity (d) for positive (empty circles) and negative (filled circles) applied biases. Solid lines are fits to the data; see main text for details.*
Source: After McGilly et al. 2015a.

Ohm's law, and the energy balance (for the details, see McGilly et al. 2015b), to obtain

$$\dot{u}(x,t) = (rc)^{-1}u''(x,t), 0 < x \neq X, \tag{13.1}$$

$$u(0,t) = V_t\,(t), \tag{13.2}$$

$$u(\infty,t) = 0, \tag{13.3}$$

$$u|_{X-} = u|_{X+} = u(X,t) = \begin{cases} V_s^-, \dot{X} < 0 \\ V \in \left(V_s^-, V_s^+\right), \dot{X} = 0, \\ V_s^+, \dot{X} > 0 \end{cases} \tag{13.4}$$

$$\left.\frac{u'}{r}\right|_{X-} - \left.\frac{u'}{r}\right|_{X+} = 2P_s w\dot{X}, \tag{13.5}$$

where the prime and the dot denote derivation with respect to coordinate (x) and time (t), respectively; $|_{X\mp}$ denotes the limit at the DW approaching from the left (domain 2) and right (domain 1), respectively; $r = \rho/wd$ is the electric resistance per unit length of the electrode; w, d, and ρ are the width, thickness, and resistivity of the electrode, respectively; $c = (\epsilon\epsilon_0\, w)/h$ is the system's capacitance per unit length; $\epsilon_0 = 8.85 \times 10^{-12}$ F/m is the electric permittivity of vacuum; ϵ is the relative permittivity of the ferroelectric material; h is the thickness of the ferroelectric layer; P_s is the spontaneous polarization; V_t is the tip voltage; and switching voltages $V_s^{\pm}$ are introduced, where "+" and "−" indicate the growing and shrinking modes of domain 2, respectively. For the experimentally relevant parameters, $V_s^{\pm}$ does not depend on the DW velocity $\dot{X}$ and are determined in the framework of the "dry friction" concept; for details, see McGilly et al. 2015b:

$$V_s^{\pm} = \frac{\tau h}{wP_S} \pm \frac{P_f^{max} h}{2P_S},$$

(13.6)

where τ is the DW energy per unit area, $P_f = P_f^{max} \cdot \frac{\dot{X}}{|\dot{X}|}$ is the pressure on the moving DW exerted as a result of different dissipative processes, related to overcoming Peierls barriers, defect pinning, etc.

Analysis of Equations (13.1)–(13.5) reveals a close similarity of the problem with the classical Stefan problem (Gupta 2003) for heat. Thus, the existing solutions can be used with respect to the substitutions indicated by McGilly et al. (2015b). An important parameter involved here is the dimensionless Stefan number; classically, this is the ratio of the separate amount of heat needed for melting to that needed for heating. The analogous Stefan number in our system is $S_2 = [c(V_t - V_s)]/(2P_s w)$, that is, the ratio of the charge accumulated in the capacitance of the system to the screening charge related to the DW motion. An analysis of the physical quantities in our problem shows that we are dealing with small Stefan numbers. The solution to the Stefan problem for DW displacement is then

$$X(t) = \sqrt{\frac{(V_t - V_s)\,t}{rP_s w}},$$

(13.7)

which perfectly explains the experimental observations (Figures 13.3(c,d)). The associated voltage distribution is shown in Figure 13.4(b). It obeys the linear law $V_t - x(V_t - V_s)/X$ in domain 2, while for the voltage in domain 1 the exact expression involving special functions is given by McGilly et al. (2015b). This demonstrates that initially the voltage is locally confined close to the tip and decays rapidly, excluding the possibility of second nucleation far from the tip position. The dependence of DW displacement on the applied tip voltage was tested in a series of experiments where pulses of same duration but different amplitude were applied (Figure 13.3(d)). Along with the "positive switching" experiments considered above, "negative switching" experiments were also conducted, where in its initial state the film contained domain 2 under the entire area of the line electrode. Using a negatively biased PFM tip in contact with the edge

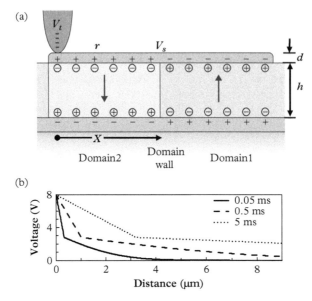

Figure 13.4 *Theoretical interpretation of domain wall (DW) motion under linear electrodes. (a) A cross-sectional view of the experimental setup showing partial switching of the capacitor structure. Polarization is given by arrows, free charges are pluses and minuses, whilst bound charges are circled. (b) The calculated voltage distribution according to the Stefan model for a moving DW 0.05 ms after voltage application (solid), 0.5 ms (dashed) and 5 ms (dotted).*
Source: After McGilly et al. 2015a, 2015b.

of the electrode, domain 2 was gradually removed, turning the film to the monodomain 1 state. The results are summarized in Figure 13.3(d), where good agreement with the theoretical prediction of Equation (13.7) is achieved. From the theoretical fit to the data (Figure 13.3(d)), V_s can be extracted and is found to be -2.70 ± 0.05 V and $+2.88 \pm 0.05$ V for negative and positive switching, respectively. Note that V_s defined in the Stefan model for the existing DW was measured to be lower than the switching voltage in the hysteresis loop (Figure 13.2(e)), the latter being controlled by the nucleation process. Interestingly, the discrepancy in the velocities for positive and negative switching can be explained by a polarization-sensitive resistance $r(P_s)$ due to different electron scattering conditions at the interface dependent on the species of the bound and free charge present. This feature can be used to detect position of the DW; see Section 13.6.

The agreement of theory with experiment is highly reproducible over a wide range of parameters; see additional data in McGilly et al. 2015a, 2015b, 2016. Reproducibility tests for 10-μm-long line electrodes have demonstrated a precision in DW positioning of ~200 nm, which means a possibility of 50 distinct locations for an individual DW along the electrode.

Expression (13.7) was also used to determine the resistance r of the electrodes to be $(5.5 \pm 1.2) \times 10^{15}$ Ωm^{-1} (here the sensitivity of the resistance to the polarization sign is included in the error). Based on an approximation of the electrode shape (Figure 13.2(c)) with a triangle, the corresponding resistivity $\rho_{eff} = rwd/2$ was found to be in the range 27 ± 7 Ωm. These values are considerably higher than that for bulk Pt and values reported for similar EBID Pt (Botman et al. 2006); however, in that case, the electrode thickness was an order of magnitude larger (150 nm) than in our case (~10 nm). Resistivities are known to sharply depend on electrode thicknesses due to electron scattering from the surfaces (Hoffman and Fischer 1976; Fischer et al. 1980). This hints that varying the resistivity by simply modifying the thickness of the electrode is an attractive method to significantly control DW motion.

13.5 Control of DW Velocity via Thickness-Dependent Resistivity of Pt Electrodes

To study the dependence of the DW velocity on the electrode thickness, EBID Pt top electrodes each of length 10 μm and width 200 nm but of different thickness in the range 4–100 nm were deposited as can be seen in the topography image in Figure 13.5(a).

The same stroboscopic approach introduced in Figure 13.2(d) and used in Figure 13.3 was here applied to produce DW displacement as a function of time for each line electrode thickness as shown in Figure 13.5(b). From the accompanying DW velocity as a function of the position along the line (Figure 13.5(c)), we can clearly see a large spread in values, wherein the DWs move faster for thicker top electrodes than in the case of the thinner ones.

From Equation (13.7), the DW velocity may be presented in terms of either time or displacement as

$$v = \frac{1}{2}\sqrt{\frac{(V_t - V_s)}{rP_s wt}} = \frac{(V_t - V_s)}{2rP_s w} \cdot \frac{1}{X} \tag{13.8}$$

Equation (13.8) was used to fit the experimental data, as shown in Figure 13.5(c), and gives good agreement. The obtained data are summarized in the form of a velocity–thickness dependence plot (Figure 13.5(d)), where the fitted model values were taken for each thickness to evaluate the DW velocity at a displacement of 5 μm. It can be seen in Figure 13.5(d) that velocities extend over at least seven orders of magnitude, which turns our expectations into a remarkable observation. It is clear from Equation (13.8) that the velocity only depends linearly on the electrode thickness d and as such is unable to describe the orders of magnitude changes seen. In this context, the dependence of the effective electrode resistivity ρ_{eff} on its thickness d deserves a more careful look.

In order to quantify the resistivity of the line EBID Pt top electrodes as a function of the thickness, we performed four-point probe measurements using the experimental setup illustrated in Figure 13.6(a). The electrodes were deposited across the four contact pads, and the two outer ones served to drive the current while the two inner ones allowed measurement of the potential difference with a Keithley 2636A SourceMeter. These

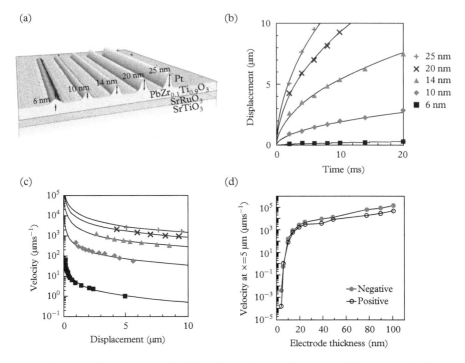

Figure 13.5 *Controlling domain wall (DW) velocity through electrode modification. (a) A 3D topography image of Pt line electrodes of different thicknesses. (b) The displacement as a function of time for the electrodes shown in (a) and (c) the associated velocity–displacement relations. (d) Velocity measured at a displacement of 5 μm as a function of electrode thickness.*
Source: After McGilly et al. 2016.

measurements, combined with knowledge of the length and the cross-sectional shape of the electrode (see Figure 13.2(c)), allowed calculation of the resistance per length r and the effective electrode resistivity ρ_{eff}. The results for ρ_{eff} are displayed in Figure 13.6(b). The uncertainty associated with the cross-sectional non-uniform thickness has been included in the data point error bars. From a comparison with the calculated resistivities from the Stefan model for DW motion (Equation (13.8)), we see that the trend is similar to the real resistivity measured by the four-point probe method. This justifies the approach where the electrode resistivity is evaluated from the fit to the Stefan model. The transition from a moderate slope for relatively thick electrodes in Figure 13.6(b) to a sharp increase in the resistivity for very thin electrodes supports the theoretical interpretation based on charge carrier scattering from the electrode surfaces. The scattering effects are expected to become much more pronounced where the electrode thickness becomes comparable to, and then falls below, the electron mean free path (Hoffman and Fischer 1976), which is estimated to be ~12 nm in Pt thin films (Fischer et al. 1980). It is noteworthy here that the resistivities reported for the

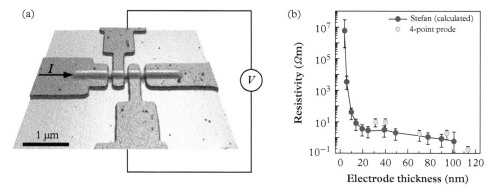

Figure 13.6 *Electric characterization of electron-beam-induced deposition (EBID) Pt. (a) Topography image of the experimental setup for four-point probe measurements of the EBID Pt. The large Ti/Au contact pads appear darker than the STO substrate. The EBID Pt stripe electrode is shown with the arrow. (b) The dependence of the resistivity on the electrode thickness calculated by the Stefan model (dark points) and as measured by the four-point probe method (bright points).*
Source: After McGilly et al. 2016.

thickest EBID Pt electrodes, although consistent with other reported values (i.e. 0.1 Ωm for thickness >100 nm with similar deposition parameters (Hoffman and Fischer 1976)), are still four orders of magnitude higher than those of the thin and ultra-thin epitaxial SRO thin films (Schultz et al. 2009) commonly used for lower electrodes in ferroelectric films. Therefore, these results are obtained in a "high-resistivity electrode regime."

In summary, we have shown that 180° DW velocities in thin films of PZT depend strongly on the EBID Pt top electrode resistivity as represented by the proxy parameter of electrode thickness. This can serve as a method to tune the DW velocity and therefore switching characteristics in ferroelectric thin films when working in the regime of "high-resistivity" electrodes.

13.6 Charge-Controlled 2D DW Motion

Earlier we considered 1D DW motion under linear electrodes. The EBID technology allows deposition of 2D electrode structures where the limited electrode resistivity, now per unit area, is a function of the two surface coordinates. This opens further possibilities for DW motion control. The theoretical description of such DW motion can be readily developed within the same approach.

In the 2D case, we assume that a single DW is nucleated under the tip, and that during the domain growth it encircles the tip and maintains a convex shape. It is convenient to use polar coordinates centered at the point of contact of the PFM tip with the electrode. The mathematical formulation for the problem is analogical to the 1D case, but now the evolution of the voltage $u(l, \varphi, t)$ and the DW displacement $L(\varphi, t)$ are in the general case functions of the azimuthal angle, φ. Here, l is the radius (distance to the tip center). Inside domains 1 and 2, we solve the equations of charge conservation and Ohm's law

$$\text{div} \left(\overrightarrow{j} \right) + \gamma \dot{u} = 0, \tag{13.9}$$

$$\rho_{(2D)} \overrightarrow{j} = -\overrightarrow{\text{grad}}(u), \tag{13.10}$$

where $\overrightarrow{j}$ is the surface current density, γ is the capacitance per unit area of the system, $\rho_{(2D)}$ is the electric resistance of the unit area of the system, and the operators *div* and $\overrightarrow{\text{grad}}$ represent the two-dimensional divergence and gradient, respectively. On the moving domain boundary, the charge conservation yields

$$v \equiv \overrightarrow{v} \cdot \overrightarrow{n} = \frac{1}{2P_s} \left[\left[\overrightarrow{j} \right] \right] \overrightarrow{n}, \tag{13.11}$$

where $\overrightarrow{v}$ is the velocity of the DW, $\overrightarrow{n}$ is the vector normal to the wall directed outside, [[a]] denotes the jump of the variable a at the wall. The force balance yields the voltage on the DW

$$V_{s(2D)} \equiv u\,(L,\varphi,t) = -\frac{\tau h}{2RP_s} - \frac{P_f h}{2P_s}, \tag{13.12}$$

where R is the local radius of curvature of the DW. To describe the charge injection from the PFM tip, we cut out a circle C_t of radius L_t (the effective contact radius) and apply the boundary condition $u|_{C_t} = V_t$. The solution of the 2D problem in the general case is not feasible analytically. However, in each particular case, a numerical solution may be obtained with very little computational effort. Here, aiming at an analytical solution, we address both theoretically and experimentally the growth of a circular domain. Theoretically, the system of Equations (9)–(12) is solved with initial conditions $u|_{t=0} = 0$, $L|_{t=0} = L_t$, $u|_{C_t,t>0} = V_t$. To perform the corresponding experiment, 5×5 μm^2 EBID Pt electrodes of thickness 5.4 ± 0.4 nm were deposited on top of the 70-nm-thick PZT thin film. Then, using a PFM tip in contact with the top electrode (see Figure 13.7(a)), voltage pulses of $V_t = 6$ V and different time durations were applied to nucleate and advance the DW as used in the previously described experiments (Figure 13.2). The PFM images (Figure 13.7(b)) show a single nucleated circular domain that grows as a function of time. The dependence of the domain radius on the switching time forms a smooth curve, rising to the micrometer scale within hundreds of milliseconds. As described by Equation (13.12), the larger the domain radius L, the smaller is the contribution of the DW surface energy to the switching voltage. The pinning-related contribution is also expected to be weak in the high-purity film. That is why we assume here for simplicity that $V_t - V_{s(2D)} \gg V_{s(2D)}$, implying applicability of the so-called one-phase Stefan problem (Gupta 2003), where the current flowing inside domain 1 is neglected when determining the DW position from charge balance. This simpler version has an analytical solution for the time needed for the circular DW to reach radius L:

$$t = A\left(1 - \xi^2 + 2\xi^2 Ln(\xi)\right),$$ (13.13)

where

$$\xi = \frac{L}{L_t},$$ (13.14)

and

$$A = \frac{P_s \rho_{2D} L_t^2}{2\left(V_t - V_{s(2D)}\right)}.$$ (13.15)

Equations (13.13)–(13.15) determine the inverse function for the time dependence of the domain radius. Note that in the theory the time is measured from the moment when the domain already has its initial radius equal to the tip contact radius. As determined from experiment, the nucleation time $<<1$ ms, which is negligible compared to the total time for switching, ~100 ms, and so validates this theoretical approach. The theoretical and experimental results are compared in Figures 13.7(b,d), where we selected the appropriate parameter A to fit the experimental data. Their good agreement further confirms that the Stefan model is applicable for the description of a wide range of processes involving DW motion under electrodes with limited conductivity.

The one-phase Stefan model assumes that at each moment of time the capacitance between the electrodes is charged to voltage $\sim V_{s(2D)}$ in the surrounding domain 1 to a distance greatly exceeding the domain radius. The evolution of the voltage inside domain 2 is then described as

$$u = V_t - \frac{\left(V_t - V_{s(2D)}\right) Ln\left(\frac{l}{L_t}\right)}{Ln\left(\frac{L}{L_t}\right)}.$$ (13.16)

The resulting voltage distributions for three different moments of time are plotted in Figure 13.7(c).

It is very interesting to compare the resistivities of the Pt electrodes obtained in 1D and 2D experiments. Taking for the radius of contact $L_t = 10$ nm and for switching voltage $V_{s(2D)} = 1V$, we obtain $\rho_{(2D)} = 1.4 \times 10^{14}$ Ω. The corresponding resistivity, $\rho = \rho_{(2D)} d = 7.6 \times 10^5$ Ωm, is in perfect agreement with the 1D data (Figure 13.6(b)). Note that the obtained result for $\rho_{(2D)}$ is only very weakly dependent on the parameters L_t and $V_{s(2D)}$. This can be explained with the approximate expression, obtained for large enough domains $\left(1 << \xi^2\right)$ from Equations (13.13)–(13.15):

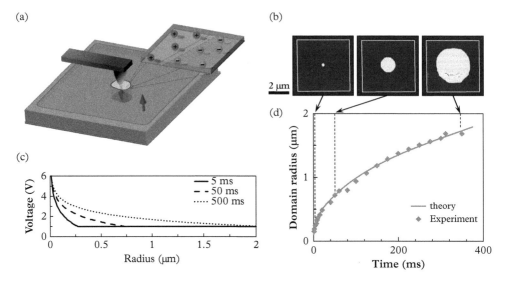

Figure 13.7 *Growth of a circular domain. (a) Schematic drawing for the experimental configuration; screening charges and their motion accompanying the growth of the domain are shown in the magnified image. (b) Piezoresponse force microscopy (PFM) phase images for three selected pulse durations of 1, 50, and 350 ms (left to right). (c) The calculated voltage distribution according to the 2D axisymmetric one-phase Stefan model 5 ms after voltage application (solid), 50 ms (dashed), and 500 ms (dotted). (d) Domain radius time dependence: experimental averaged radius (rhombs) and theoretical prediction, Equation (13.13) (solid line).*

$$t = \frac{\rho_{(2D)} P_s L^2}{2 \left(V_t - V_{s(2D)} \right)} \left(2Ln \left(L/L_t \right) - 1 \right) \tag{13.17}$$

The good agreement between the theory and experiment in both 1D linear and 2D axisymmetric cases demonstrates the high predictive power of the Stefan model for describing DW motion in thin films under controlled charge injection using EBID electrodes. Numerical simulations of Equations (13.9)–(13.12) may be performed to describe more complicated systems. Keeping in mind some particular application, one can readily incorporate arbitrary distributions of electrode thickness, multiple charge injection spots, multiple DWs, etc. Correspondingly, the predicted behavior may be tested experimentally. Some experimental tests for electrodes with complex geometries and multiple DWs were already demonstrated by McGilly et al. (2015a, 2015b, 2016). In particular, it was demonstrated that insertion of a section with poorer resistivity into a linear electrode serves as a "speed limit" for DW motion under the part of the electrode remaining after such insertion (McGilly et al. 2016). Insertion of a section with increased width also resulted in the DW slowing down, but in that case the DW speed was recovered after the end of the section (McGilly et al. 2016). Management of multiple DWs and DW "splitting" by bifurcating electrodes was demonstrated by McGilly et al. (2015a).

13.7   Applications and Outlook

Precise and accurate positioning of DWs could be useful for future applications taking advantage of well-defined switched fractions of a capacitor for, e.g. memristors (Chanthbouala et al. 2012), multi-level data storage, or using the DWs themselves as compliant electrical connectors between two electrodes (Seidel 2012; Seidel et al. 2009). As mentioned previously, the resistance of the EBID electrode appears to depend on the switched fraction of polarization, i.e. $r(P_s) \equiv r(X)$. Thus, the position of the DW determines the top electrode's resistance state. In this way, a certain memristive behavior is expected in which the readout of the current through the top electrode depends on the history of the current flow through the device in a continuous manner. However, this behavior is predicated on being able to control DW positions and to be able to do so repeatedly. This study shows that it is in fact feasible to do this for individual DWs and so highlights a pathway toward DW memristor devices. While the DWs shown here do not exhibit strongly enhanced conduction relative to the surrounding domains, we see no immediate reason why this approach could not be extended to other systems such as charged DWs in $BiFeO_3$. This would open the possibility of using ferroelectric walls as rewritable interconnectors in reconfigurable nanocircuits where, of course, precise positioning of the DWs would be required.

From a fundamental point of view, the ability to nucleate and isolate an individual propagating DW could be useful for basic research. There are still a considerable amount of unknowns in certain material systems with regard to DW motion, especially as researchers approach the 2D limit. In the extreme case, layered van der Waals ferroelectrics such as In_2Se_3 (Cui et al. 2018), SnTe (Chang et al. 2016), and WTe_2 (Fei et al. 2018) can be exfoliated down to few layers and still possess a sizable polarization. However, the DW mobility is almost entirely unexplored, and so this is likely to be a fertile playground in years to come as interest in such materials increases. From an applications standpoint, layered ferroelectrics represent the ultimate in downsizing and an absolute scaling limit due to the material being the thinnest possible at only a few atoms thick. Furthermore, as much has been made of the 2D nature of traditional ferroic walls, it worth considering that when the ferroic material itself is two dimensional, then by necessity the associated walls will be one-dimensional entities and therefore likely to have unique properties and dynamics.

Although the velocities measured here are lower than those of similar spin-current driven magnetic DWs (Lewis et al. 2010; Yang et al. 2010), the unique properties of ferroelectric DWs, i.e. conduction, could still merit their use. Strategies that could help optimize DW velocities would be to further decrease the EBID top electrode resistivity. However, to achieve this through ever-increasing electrode thicknesses is not feasible. Indeed, DW velocities of 10^1–10^2 ms^{-1} would allow for nanosecond operation but would require a top electrode of thickness >100 nm. Here, a promising method is deposition of pure Pt nanostructures under an O_2 gas flux (Villamor et al. 2015) or by using other metals including Au, W, Ir, Al, Pd, Co, and Fe which can be deposited by the EBID method (Botman et al. 2009). However, the applicability of the theoretical description provided here, first of all the "dry friction" concept, needs to be tested for much higher DW velocities. Ultimately, at some point, the increase in conductivity will bring the system back to "normal" switching conditions; that is, capacitor charging is fast compared to nucleation times, and reverse domains will form randomly at defects rather than beneath the tip/charge source.

The present study only touches on the 2D control of DW motion using ferroelectric films with EBID electrodes. The full potential of these systems is still to be explored. As demonstrated here, the theoretical predictions using the Stefan model are a viable method of guiding experiments, not just those involving EBID electrodes but in other real systems, especially those with low-quality electrodes. Here, we have limited ourselves to simple cases; more sophisticated experiments may be guided by numerical simulations. The interrelations of the DW speed and shape with current and voltage in the circuit are of great interest for nanoelectronic applications and should be explored further.

In conclusion, ferroelectric DW motion can be used within devices to potentially bring new functionalities as long as nucleation and propagation are controlled. Here, we have shown that this can be achieved through the use of PZT thin films with EBID Pt electrodes with limited conductivity. An analogy with the classical Stefan problem is presented for 1D and 2D cases and shows excellent agreement with experiment. We believe that the Stefan problem analogy can be extended to many ferroelectric systems to describe DW motion where a limited charge source is present. The demonstrated precise coupling of DW motion to electric currents in the electrodes is highly attractive for potential future nanoelectronic applications.

· ·

REFERENCES

Botman A., Mulders J. J. L., Hagen C. W., "Creating pure nanostructures from electron-beam-induced deposition using purification techniques: a technology perspective," *Nanotechnology* 20, 372001 (2009).

Botman A., Mulders J. J. L., Weemaes R., Mentink S., "Purification of platinum and gold structures after electron-beam-induced deposition," *Nanotechnology* 17, 3779–3785 (2006).

Chang K. Liu J., Lin H., Wang N., Zhao K., Zhang A., Jin F., Zhong Y., Hu X., Duan W., Zhang Q., Fu L., Xue Q. K., Chen X., Ji S. H., "Discovery of robust in-plane ferroelectricity in atomic-thick SnTe," *Science* 353, 274–278 (2016).

Chanthbouala A., Garcia V., Cherifi R. O., Bouzehouane K., Fusil S., Moya X., Xavier S., Yamada H., Deranlot C., Mathur N. D., Bibes M., Barthélémy A., Grollier J., "A ferroelectric memristor," *Nat. Mater.* 11, 860–864 (2012).

Cui C. W., Hu W. J., Yan X., Addiego C., Gao W., Wang Y., Wang Z., Li L., Cheng Y., Li P., Zhang X., Alshareef H. N., Wu T., Zhu W., Pan X., Li L. J., "Intercorrelated in-plane and out-of-plane ferroelectricity in ultrathin two-dimensional layered semiconductor In_2Se_3," *Nano Lett.* 18, 1253–1258 (2018).

Fei Z., Zhao W., Palomaki T. A., Sun B., Miller M. K., Zhao Z., "Ferroelectric switching of a two-dimensional metal," *Nature* 560, 336 (2018).

Fischer G., Hoffmann H., Vancea J., "Mean free path and density of conductance electrons in platinum determined by the size effect in extremely thin films," *Phys. Rev. B* 22, 6065 (1980).

Grigoriev A., Do D.-H., Kim D. M., Eom C.-B., Adams B., Dufresne E. M., Evans P. G., "Nanosecond domain wall dynamics in ferroelectric Pb(Zr,Ti)O₃ thin films," *Phys. Rev. Lett.* 96, 187601 (2006).

Gruverman A.Rodriguez B. J., Dehoff C., Waldrep J. D., Kingon A. I., Nemanich R. J., "Direct studies of domain switching dynamics in thin film ferroelectric capacitors," *Appl. Phys. Lett.* 87, 082902 (2005).

Gruverman A., Wu D., Scott J. F., "Piezoresponse force microscopy studies of switching behavior of ferroelectric capacitors on a 100-ns time scale," *Phys. Rev. Lett.* 100, 097601 (2008).

Gupta S. C., *The Classical Stefan Problem: Basic Concepts, Modelling and Analysis* (Elsevier, Amsterdam, 2003).

Hoffmann H., Fischer G., "Electrical conductivity in thin and very thin platinum films," *Thin Solid Films* 36, 25–28 (1976).

Kasuya T., Shimoda N., "Stefan problem for a finite liquid phase and its application to laser or electron beam welding," *J. Appl. Phys.* 82, 3672–3678 (1997).

Lewis E. R.Petit D., O'Brien L., Fernandez-Pacheco A., Sampaio J., Jausovec A-V., Zeng H. T., Read D. E., Cowburn R. P., "Fast domain wall motion in magnetic comb structures," *Nat. Mater.* 9, 980 (2010).

Li J., Nagaraj B., Liang H., Cao W., Lee C. H., Ramesh R., "Ultrafast polarization switching in thin-film ferroelectrics," *Appl. Phys. Lett.* 84, 1174 (2004).

Lin Z., "A free boundary problem for a predator–prey model," *Nonlinearity* 20, 1883 (2007).

McGilly L. J., Feigl L., Sluka T., Yudin P., Tagantsev A. K., Setter N., "Velocity control of 180 domain walls in ferroelectric thin films by electrode modification," *Nano Lett.* 16, 68–73 (2016).

McGilly L. J., Yudin P., Feigl L., Tagantsev A. K., Setter N., "Controlling domain wall motion in ferroelectric thin films," *Nat. Nanotechnol.* 10, 145 (2015a).

Meng Q. P., Han M.-G., Tao J., Xu G., Welch D. O., Zhu Y., "Velocity of domain-wall motion during polarization reversal in ferroelectric thin films: beyond Merz's law," *Phys. Rev. B* 91, 054104 (2015).

Peskir G., "On the American option problem," *Mathematical Finance: An International Journal of Mathematics, Statistics and Financial Economics* 15, 169–181 (2005).

Rodriguez B. J., Nemanich R. J., Kingon A., Gruverman A., "Domain growth kinetics in lithium niobate single crystals studied by piezoresponse force microscopy," *Appl. Phys. Lett.* 86, 012906 (2005).

Schultz M., Levy S., Reiner J. W., Klein L., "Magnetic and transport properties of epitaxial films of SrRuO₃ in the ultrathin limit," *Phys. Rev. B* 79, 125444 (2009).

Seidel J., "Domain walls as nanoscale functional elements," *J. Phys. Chem. Lett.* 3, 2905–2909 (2012).

Seidel J., Martin L. W., He Q., Zhan Q., Chu Y. H., Rother A., Hawkridge M. E., Maksymovych P., Yu P., Gajek M., Balke N., Kalinin S. V., Gemming S., Wang F, Catalan G., Scott J. F., Spaldin N. A., Orenstein J., Ramesh R., "Conduction at domain walls in oxide multiferroics," *Nat. Mater.* 8, 229–234 (2009).

Stefan J., "Versuche über die scheinbare Adhäsion," *Ann. Phys.* 230, 316–318 (1875).

Stefan J., "Über die Verdampfung und die Auflösung als Vorgänge der Diffusion," *Sitzungsber. kais. Akad. Wiss. Wien. Math. Naturwiss. Kl.,* Abt. II a, 98, 1418–1442 (1889).

Supplementary Materials

McGilly L. J., Yudin P., Feigl L., Tagantsev A. K., Setter N., "Controlling domain wall motion in ferroelectric thin films," *Nature Nanotechnol.* 10, 145 (2015b).

Tybell T., Paruch P., Giamarchi T., Triscone J.-M., "Domain wall creep in epitaxial ferroelectric $Pb(Zr_{0.2}Ti_{0.8})O_3$ thin films," *Phys. Rev. Lett.* 89, 097601 (2001).

Villamor E., Casanova F., Trompenaars P. H. F., Mulders J. J. L., "Embedded purification for electron beam induced Pt deposition using MeCpPtMe3," *Nanotechnology* 26, 095303 (2015).

Whyte J. R., McQuaid R. G., Sharma P., Canalias C., Scott J. F., Gruverman A., Gregg J. M., "Ferroelectric domain wall injection," *Adv. Mater.* 26, 293 (2014a).

Whyte J. R., McQuaid R. G. P., Ashcroft C. M., Einsle J. F., Canalias C., Gruverman A., Gregg J. M., "Sequential injection of domain walls into ferroelectrics at different bias voltages: Paving the way for 'domain wall memristors,'" *J. Appl. Phys.* 116, 066813 (2014b).

Yang S.-H., Ryu K.-S., Parkin S., "Memory on the racetrack," *Nat. Nanotechnol.* 10, 221 (2010).

14

Multiscale Simulations of Domains in Ferroelectrics

S. Liu[1], I. Grinberg[2], and A. M. Rappe[3]

[1] *School of Science, Westlake University, Hangzhou, China*
[2] *Department of Chemistry, Bar-Ilan University, Ramat Gan, Israel*
[3] *Department of Chemistry, University of Pennsylvania, Philadelphia, United States*

14.1 Introduction

Ferroelectric materials characterized by switchable polarization are the best dielectric, piezoelectric, and pyroelectric materials known today, enabling a plethora of applications including capacitors, non-volatile memory, sensors, resonators, actuators, and transducers (Lines and Glass 1977; Scott 2007; Liu et al. 2015). Technologically important ferroelectric materials such as $Pb(Zr,Ti)O_3$ (PZT) and $Ba(Sr,Ti)O_3$ (BST) are often processed in the form of ceramics, which are polycrystalline materials composed of different grains (Figure 14.1). Each grain may have different regions, each possessing a uniform polarization orientation. Each homogeneously polarized volume is called a ferroelectric domain. The transition region separating domains of different polarization directions is called a domain wall and is characterized by the gradual change in polarization orientations and magnitudes. The occurrence of domains and domain walls in polycrystalline ferroelectrics is inevitable and is driven by minimizing the depolarization energy associated with surface charges and elastic residual stresses governed by the mechanical boundary conditions and the thermal/electrical history (Rabe et al. 2007; Pramanick et al. 2012). The functionalities of ferroelectrics rely heavily on the interactions between the polarization and the external stimuli, which are intimately related to the motion of domain walls that grow the stimuli-favored ferroelectric domains at the expense of energetically less favorable domains.

   Switching processes in polycrystalline ferroelectrics often go through a series of complex, hierarchically organized sequences of physical processes on time scales ranging from femtoseconds (e.g. dipole flipping) to picoseconds (e.g. domain nucleation and growth) and to a few microseconds (e.g. polarization reversal). The inevitable presence of chemical inhomogeneity on length scales ranging from angstroms and nanometers

S. Liu, I. Grinberg, and A. M. Rappe, *Multiscale Simulations of Domains in Ferroelectrics* In: *Domain Walls: From Fundamental Properties to Nanotechnology Concepts*. Edited by: Dennis Meier, Jan Seidel, Marty Gregg, and Ramamoorthy Ramesh, Oxford University Press (2020).
© S. Liu, I. Grinber, and A. M. Rappe.
DOI: 10.1093/oso/9780198862499.003.0014

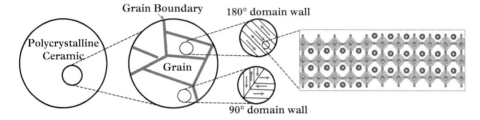

Figure 14.1 *Schematic of structural hierarchy in polycrystalline ferroelectrics.*

(e.g. defects and dislocations) to micrometers (e.g. grain boundaries) introduces additional complexity. The macroscopic properties of ferroelectrics are therefore truly *emergent functional properties* arising from the complex interactions between various intrinsic and extrinsic effects across many different scales in space and time. In order to achieve rational design and optimization of ferroelectrics, it is of great importance to understand the fundamental physics and chemistry that involve all scales, which cannot be realized by a single theory or method. A complete understanding of a complex multicomponent material system requires a multiscale approach that progressively bridges methods at different scales, enabling accurate and efficient predictions of materials properties from the atomistic level over the mesoscale and eventually up to the macroscale accessible to experimental measurements.

The multiscale approach discussed here consists of three levels of methods: quantum mechanical methods such as density functional theory (DFT), classical statistical methods such as molecular dynamics (MD) simulations, and phenomenological continuum methods such as the phase-field model (PFM). The first-principles DFT method is a powerful predictive method requiring essentially only the positions of atoms as the inputs. The development of the modern theory of polarization (King-Smith and Vanderbilt 1993, 1994; Resta 1993, 1994; Rabe et al. 2007) allows for accurate predictions of ferroelectric polarization and derived quantities such as piezoelectric and pyroelectric coefficients (de Gironcoli et al. 1989; Sághi-Szabó et al. 1998; Bellaiche et al. 2000; Liu et al. 2016a; Tan et al. 2018). The combination of high-throughput DFT calculations and the tools of data science for automated property calculations (Curtarolo et al. 2012; Jain et al. 2013; Garrity et al. 2014) have led to the establishment of a DFT database of piezoelectrics (de Jong et al. 2015), helping to identify novel piezoelectrics (Liu and Cohen 2017a). However, the relatively high computational cost of DFT limits the size of the model system to a few hundred atoms, and in many cases DFT methods can only estimate the static intrinsic bulk properties of simple chemical compositions under ideal conditions (such as at zero Kelvin). The study of the finite-temperature dynamical properties of multicomponent ferroelectric solid solutions involving domain-wall mobility is still beyond the scope of conventional DFT studies. Statistical methods such as MD simulations are ideal techniques for investigating dynamic processes on larger length/time scales, but their applications are often hindered by the limited availability and accuracy of classical force fields. One approach to address this issue is to fit classical force fields to first-principles results, thus combining the efficiency of MD and the accuracy of DFT. Using DFT and MD calculations, it is possible to further parameterize a free energy functional that depends on well-defined order parameters (e.g. polarization) (Völker et al. 2011), which can be used for phase-field simulations (Nambu and Sagala 1994; Hu and Chen 1998; Semenovskaya and Khachaturyan 1998; Chen 2008) to model evolution of large ensembles of ferroelectric domains on the mesoscopic length scale within the framework of Ginzburg–Landau theory.

In this chapter, we will mainly focus on our recent studies of ferroelectrics (Shin et al. 2007; Takenaka et al. 2013, 2017; Liu et al. 2016b), where large-scale MD simulations using first-principles-based force fields played a central role in revealing important physics inaccessible to direct DFT calculations but critical for developing the physically based free energy functional for coarse-grained phase-field-type simulations. The outline is as follows: after reviewing typical atomistic potentials of ferroelectrics for MD simulations, we describe a progressive theoretical framework that combines DFT, MD, and a mean-field theory, which eventually allows for the first time an efficient and accurate estimation of coercive fields for a broad range of ferroelectrics. We then focus on relaxor ferroelectrics, a type of functional materials characterized by diffuse phase transitions and possessing giant dielectric and piezoelectric responses. By examining the spatial and temporal polarization correlations in prototypical relaxor ferroelectrics with million-atom MD simulations and novel analysis techniques such as the dynamic pair-distribution function, we show that the widely accepted model of polar nanoregions embedded in a nonpolar matrix is incorrect for Pb-based relaxors. Rather, the unusual properties of theses relaxor ferroelectrics stem from the presence of a multi-domain state with extremely small domain sizes (2–10 nm), giving rise to greater flexibility for polarization rotations and ultrahigh dielectric and piezoelectric responses. Finally, we discuss the challenges and opportunities for multiscale simulations of ferroelectric materials.

14.2 Atomistic Potentials of Ferroelectrics

MD is a statistical method that calculates the time-dependent behavior of an ensemble of atoms by integrating Newton's equations of motion (Metropolis et al. 1953; Alder and Wainwright 1959; Rahman 1964; Frenkel and Smit 2001). All atoms are treated as structureless classical particles, experiencing an interatomic potential (often referred to as force field) described by an analytical energy function that approximates the true potential energy surface. In practice, the full interaction is often decomposed into two-body and many-body terms, long-range and short-range terms, and electrostatic and non-electrostatic interactions, which are represented by suitable functional forms depending only on the atomic positions. In most cases, the force-field parameters are not derived (calculated) directly from first principles but are obtained by fitting to quantum mechanical results such as energies and forces, and sometimes experimental properties such as density, vibrational frequency, and phase transition temperatures.

For typical ferroelectrics such as $PbTiO_3$ and $BaTiO_3$, two types of force fields have been developed and are widely used in general. One is the shell model (Mitchell and Fincham 1993; Sepliarsky et al. 2005), where each atom is described as two coupled particles: a higher mass core and a lower mass or massless shell. The core and shell are connected by a nonlinear spring. The model also includes Coulomb interactions among cores and shells of different atoms, and short-range interactions between the shells. The shell model parameters for a wide range of ferroelectrics have been developed by fitting to first-principles results. Examples include $PbTiO_3$ (Sepliarsky and Cohen 2002), $BaTiO_3$ (Tinte et al. 1999; Vielma and Schneider 2013), $KNbO_3$ (Sepliarsky et al. 1995), and solid solutions such as $Ba(Ti,Sr)O_3$ (Tinte et al. 2004), $Pb(Zr,Ti)O_3$ (Gindele et al. 2015), and $(1-x)Pb(Mg_{1/3}Nb_{2/3}O_3)-xPbTiO_3$ (Sepliarsky and Cohen 2011).

We have recently developed a new model potential based on the bond valence theory (Liu et al. 2013a, 2013b). Bond valence (BV) is a useful concept describing the bonding strength and was first developed by X-ray crystallographers for crystal structure analysis (Brown and Shannon 1973; Brown and Wu 1976). The BV V_{ij} between atom i and atom j can be calculated with the following formula:

$$V_{ij} = \left(\frac{r_{0,ij}}{r_{ij}} \right)^{C_{ij}},$$
(14.1)

where r_{ij} is the bond length, $r_{0,ij}$, and C_{ij} are Brown's empirical parameters that are readily available for many atomic pairs (Brese and OKeeffe 1991; Brown 2009).

We construct two energy terms based on bond-valence sum (BVS) conservation and bond-valence vector sum (BVVS) conservation (Harvey et al. 2006; Brown 2009). BVS conservation states that each atom i prefers to have the sum of the bond valences of all its bonds, $V_i = \sum_i V_{ij}$, equal to its atomic valence, $V_{0,i}$. This leads to the BVS energy, $E_{BVS} = \sum_i S_i (V_i - V_{0,i})^2$, which indicates that any instantaneous V_i deviating from the desired atomic valence $V_{0,i}$ will cause an energy penalty. The bond-valence vector (BVV), $\mathbf{W}_{ij}$, is defined as a vector directed along the bond line with its magnitude equal to the scalar bond valence. The BVVS, $\mathbf{W}_{0,i}$, is the vector sum of individual $\mathbf{W}_{ij}$ (Harvey et al. 2006). We propose that each atom assumes a desired length of $|\mathbf{W}_{0,i}|$ and construct the BVVS energy as $E_{BVVS} = \sum_i D_i \left(\mathbf{W}_i^2 - \mathbf{W}_{0,i}^2 \right)^2$. The general form of the force field is given as

$$E = E_c + E_r + E_{BVS} + E_{BVVS},$$
(14.2)

where E_c is the Coulomb energy, and E_r is the short-range repulsion. We have proved that E_{BVS} and E_{BVVS} can be justified quantum mechanically in a framework of a tight-bonding model (Horsfield et al. 1996): the BVS energy term is formally equivalent to a second-moment bond order potential (Pettifor and Oleinik 1999; Pettifor et al. 2002), and the BVVS term is linked to the fourth-moment of the local density of states (Sutton 1993; Liu et al. 2013b).

14.3   Predicting Coercive Fields from First Principles

The polarization-electric field (P—E) hysteresis loop of a ferroelectric material contains critical information that relates microscopic structures to macroscopic properties (Jin et al. 2014). Two important quantities, the spontaneous remnant polarization (P_s) and coercive field (E_c, the electric field required to reverse the polarization), can be directly extracted from the P—E loop. Thanks to the development of the modern theory of polarization (King-Smith and Vanderbilt 1993, 1994; Resta 1993, 1994), we can now accurately predict the magnitude of the intrinsic remnant polarization with quantum mechanical methods.

The theoretical prediction of coercive fields is more challenging because it is not an intrinsic property. The polarization reversal in most cases is realized through domain-wall motions driven by external electric fields, which show a strong dependence on the strength and frequency of the applied fields. Efforts have been made to estimate coercive fields with phenomenological free energy theory (Kim et al. 2002; Beckman et al. 2009) based on the DFT energy difference between the polar and nonpolar phases. However, theoretical predictions of coercive fields are often orders of magnitude higher than the experimentally observed values. For example, Kim et al. calculated the theoretical coercive fields as 2.75 MV/cm for $LiTaO_3$ and 5.42 MV/cm for $LiNbO_3$ (Kim et al. 2002). Experimentally, the coercive field for near-stoichiometric $LiTaO_3$ is $\approx$ 17 kV/cm, for near-stoichiometric $LiNbO_3$ it is $\approx$ 40 kV/cm, and it is $\approx$210 kV/cm for congruent $LiNbO_3$ [$Li/(Li + Nb) = 0.485$] (Gopalan et al. 1998; Kim et al. 2001). Similarly, the ideal intrinsic coercive field values for PZT thin films estimated with DFT methods ranged from 2.5 MV/cm to 10 MV/cm (Beckman et al. 2009), which are 2–3 orders of magnitude higher than the experimental values (10 kV/cm) (Kong and Ma 2001; Lente and Eiras 2001; Lente et al. 2004; Shrout and Zhang 2007).

It is evident that an accurate prediction of the coercive field requires a quantitative understanding of the relationship between domain-wall mobility, temperature, and the driving electric field. In this section, we describe a multiscale approach for first-principles-based predictions of "intrinsic" coercive fields in the absence of defects. It is noted that the term "intrinsic" coercive field was often used to refer to the value of electric field predicted by the Landau–Ginzburg mean-field theory (Ginzburg 1945), where the polarization reversal is realized through nearly simultaneous, collective flipping of dipoles of an entire crystal without the domain nucleation-and-growth process. Such "intrinsic" ferroelectric switching was reported in ultrathin films of ferroelectric vinylidene fluoride copolymer (Bune et al. 1998), in which the nucleation was suggested to be inhibited by finite-size effects (Ducharme et al. 2000). Here, the term "intrinsic" coercive field stands for the field which reverses the polarization by domain-wall motions in the absence of "extrinsic" defects.

In the multiscale approach, large-scale bond-valence MD simulations are first used to understand the mechanism of domain-wall motion. The insights gained from MD simulations are then incorporated into a continuum model that quantifies the dynamics of many types of domain walls in various ferroelectrics, which eventually

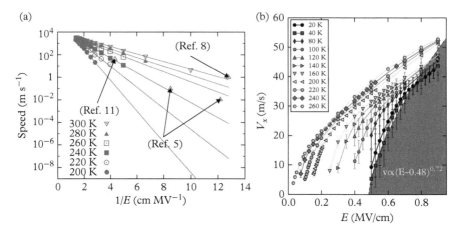

Figure 14.2 *Temperature- and field-dependent domain-wall speeds from molecular dynamics simulations. (a) The 180° domain-wall speed as a function of temperature and electric field plotted as ln v versus 1/E. The linear relationship conforms to Merz's law. (b) The 90° domain-wall mobility as a function of electric field under different temperatures. Temperature- and field-dependent data suggest an intrinsic creep-depinning transition. The domain-wall velocity data at 40 K are in the flow region (shaded area) and are fitted to Equation (14.4) giving θ = 0.72 and E_{C0} = 0.482 MV/m.*

Source: (a) Reprinted from Shin et al. 2007. Copyright © 2007 Springer Nature. (b) Reprinted from Liu et al. 2016b. Copyright © 2016 Springer Nature.

allows for coarse-grained simulation of the *P—E* hysteresis loop and the prediction of coercive fields.

14.3.1 Domain-Wall Mobility from MD Simulations

A quantitative estimation of the speed of domain-wall motion driven by a given electric field is critical for a complete understanding of domain-wall-related dynamical properties. Within the framework of statistical physics, the domain-wall motion is usually understood as an elastic interface moving in a pinning potential created by defects (Ioffe and Vinokur 1987; Chauve et al. 2000). The propagation of domain walls under weak electric fields can be described with a creep process (Tybell et al. 2002; Jo et al. 2009), and the temperature (T) and field (E) dependence of domain-wall velocity v is given as

$$v \propto \exp\left[-\frac{U}{k_B T}\left(\frac{E_{C0}}{E}\right)^{\mu}\right], \tag{14.3}$$

where U is a characteristic energy barrier, k_B is Boltzmann's constant, E_{C0} is a critical field at which depinning occurs at 0 K, and μ is the dynamical exponent. The empirical Merz's law $v = v_0 \exp\left(-E_a/E\right)$ is a reformulation of Equation (14.3) with $\mu = 1$ and $E_a = UE_{C0}/k_B T$. The dynamical exponent μ is used to distinguish the defect type: $\mu = 1$ is due to the random field defects which break the symmetry of the ferroelectric double-well potential (Tybell et al. 2002; Jo et al. 2009); $\mu = 0.5$ suggests the presence of random bond disorder which modifies the double-well potential depth while preserving the symmetry (Paruch et al. 2006; Pertsev et al. 2008). The presence of defects is implicitly assumed to be responsible for the creep behavior. When the electric field becomes larger than the crossing field E_{C0}, the wall experiences a pinning-depinning transition (Jo et al. 2009) with the velocity becoming temperature-independent and given by

$$v \propto (E - E_{C0})^\theta, \tag{14.4}$$

where θ is the velocity exponent that reflects the dimensionality (D) of the wall.

The very first effort of estimating the domain-wall velocity with MD simulations focused on 180° domain walls in $PbTiO_3$ (Shin et al. 2007). By running MD simulations at different temperatures and electric fields, Shin et al. determined the domain-wall speeds as a function of T and E and found that v follows Merz's law, $v = v_0 \exp(-E_a/E)$ (Figure 14.2(a)). However, MD simulations were carried out in the absence of any defects, suggesting an intrinsic origin of $\mu = 1$.

In a later study on 90° domain walls in defect-free $PbTiO_3$, the domain-wall velocity was determined over a broad range of temperatures and electric fields using bond-valence MD simulations with a $40 \times 40 \times 40$ 320,000 -atom supercell (Liu et al. 2016b). As shown in Figure 14.2(b), the v—E—T plot reveals two regions. In the low-field region, the domain-wall speed has a strong temperature dependence and $\ln v$ depends on $1/E$ linearly, conforming to Merz's law and indicating a creep behavior in the absence of defects, similar to the previous findings for the 180° walls. In the high-field region, the temperature dependence becomes much weaker, which is a typical depinning signature. These findings suggest a universal intrinsic response of ferroelectric domain walls to low driving fields characterized by $\mu = 1$. Moreover, it demonstrates that the presence of defects is not necessary for the intrinsic creep-depinning transition.

We propose that the nucleation-and-growth mechanism at the domain wall is responsible for this intrinsic transition. In the presence of an electric field, a nucleus with the dipoles aligned with the electric field will form at the domain wall. The size of the nucleus and thus the nucleation barrier depends inversely on the field strength. At low fields, the large nucleation barrier relative to thermal fluctuations makes nucleation the rate limiting step, giving rise to an Arrhenius dependence of the velocity on the driving force. This is similar to the temperature dependence of reaction rates, where the molecules in a chemical reaction rely on thermal fluctuations to overcome a reaction barrier. At high fields, the nucleus size and nucleation barrier approach zero, and applying a higher electric field essentially generates nuclei more rapidly, resulting in a near-linear dependence on the electric field and a weak temperature dependence.

Results from these simulations suggest that the creep-like behavior does not necessarily result from the presence of defects. It is equally important to point out that the presence of defects will certainly impact the domain-wall dynamics. Experimentally, non-unitary values of the dynamical exponent have been found in epitaxial thin films of PZT, $BaTiO_3$, and $(K,Na)NbO_3$ (Paruch et al. 2006; Pertsev et al. 2008; Men et al. 2017), and ferroelectric polymer (Xiao et al. 2013), suggesting the presence of a large amount of defects and a strong pinning effect (Paruch and Guyonnet 2013). Understanding the effects of defects on the structural and electronic properties of domain walls (Jiang et al. 2017; Körbel et al. 2018) and their consequences for the functional properties of ferroelectrics remain an active research topic (Kalinin et al. 2010).

MD simulations reveal that the domain-wall mobility is fundamentally determined by the nucleation mechanism at the wall. Constructing a physically based continuum model that goes beyond MD simulations therefore requires a detailed understanding of the mechanism and the relative importance of different energetic terms involved in the nucleation process.

14.3.2   Nucleation at Domain Walls

We begin this section by giving a brief summary of the classic nucleation model developed by Miller and Weinreich and then discuss the fundamental problem of this model which we addressed by using the insights gained from MD simulations.

In the seminal work of Miller and Weinreich, a nucleation-and-growth mechanism was proposed to explain the electric field dependence of the domain-wall velocity $v(E)$ in $BaTiO_3$ (Miller and Weinreich 1960). The Miller–Weinreich (MW) nucleation model (illustrated in Figure 14.3(a)) is based on following assumptions: (1) The nucleus boundary is sharp and is oriented orthogonally to the original domain wall; (2) The nucleus originates from the surface of the material, and the total boundary charge is nonzero ($\rho_1 + \rho_2 > 0$); (3) The boundary of the nucleus will have the same interface energy as that of the planar domain wall (σ_w) on which the nucleus is located. The energy of a triangular-shaped (l and a labeled in Figure 14.3(a)) nucleus, U_{nuc}, is then derived as

$$U_{nuc} = -2P_s Ealc + 2\sigma_w c\left(a^2 + l^2\right)^{1/2} + U_d. \tag{14.5}$$

The first term on the right-hand side of the equation is the energy gain for the polarization (P_s) to be aligned with the electric field E within a nucleus of volume of alc. U_d is the depolarization energy due to repulsions of the boundary charges on the two edges of the nucleus:

$$U_d = \frac{8P_s^2 c^2}{\varepsilon}\frac{a^2}{l}\ln\left(\frac{2a}{eb}\right) = 2\sigma_p ca^2/l, \tag{14.6}$$

where ε is the dielectric constant, b is the lower limit of charge separation, and σ_p is introduced to represent the depolarization-contributed domain-wall energy:

$$\sigma_p = \frac{4P_s^2 c}{\epsilon}\ln\left(\frac{2a}{eb}\right). \tag{14.7}$$

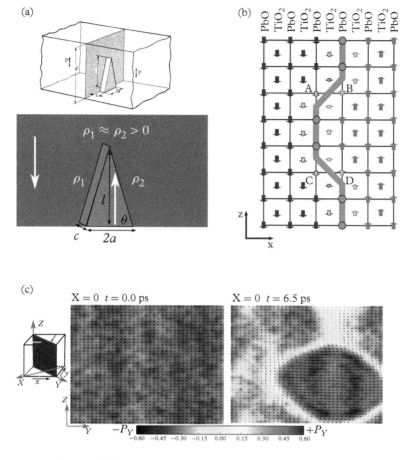

Figure 14.3 *(a) Miller and Weinreich nucleation model at a 180° domain wall. (b) Beveled-shape nucleus with diffusive boundary at a 180° domain wall suggested by MD simulations. (c) Nucleation process at a 90° domain wall. In the presence of an electric field, a diamond-like nucleus forms at t = 6.5 ps.*

Source: (a) Reprinted from Miller and Weinreich 1960. Copyright © 1960 American Physical Society. (b) Reprinted from Shin et al. 2007. Copyright © 2007 Springer Nature. (c) Adapted from Liu et al. 2016b. Copyright © 2016 Springer Nature.

The dimensions of the critical nucleus a^* and l^* determined by the conditions of $\partial U_{\mathrm{nuc}}/\partial a = 0$ and $\partial U_{\mathrm{nuc}}/\partial l = 0$ are given as

$$a^* = \frac{\sigma_w}{P_s E} \frac{\sigma_w + 2\sigma_p}{\sigma_w + 3\sigma_p} \tag{14.8}$$

and

$$l^* = \frac{\sigma_w^{\frac{1}{2}}}{P_s E} \frac{\sigma_w + 2\sigma_p}{(\sigma_w + 3\sigma_p)^{\frac{1}{2}}} \qquad (14.9)$$

The ratio of σ_ρ to σ_w determines the aspect ratio of the critical nucleus: $l^*/a^* = \sqrt{(\sigma_w + 3\sigma_p)}/\sigma_w$. The assumption that the shape of the nucleus is a triangle with $l^* > a^*$ means that $\sigma_p > \sigma_w$.

Miller and Weinreich used the nucleation model to *parameterize* the domain-wall energy with the then available domain-wall velocity data, which allowed the fitting of v–E relationships in many experiments. Despite this success, Tybell et al. later found that first-principles values of σ_w are dramatically higher than the fitted values, whereas using the accurate DFT values (Meyer and Vanderbilt 2002) in the MW model results in velocities dramatically lower than those observed experimentally (Tybell et al. 2002). This poses an important question: what is the relative importance of σ_w and σ_p?

This problem was first addressed in the work of Shin et al. on 180° domain walls in PbTiO$_3$ with MD simulations discussed in the previous section. It was found that the critical nucleus at the 180° domain wall is a diffuse beveled square (Shin et al. 2007), not a tall narrow sharp triangle assumed by MW. This is significant because the beveled shape of the nucleus effectively reduces the *effective domain-wall area* or alternatively the *effective* local domain-wall energy $\sigma_{\text{eff},w}$ for a given nucleus of width a, as explained in Figure 14.3(b).

Specifically, the magnitude of a^* is determined by the σ_w/PE term in both the limits of $\sigma_p > \sigma_w$ and $\sigma_w > \sigma_p$, as derived from Equation (14.8). Thus, a smaller σ_w leads to a smaller critical width a^* and a reduced σ_p^* term as well because $\sigma_p \propto \ln a$ (see Equation (14.7)). Therefore, a decrease in σ_w due to the bevel shape of the nucleus that favors a smaller critical nucleus will significantly decreases the magnitude of σ_p^*.

Recent MD simulations of ferroelastic 90° domain walls in PbTiO$_3$ reveal a more detailed atomistic picture of nucleation and growth. A diamond-like nucleus with substantial diffuseness at the boundary characterized by a gradual polarization change was observed (Figure 14.3(c)), with three other factors contributing to the reduced role of σ_p.

First, as confirmed by MD simulations, the dielectric constant ε at the 90° domain wall is enhanced by a factor of two relative to the bulk value (Xu et al. 2014; Liu and Cohen 2017b; Gu et al. 2018). Because $\sigma_p \propto 1/\varepsilon$, the actual σ_p value is then reduced by a factor of two relative to the original MW estimate. Second, the diamond shape of the nucleus acquires an *interaction cancelation* effect: the repulsive energy penalty due to the interaction between ρ_1 and ρ_2, and between $-\rho_3$ and $-\rho_4$ is canceled by the attractive energy gain of the interaction between ρ_1 and $-\rho_3$, and between ρ_2 and $-\rho_4$. This changes the dependence of σ_p on a from $\ln(2a/eb)$ to $\ln(a/eb)$ (the ln 2 term arises from the repulsive interaction between the charges on the two opposite sides of the triangle in MW model). Finally, the nucleus boundary has a much smaller depolarization charge because of the smaller polarization change across the *diffusive* domain wall. The magnitude of the dipoles at the domain-wall layer is ≈ 2 times smaller than the bulk value. It is this domain-wall layer that undergoes the nucleation-and-growth process. Therefore, the appropriate value of P to be used in Equation (14.7) is much smaller than the bulk value P_s used by MW. The smaller boundary charge decreases the strength of electrostatic interactions and σ_p by a further factor of ≈ 4.

In summary, rather than the $\sigma_p > \sigma_w$ limit assumed by MW, the actual nucleation takes place in the $\sigma_w > \sigma_p$ limit, with the local interface energy playing the dominant role and governing the energetics of nucleation and growth. This insight is critical for the development of a simple and accurate analytical model that captures the most important energy terms.

14.3.3   Continuum Nucleation Model

The study of ferroelectric domains in the framework of Landau–Ginzburg–Devonshire (LGD) theory has a long history. A few examples include the study of ferroelectric phase transitions in $BaTiO_3$ (Devonshire 1954), derivation of the electric field–temperature phase diagrams of single-crystal $BaTiO_3$ (Bell 2001), development of a phenomenological theory for the 90° domain wall (Ishibashi 1993; Ishibashi and Salje 2002), investigation of the flexoelectric effect of a 180° domain wall (Gu et al. 2014), determination of the domain structures in confined nanoferroelectrics (Slutsker et al. 2008), estimation of the domain-wall free energy as a function of temperature (Kumar and Waghmare 2010), and quantification of the threshold field for the nanodomain formation induced by the biased scanning probe microscopy tip (Morozovska et al. 2016).

Building upon the nucleation physics revealed by a careful analysis of MD results, we develop a universal nucleation-and-growth-based analytical model that quantifies the nucleation kinetics for many types of domain walls. Taking the 90° domain wall in *x-y* coordinates as an example, we can perform a simple coordinate transformation and convert it to a 180° domain wall: the polarization component parallel to the domain wall (P_Y) is reversed by 180° across the boundary, while the polarization component perpendicular to the domain wall (P_X) remains almost unchanged (Figure 14.4(a)). This mapping scheme allows us to treat all types of non-180° domain walls as a generalized 180° domain wall.

Our analytical model is based on LGD theory. The nucleation energy U_{nuc} is given as

$$U_{nuc} = \Delta U_E + \Delta U_i, \tag{14.10}$$

where

$$\Delta U_E = -E\int_{-\infty}^{\infty}dX\int_{-\infty}^{\infty}dY\int_{-\infty}^{\infty}dZ\,(P_{nuc}(X,Y,Z) - P_{DW}(X,Y,Z)),$$

$$\Delta U_i = \int_{-\infty}^{\infty}dX\int_{-\infty}^{\infty}dY\int_{-\infty}^{\infty}dZ\,\{[U_g(P_{nuc}) + U_{loc}(P_{nuc})] - [U_g(P_{DW}) + U_{loc}(P_{DW})]\}. \tag{14.11}$$

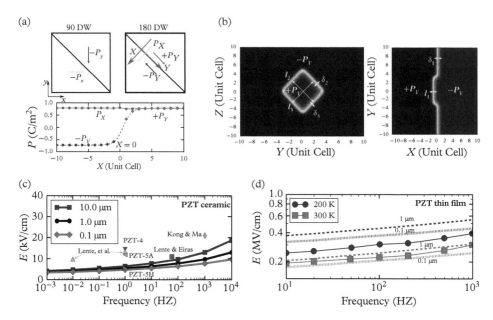

Figure 14.4 *(a) Schematic of mapping a 90° domain wall in x –y plane to a generalized 180° domain wall in the X—Y plane through coordinate transformation. (b) Polarization profile of a nucleus generated by Equation (14.12). (c) Simulated frequency dependence of coercive fields E_c for PZT ceramics for various domain sizes (see legend) at 300 K. (d) Simulated frequency- and temperature-dependent coercive fields for PZT thin films.*
Source: Adapted from Liu et al. 2016b. Copyright © 2018 Springer Nature.

Here, $P_{nuc}(X, Y, Z)$ and $P_{DW}(X, Y, Z)$ are the polarization profiles of a domain wall with and without the nucleus, respectively. ΔU_E is the electrostatic energy gain due to the alignment of dipoles inside the nucleus with the external electric field E. U_i is the interface energy and has two contributions. U_{loc} is the local energy penalty due to the deviation of the local polarization from the bulk value (P_s) and is given by $U_{loc}(P) = A_{loc}\left[1 - (P/P_s)^2\right]^2$, where A_{loc} is the energy difference between the ferroelectric phase and the paraelectric phase. U_g is the gradient energy due to the polarization gradient $(\partial_j P_i)$ and is given by $U_g(P_i) = \sum_j g_{ij}(\partial_j P_i)^2$, where g_{ij} is the coefficient of the gradient of the i-th component of P along direction j. The value of g_{ij} can be derived from the energy and diffusiveness of the domain wall. The formation of a quasi-two-dimensional nucleus within the domain wall creates new interfaces, and thus results in ΔU_i. Unlike the MW model, we do not consider the depolarization energy based on the discussion in the previous section.

At the lowest approximation, the nucleation energy depends only on the spatial variation of P_Y assuming that both P_X and P_Z remain unchanged across the domain wall. MD simulations reveal a diamond-shaped nucleus, and thus the profile of P_Y for a domain wall with a nucleus of size $l_1 \times l_2 \times l_3$ is defined as

$$P_Y = \frac{2P_s}{\sqrt{2}} f(X, l_1, \delta_1) f\left(Y + Z, \sqrt{2}l_2, \delta_2\right) f\left(Y - Z, \sqrt{2}l_3, \delta_3\right) + \frac{P_s}{\sqrt{2}} g(X, l_1, \delta_1),$$

$$(14.12)$$

where $f(x, l, \delta) = \frac{1}{2}\left[\tanh\left(\frac{x+l/2}{\delta/2}\right) - \tanh\left(\frac{x-l/2}{\delta/2}\right)\right]$, $g(x, l, \delta) = \tanh\left(\frac{x-l/2}{\delta/2}\right)$, and δ_i characterizes the diffuseness of the nucleus along direction i. It is easy to show that the value
of $f(x, l, \delta)$ is equal to 1 when $-l/2 < x < l/2$ and becomes zero elsewhere, ideal for
approximating a nucleus of dimension l with a diffuseness of δ. Figure 14.4(b) shows
the polarization profile in the Y-Z and X-Y planes generated by Equation (14.12). With
this analytical nucleation model, we can easily map out the energy of the nucleus as a
function of nucleus size ($l_1 \times l_2$), identify the critical nucleus, and estimate the nucleation
activation energy (ΔU_{nuc}). We then use the Avrami theory of transformation kinetics
to relate ΔU_{nuc} to the activation field E_a in Merz's law, which gives $E_a \approx \frac{1}{d+1} \frac{\Delta U_{nuc}}{k_B T} E$,
where $d = 2$ is the dimensionality (Shin et al. 2007). It turns out that E_a estimated with
the analytical model agrees well with the values obtained directly from MD simulations
(Liu et al. 2016b).

For a given domain wall, all input parameters for the analytical model (e.g. P_s, A_{loc},
and g_i) can be directly estimated from DFT calculations, which enables rapid estimation
of activation fields without the much more time-consuming MD simulations.

14.3.4 Coarse-Grained Simulation of *P—E* Hysteresis Loop

The ultimate goal is to predict the coercive field by simulating the *P—E* hysteresis
loop. We consider the polarization reversal achieved via electric-field-driven domain-wall
motions: the polarization evolution depends on the distance traversed by the domain-
wall. Following the experimental setup used in most hysteresis loop measurements, a
triangular electric field, $E(t)$, with frequency f and maximum magnitude E_0, is used in
the simulation:

$$E(t) = \begin{cases} 4f\, E_0 t & 0 < t < \frac{1}{4f} \\ -4f\, E_0 t + 2E_0 & \frac{1}{4f} < t < \frac{3}{4f} \\ 4f\, E_0 t - 4E_0 & \frac{3}{4f} < t < \frac{1}{f} \end{cases} \qquad (14.13)$$

At $t = 0$, the domain of size d is fully poled with remnant polarization $-P_s$. The
polarization at t can be calculated with

$$P(t) = -P_s + \frac{\int_0^{1/f} v(t)dt}{d} P_s, \qquad (14.14)$$

where $v(t)$ is the domain-wall velocity at time t and is calculated with Merz's law: $v(t) = v_0 \exp(-E_a/E(t))$. When the value of $P(t)$ obtained from Equation (14.14) is larger than
P_s (the domain is fully reversed), $P(t)$ is set to P_s. Plotting $P(t)$ with respect to $E(t)$

gives the hysteresis loop. The coercive field E_c is the magnitude of the electric field when $P(t) = 0$.

To test our simple continuum "macroscopic" model, we first obtain E_a for PbTiO$_3$ from the LGD model with parameters (Zhao and Truhlar 2008) estimated with DFT PBEsol calculations (Perdew et al. 2008). We use $v_0 = 300$ m/s based on MD simulations (it was found later that the coercive field is not sensitive to v_0). We then simulate hysteresis loops at 300 K and obtain the frequency dependence of E_c for varying domain sizes (Figures 14.4(b,c)). Our theoretical coercive fields using the parameters of the 90° domain wall agree well with experimental E_c values (5–20 kV/cm) over a large frequency range (Lente and Eiras 2001; Kong and Ma 2001; Lente et al. 2004; Shrout and Zhang 2007). The E_c values based on the 180° domain wall are larger, which is in agreement with thin-film results (Yang et al. 2010). This suggests that the polarization reversal in ceramics is more likely to go through successive 90° rotations (Lente et al. 2004) because of the smaller nucleation barrier at the 90° domain wall. The model has also been applied successfully to predict the coercive fields of BaTiO$_3$ and BiFeO$_3$.

The good theory–experiment agreement for the E_c values suggests that the coercive field is largely determined by the intrinsic properties of the material, with the nucleation barrier at the domain wall controlling the kinetics of polarization reversal.

14.4 Dynamics in Single-Crystal Relaxors

Relaxor ferroelectrics characterized by a diffuse phase transition with a frequency-dependent dielectric peak were first discovered in the late 1950s. Since then, they have been the focus of intense research due to their fascinating physical properties; an additional impetus to the study of relaxors was provided by the demonstration of the giant dielectric and piezoelectric response of single-crystal relaxors in the seminal work of Park and Shrout (1997). The common feature of ABO_3 perovskite-type relaxors is the intrinsic structural and chemical disorder due to the random arrangement of ions on crystallographically equivalent sites. For example, the typical ferroelectric relaxor Pb(Mg$_{1/3}$Nb$_{2/3}$)O$_3$ (PMN) has Mg^{2+} and Nb^{5+} randomly distributed at the B-sites. To explain the phase transitions of relaxor ferroelectrics, the polar nano-region (PNR) model was proposed in 1983 with Burns T_b, intermediate T^*, and freezing T_f phase transition temperatures related to the appearance, growth, and freezing-in of the nanoscale polar domains in a nonpolar matrix (Burns and Dacol 1983). There have been numerous studies, both experimentally and theoretically, aiming to fully resolve the relationship among the PNR model, diffuse phase transition, and various characteristic transition temperatures (Hirota et al. 2002; Kamba et al. 2005; Dec et al. 2006; Nikolaeva et al. 2006; Burton et al. 2008; Xu et al. 2008; Dkhil et al. 2009; Gehring et al. 2009; Ganesh et al. 2010; Paściak et al. 2012; Hehlen et al. 2016). However, the PNR model is more of a post hoc justification of the experimental results (Hlinka 2012) and lacks the predictive capability to guide the optimization of materials properties (Krogstad et al. 2018).

In this section, we describe our studies of the relaxor ferroelectric $Pb(Mg_{1/3}Nb_{2/3})O_3$–$PbTiO_3$ (PMN–PT) with bond-valence MD simulations (Takenaka et al. 2013, 2017). We demonstrate that the complex phase transitions of relaxors are characterized by well-defined and observable local order parameters. We suggest replacing the PNR model where the relaxor state is assumed to be inhomogeneous and consisting of static polar domains in a nonpolar matrix with a model of a hydrogen-bond-like network of dipoles where the relaxor phase is a multi-domain state with extremely small domain sizes without the nonpolar matrix.

14.4.1 Order Parameters Characterizing Relaxor Phase Transitions

A detailed atomistic understanding of the dynamical correlations of local cation displacements in time and length scales is critical for the construction of a universal model for relaxor ferroelectrics. We employ the technique of the dynamic pair-distribution function (DPDF) to analyze the size and directions of the displacement correlations (Dmowski et al. 2008). The generalized pair-distribution function for an atomic pair, $g(r, t)$, is the ensemble-averaged probability of finding an atom r distance away at a time delay t. $g(r, t)$ is defined as

$$g(r,t) = \frac{1}{N < b^2 >} \sum_{v,\mu=1}^{N} \frac{1}{\sqrt{\pi \left(\sigma_v + \sigma_\mu\right)}} b_v b_\mu \frac{1}{4\pi r^2} \times \int \exp\left[-\frac{\left(r - |R_v(t') - R_\mu(t+t')|\right)^2}{\sigma_v + \sigma_\mu}\right] dt'$$

$$(14.15)$$

where N is the number of atoms, b is the neutron scattering length, $\langle b^2 \rangle$ is the averaged scattering length, σ is the Gaussian smearing width, μ and v are species index, and R is the atomic position. The Fourier transform of $g(r,t)$ gives the frequency spectrum $G(r,\omega)$, which can be measured by inelastic neutron scattering experiments (Dmowski et al. 2008, 2017). The species-pair-dependent PDF, $g_{AB}(r,t)$, measures the probability of an atom of species B located at a distance r away from an atom of species A after a time delay t. Similarly, we can evaluate the Fourier transform of $g_{AB}(r,t)$ to obtain the frequency spectrum $G_{AB}(r,\omega)$.

The instantaneous PDF $g_{AB}(r,t=0)$ is a measure of the instantaneous structural ordering. The peak position of $g_{AB}(r,t=0)$ corresponds to the minimum-energy distance r_0 and the peak broadens because of structural variations and out-of-phase vibrations of *A-B* pairs. The in-phase vibrations preserving the preferred distance r_0 will not broaden the peak. Similarly, the peak of $G_{AB}(r,\omega=0)$, namely, the time-averaged PDF, is also broadened by structural variations and by out-of-phase vibrations of *A-B* pairs around r_0. In addition, it is also broadened due to the in-phase vibrations of *A-B* pairs. Therefore, the difference between $g_{AB}(r,t=0)$ and $G_{AB}(r,\omega=0)$ represents the amplitude and the intensity of the in-phase vibrations of *A-B* pairs. We use the DPDF method to identify the frequencies at which a particular type of atomic pair is vibrating in-phase at a particular separation.

We examine the cation-oxygen instantaneous PDF and time-averaged PDF to investigate the structural dynamics as a function of temperature T through the phase transitions in 0.75PMN–0.25PT. Local cation displacements with respect to the surrounding oxygen cages are obtained simply by subtracting the ideal shortest cation-oxygen distances in the nonpolar high-symmetry perovskite structure from the first peak positions of the PDFs. In Figure 14.5(c), we show $D^{\mathrm{Pb}}_{\mathrm{inst}}$, the instantaneous local Pb displacements; $D^{\mathrm{Pb}}_{\mathrm{static}}$, time-averaged static local displacements; and $D^{\mathrm{Pb}}_{\mathrm{dyn}}$, the dynamic component defined as the difference between $D^{\mathrm{Pb}}_{\mathrm{inst}}$ and $D^{\mathrm{Pb}}_{\mathrm{static}}$.

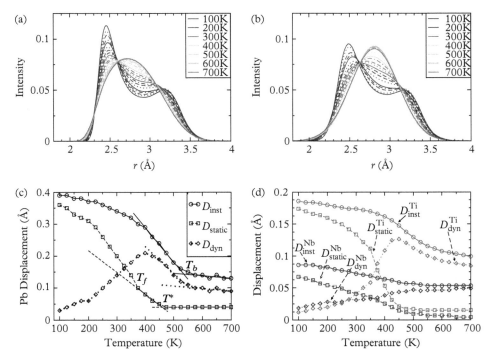

Figure 14.5 *Temperature dependence of instantaneous and time-averaged PDFs for cation-O atomic pairs. (a) Instantaneous and (b) time-averaged Pb-O partial PDFs for the first Pb-O peak. The instantaneous Pb-O PDFs are asymmetric for all temperatures and are positively skewed above $T = 550$ K, suggesting Pb atoms are displaced away from the center of their O_{12} cages even in the high-temperature paraelectric phase. The time-averaged Pb-O PDFs are almost symmetric for $T > 475$ K, indicating small and temperature-independent time-averaged local Pb displacements. The magnitude of the instantaneous cation displacement and its static and dynamic components versus T are shown for Pb in c and for Nb and Ti in (d).*

Source: Reprinted from Takenaka et al. 2013. Copyright © 2013 American Physical Society.

$D_{\text{inst}}^{\text{Pb}}$ characterizes the transition from the paraelectric phase to the dynamic relaxor phase. We find that for $T > T_b = 550$ K, both $D_{\text{inst}}^{\text{Pb}}$ and $D_{\text{static}}^{\text{Pb}}$ are small and temperature insensitive. However, below T_b, $D_{\text{inst}}^{\text{Pb}}$ increases rapidly with decreasing temperature, while $D_{\text{static}}^{\text{Pb}}$ remains small. Equivalently, the rapid increase of $D_{\text{dyn}}^{\text{Pb}}$ indicates the onset of the dynamic relaxor phase below T_b.

The $T = T^* = 450$ K transition temperature is characterized by the onset of $D_{\text{static}}^{\text{Pb}}$, which deviates substantially from its high-temperature value when $T < T^*$. Finally, at $T = T_f = 375$ K, another transition takes place with $D_{\text{static}}^{\text{Pb}}$ increasing rapidly and $D_{\text{dyn}}^{\text{Pb}}$ decreasing as the system undergoes a transition into the frozen phase.

Our study suggests that two quantities are required to characterize the phase transitions of relaxor ferroelectrics. The instantaneous information such as $D_{\text{inst}}^{\text{Pb}}$ describes the transition from the paraelectric phase to the dynamic relaxor phase, while the temperature dependence of the static displacements $D_{\text{static}}^{\text{Pb}}$ reflects the transitions at lower temperatures.

14.4.2 Slush-Like Polar Structures in Single-Crystal Relaxors

To further reveal the structural dynamics in single-crystal relaxors, we carry out MD simulations in a $72 \times 72 \times 72$ supercell of 0.75PMN–0.25PT (Takenaka et al. 2017). Using the trajectories obtained in these simulations, we are able to reproduce the experimental butterfly shape of the diffuse scattering patterns in the relaxor phase. This butterfly shape was long considered to be a structural signature of PNR in a nonpolar matrix. However, a detailed analysis of the correlations between the local Pb displacements reveals that the butterfly shape arises from a multi-domain state rather than from the PNR/nonpolar matrix state.

To quantify the correlations of Pb displacements, we use the angle-correlation distribution function $\Pi(\theta)$ for the pairs of dipoles generated by the displacements of the Pb atoms from the centers of their O_{12} cages separated by n unit cells, which is defined as

$$\Pi^{nv}(\theta) = \frac{1}{N_c N_v \sqrt{\pi \sigma}} \sum_{i=1}^{N_c} \sum_{j=1}^{N_v} \int \int \exp\left[-\frac{[\theta - \arccos(\hat{D}_i(t) \cdot \hat{D}_j(t+t'))]^2}{\sigma} \right] dt\,dt'$$

$$(14.16)$$

where θ is a given angle, v represents a high-symmetry direction ($\langle 100 \rangle$, $\langle 110 \rangle$, or $\langle 111 \rangle$), N_c is the number of Pb atoms in PMN–PT, N_v is 6, 12, and 8 for $v = \langle n00 \rangle$, $\langle nn0 \rangle$, and $\langle nnn \rangle$, respectively, σ is the Gaussian width, $\hat{D}_i(t)$ is the unit vector of a Pb displacement vector at time t, and t' is a time delay. $\Pi^{nv}(\theta)$ essentially measures the time-delay-averaged correlation strength between dipole $\hat{D}_i$ and dipole $\hat{D}_j$ which are n unit cells apart along direction v. For a given n, the peak position of $\Pi^{nv}(\theta)$ is the most probable angle between two dipoles, a value of 0° means perfect correlation, and 90° means no correlation.

We plot the most probable angle θ_p versus distance along the $\langle 100 \rangle$, $\langle 110 \rangle$, and $\langle 111 \rangle$ directions, for several temperatures T (Figure 14.6(a)). The flat curves at around 90° for $T = 600$ K and $T = 700$ K in the paraelectric phase is an indication of no correlations at high temperatures. At $T = 500$ K $< T_b$, a weak distance dependence of θ_p is found.

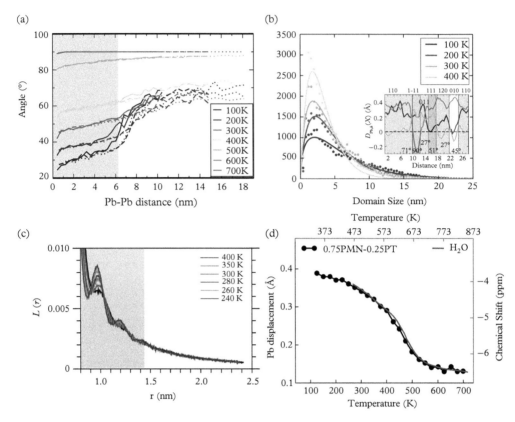

Figure 14.6 *(a) Most probable angles of the time-delay-averaged Pb-Pb angle correlations versus r for various temperatures T. (b) Domain-size distribution from molecular dynamics simulations of PMN–PT. (c) Longitudinal dipole-dipole correlation function for water. (d) Temperature evolution of the magnitude of the instantaneous local displacement of Pb atoms (black) and the ^1H NMR chemical shift.*

Source: *Reprinted from Takenaka et al. 2017. Copyright © 2017 Springer Nature.*

Most notably, it does not converge to 90° even at 18 nm, suggesting a weak but long-range correlation. This is in sharp contrast to the picture of PNR, where a polar domain $\left(\theta_p < 90°\right)$ is embedded in a nonpolar matrix $\left(\theta_p \approx 90°\right)$. At $T = 400$ K $< T^*$, the local dipoles start to form domains, as revealed by the stronger correlations for distances smaller than about 6 nm. At lower T, there is a clear distinction between the strong-correlation region for distances less than about 6 nm and the weak-correlation region for larger distances. Thus, contrary to the previous assumption of PNR growth that takes over some of the nonpolar matrix upon cooling, our results show that the size of the correlated domains is largely independent of the temperature, and it is the correlation inside the domain that increases strongly upon cooling through the phase transitions.

These results clearly demonstrate that PMN–PT adopts a multi-domain state with various low-angle domain walls separating small polar domains, without a nonpolar matrix (Figure 14.6(b)).

The insights gained from large-scale MD simulations allow us to construct a new conceptual model for relaxor ferroelectrics. We propose that the structure and dynamics in relaxor ferroelectrics closely resembles those in water (Takenaka et al. 2013, 2017). By comparing the MD results of relaxor ferroelectrics and H_2O (Hoffmann and Conradi 1997; Kumar et al. 2009; Zhang and Galli 2014; Guardia et al. 2015), we find that for both materials systems the transitions upon cooling from $T > T_b$ are characterized by the following three features: (1) at $T = T_b$ for PMN–PT and $T = 350$ K for water, a correlated region appears, but its size remains temperature insensitive upon cooling. (2) On cooling PMN–PT from T^* or water from $T = 300$ K, the increased structural ordering comes mostly from the increased correlation *within* the correlated region (Figures 14.6(a,c)). (3) The magnitude of the local order parameter (D_{inst}^{Pb} discussed in the last section) closely resembles that of the ^1H NMR chemical shift in water under pressure (Figure 14.6(d)). The ^1H NMR chemical shift correlates with the average number of hydrogen bonds formed by water molecules and is a measure of the local structural ordering.

The relaxor–water analogy can be further extended to describe the phase transitions across a broad temperature range (Figure 14.7). In the paraelectric phase ($T > T_b$), dipoles are randomly oriented and have no correlations, which is similar to the superheated water. In the dynamic relaxor phase ($T_b > T > T^*$), most Pb displacement dipoles are nonzero, giving rise to a correlated network within which a correlated region dynamically appears with changing locations and polarization orientations. This is similar to hydrogen-bonded liquid water. In the temperature range $T^* > T > T_f$, the locations of some polar domains become static, whereas the polarization inside the domain can still rotate, similar to the slush water around the melting temperature characterized by a slurry-like mixture of small ice crystals and liquid water. In this slush water state, the water molecules inside the crystalline regions have slow dynamics, whereas those in the liquid separating the ice crystals have fast dynamics; the small ice crystals can rotate with respect to each other. Below T_f, domains become static, and the correlation strength and magnitude of polarization inside each domain become stronger even though the domain size remains nearly unchanged.

The slush model of relaxor ferroelectrics explains six different physical properties of relaxors.

(1) It is well known that the ferroelectric rather than the relaxor state is obtained when a relaxor is cooled under an applied electric field. The origin of this effect is that the electric field applied during cooling biases all domains to polarize along the thermodynamically favored direction specified by the field. Thus, while the domains still exist with the polarization direction varying between the domains, their individual polarization tends to point in the field direction, until at low temperature the barrier of reorientation of an individual domain becomes high,

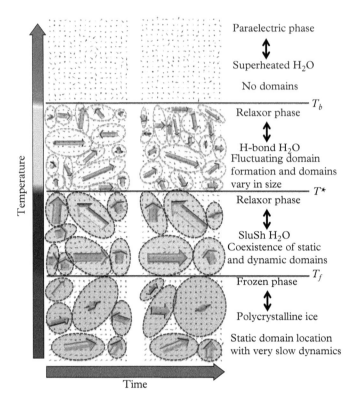

Figure 14.7 *Schematic of our slush-like model for the phase transitions in relaxors. Source: Reprinted from Takenaka et al. 2017. Copyright © 2017 Springer Nature.*

as all neighboring domains have polarization of a large component along the direction of the field. In the case of cooling without an applied electric field, the thermodynamic driving force for the domains to orient in the same direction is weak, and therefore, at low T, the domains are strongly polarized in different directions. When a field is then applied at low T, the barrier of reorientation is too large, so that the system remains in the relaxor state even under the application of a strong field.

(2) In many relaxor systems, it was observed that the addition of Ti leads to the eventual suppression of relaxor behavior, while lower Ti content increases the strength of the relaxor phase. This is due to the lesser (greater) content of the highly underbonded $Pb(Mg_{1/2}Nb_{1/2})O_3$ local environments that decouple the polarization of the neighboring unit cells and act as domain walls, which makes the domain size larger (smaller) as the Ti fraction increases (decreases) and is in agreement with the observed smaller diffuse scattering patterns at higher Ti content (Matsuura et al. 2006). Eventually, at a high enough Ti content, the increased domain size leads to the relaxor to ferroelectric phase transition.

(3) The experimentally reported signatures of the structural phase transitions such as diffuse scattering and dynamic PDFs arise from the appearance of dynamically varying domains at T_b, the nanometer-sized static domains separated by domain walls near T^*, and the strong correlations of the local dipoles within domains at T_f.

(4) Multi-domain systems with small domains and the decoupling of polarization fluctuations of each domain in relaxors naturally lead to Kohlrausch–Williams–Watts dielectric relaxation and Vögel–Fulcher characteristics of the dielectric response (Chamberlin 2014).

(5) Polarization fluctuations of small domains in the dynamic relaxor phase create strong local electric fields that smear the ferroelectric transitions, and are responsible for the diffuse phase transition and the temperature dependence of the dielectric constant (Bokov and Ye 2002).

(6) Finally, the slush structure can be related to the giant piezoelectricity of single-crystal relaxors. Experimental and theoretical work has shown that the ferroelectric domain-wall region exhibits a high permittivity that can dramatically enhance the high-frequency intrinsic dielectric response relative to a bulk single-domain material (Xu et al. 2014). The small nanoscale domains imply the presence of an extremely high density of domain walls that together with the mixture of ferroelectric- and paraelectric-like unit cells in relaxors enable greater flexibility for polarization rotations over a wide temperature range, thus favoring the giant piezoelectric response in relaxors.

14.5 Perspectives

The multiscale approach that bridges methods at different scales is a powerful tool for understanding the emergent functional properties of ferroelectrics and will play a critical role in revealing the rich physics of ferroelectric domains. The current multiscale approach is a "case-by-case" method that targets a particular material. It is necessary to develop a more general and flexible multiscale computational framework that can be readily applied to a broad range of materials in order to achieve accelerated rational design and optimization of ferroelectrics. Developing reliable and efficient classical force fields for large-scale MD simulations remains a bottleneck of the multiscale approach. Therefore, it is necessary to develop better computational techniques and infrastructure to facilitate and ultimately automate the process of force-field development to enable high-throughput multiscale simulations of ferroelectrics. This, of course, is a great challenge that is not specific to ferroelectrics but rather applies generally to the field of computational materials design. In the absence of such reliable high-throughput multiscale simulations, careful studies of prototypical ferroelectric materials systems focusing on extracting the fundamental factors that govern macroscopic materials properties will likely be a fruitful approach for obtaining better ferroelectrics. For example, the finding that the giant piezoelectricity of single-crystal relaxors arising from high-density low-angle domain walls opens up new possibilities for controlled design of new ferroelectrics by selectively pinning/depinning domain walls through doping.

In the field of ferroelectrics, much remains to be understood about emergent phenomena arising from the complex interplay between compositions, domains, domain walls, defects, heterostructures, superlattices, and various thermodynamic variables such as electric field and stress. We highlight three directions that have the greatest influence and impact. First, the effects of defects (e.g. vacancies and dopants) on the dynamics of domain walls require a deeper understanding at the atomic level. For example, it is well established experimentally that a small amount of dopants (≈ 1 atomic per cent) may drastically improve the piezoelectric properties of ferroelectrics, which was often attributed to domain pinning or/and domain-wall pinning. However, a quantitative relationship between dopant types, charge states, and domain-wall mobility is still lacking. Recent observation of giant piezoelectricity in Sm-doped $Pb(Mg_{1/3}Nb_{2/3})O_3$-$PbTiO_3$ single crystals (Li et al. 2019) suggests that we may have not yet reached the full potential of single-crystal ferroelectrics via dopant engineering. Developing an efficient and novel multiscale approach that relates dopant/defect chemistry to a colossal (or large) physical response will further facilitate the development of high-performance ferroelectrics. Second, domain-wall-related phenomena, particularly polar vortices and room-temperature skyrmions recently realized in $PbTiO_3$/$SrTiO_3$ superlattices (Yadav et al. 2016; Das et al. 2019), are likely to support novel functional properties. Recent work has demonstrated the existence of negative capacitance at the core of the polar vortices (Yadav et al. 2019). The dynamics of these polar vortices and skyrmions in the presence of external electric fields deserve detailed investigations. The potential of controlling the chirality of polar vortices may open up a whole field of research using the chirality as a new order parameter. Finally, the physics and functional properties of domain walls in novel ferroelectrics such as two-dimensional van der Waals ferroelectrics (Ding et al. 2017) and silicon-compatible ferroelectric HfO_2-based thin films (Böscke et al. 2011) remain relatively unexplored. It remains to be seen how the "nucleation-and-growth" mechanism that is intimately related to polarization switching will play out in low-dimensional ferroelectrics. We believe the powerful combination of theory, synthesis, and characterization may facilitate the harnessing of these novel ferroelectric domain-wall phenomena to drive new technologies.

Acknowledgments

SL is partly supported by the SEDD Distinguished Postdoc Fellowship at the US Army Research Laboratory and Westlake Foundation. AMR acknowledges support from the National Science Foundation, under grant DMR-1719353. IG acknowledges support from the Israel Science Foundation.

....

REFERENCES

Alder B. J., Wainwright T. E., "Studies in molecular dynamics. I. General method," *J. Chem. Phys.* 31(2), 459–466 (1959).

Beckman S. P., Wang W., Rabe K. M., Vanderbilt D., "Ideal barriers to polarization reversal and domain-wall motion in strained ferroelectric thin films," *Phys. Rev. B* 79, 144124 (2009).

Bell A. J., "Phenomenologically derived electric field-temperature phase diagrams and piezoelectric coefficients for single crystal barium titanate under fields along different axes," *J. Appl. Phys.* 89(7), 3907–3914 (2001).

Bellaiche L., Garcia A., Vanderbilt D., "Finite-temperature properties of $Pb(Zr_{1-x}Ti_x)O_3$ alloys from first principles," *Phys. Rev. Lett.* 84, 5427–5430 (2000).

Bokov A. A., Ye Z.-G., "Universal relaxor polarization in $Pb(Mg_{1/3}Nb_{2/3})O_3$ and related materials," *Phys. Rev. B* 66, 064103 (2002).

Böscke T. S., Müller J., Bräuhaus D., Schröder U., Böttger U., "Ferroelectricity in hafnium oxide thin films," *Appl. Phys. Lett.* 99(10), 102903 (2011).

Brese N. E., O'Keeffe M., "Bond-valence parameters for solids," *Acta Crystallogr. B* 47(2), 192–197 (1991).

Brown I. D., "Recent developments in the methods and applications of the bond valence model," *Chem. Rev.* 109, 6858–6919 (2009).

Brown I. D., Shannon R. D., "Empirical bond-strength–bond-length curves for oxides," *Acta Crystallogr. A* 29(3), 266–282 (1973).

Brown I. D., Wu K. K., "Empirical parameters for calculating cation–oxygen bond valences," *Acta Crystallogr. B* 32(7), 1957–1959 (1976).

Bune A. V., Fridkin V. M., Ducharme S., Blinov L. M., Palto S. P., Sorokin A. V., Yudin S. G., Zlatkin A., "Two-dimensional ferroelectric films," *Nature* 391(6670), 874–877 (1998).

Burns G., Dacol F. H., "Glassy polarization behavior in ferroelectric compounds $Pb(Mg_{13}Nb_{23})O_3$ and $Pb(Zn_{13}Nb_{23})O_3$," *Solid State Commun.* 48(10), 853–856 (1983).

Burton B. P., Cockayne E., Tinte S., Waghmare U. V., "Effect of nearest neighbor Pb-O divacancy pairs on the ferroelectric-relaxor transition in $Pb(Sc_{1/2}Nb_{1/2})O_3$," *Phys. Rev. B* 77, 144114 (2008).

Chamberlin R., "The big world of nanothermodynamics," *Entropy* 17(1), 52–73 (2014).

Chauve P., Giamarchi T., Le Doussal P., "Creep and depinning in disordered media," *Phys. Rev. B* 62(10), 6241–6267 (2000).

Chen L.-Q., "Phase-field method of phase transitions/domain structures in ferroelectric thin films: a review," *J. Am. Ceram. Soc.* 91(6), 1835–1844 (2008).

Curtarolo S., Setyawan W., Hart G. L. W., Jahnatek M., Chepulskii R. V., Taylor R. H., Wang S., Xue J., Yang K., Levy O., Mehl M. J., Stokes H. T., Demchenko D. O., Morgan D., "AFLOWLIB.ORG: a distributed materials properties repository from high-throughput ab initio calculations," *Comput. Mater. Sci.* 58, 218–226 (2012).

Das S., Tang Y. L., Hong Z., Gonçalves M. A. P., McCarter M. R., Klewe C., Nguyen K. X., Gómez-Ortiz F., Shafer P., Arenholz E., Stoica V. A., Hsu S.-L., Wang B., Ophus C., Liu J. F., Nelson C. T., Saremi S., Prasad B., Mei A. B., Schlom D. G., Íñiguez J., García-Fernández P., Muller D. A., Chen L. Q., Junquera J., Martin L. W., Ramesh R., "Observation of room-temperature polar skyrmions," *Nature* 568(7752), 368–372 (2019).

de Gironcoli S., Baroni S., Resta R., "Piezoelectric properties of III-V semiconductors from first-principles linear-response theory," *Phys. Rev. Lett.* 62, 2853–2856 (1989).

de Jong M., Chen W., Geerlings H., Asta M., Persson K. A., "A database to enable discovery and design of piezoelectric materials," *Sci. Data* 2, 150053 (2015).

Dec J., Shvartsman V. V., Kleemann W., "Domainlike precursor clusters in the paraelectric phase of the uniaxial relaxor $Sr_{0.61}Ba_{0.39}Nb_2O_6$," *Appl. Phys. Lett.* 89(21), 212901 (2006).

Devonshire A. F., "Theory of ferroelectrics," *Adv. Phys.* 3(10), 85–130 (1954).

Ding W., Zhu J., Wang Z., Gao Y., Xiao D., Gu Y., Zhang Z., Zhu W., "Prediction of intrinsic two-dimensional ferroelectrics in In_2Se_3 and other III_2-VI_3 van der Waals materials," *Nat. Commun.* 8(14956), (2017).

Dkhil B., Gemeiner P., Al-Barakaty A., Bellaiche L., Dul'kin E., Mojaev E., Roth M., "Intermediate temperature scale T* in lead-based relaxor systems," *Phys. Rev. B* 80, 064103 (2009).

Dmowski W., Diallo S. O., Lokshin K., Ehlers G., Ferré G., Boronat J., Egami T., "Observation of dynamic atom-atom correlation in liquid helium in real space," *Nat. Commun.* 8, 15294 (2017).

Dmowski W., Vakhrushev S. B., Jeong I.-K., Hehlen M. P., Trouw F., Egami T., "Local lattice dynamics and the origin of the relaxor ferroelectric behavior," *Phys. Rev. Lett.* 100, 137602 (2008).

Ducharme S., Fridkin V. M., Bune A. V., Palto S. P., Blinov L. M., Petukhova N. N., Yudin S. G., "Intrinsic ferroelectric coercive field," *Phys. Rev. Lett.* 84, 175–178 (2000).

Frenkel D., Smit B., *Understanding Molecular Simulation: From Algorithms to Applications*, Volume 1 (Academic Press, Cambridge, 2001).

Ganesh P., Cockayne E., Ahart M., Cohen R. E., Burton B., Hemley R. J., Ren Y., Yang W., Ye Z.-G., "Origin of diffuse scattering in relaxor ferroelectrics," *Phys. Rev. B* 81, 144102 (2010).

Garrity K. F., Bennett J. W., Rabe K. M., Vanderbilt D., "Pseudopotentials for high-throughput DFT calculations," *Comput. Mater. Sci.* 81, 446–452 (2014).

Gehring P. M., Hiraka H., Stock C., Lee S.-H., Chen W., Ye Z.-G., Vakhrushev S. B., Chowdhuri Z., "Reassessment of the Burns temperature and its relationship to the diffuse scattering, lattice dynamics, and thermal expansion in relaxor $Pb(Mg_{1/3}Nb_{2/3})O_3$," *Phys. Rev. B* 79, 224109 (2009).

Gindele O., Kimmel A., Cain M. G., Duffy D., "Shell model force field for lead zirconate titanate $Pb(Zr_{1-x}Ti_x)O_3$," *J. Phys. Chem. C* 119(31), 17784–17789 (2015).

Ginzburg V. L., "The dielectric properties of crystals of seignettcelectric substances and of barium titanate," *Zh. Eksp. Teor. Fiz.* 15, 739 (1945).

Gopalan V., Mitchell T. E., Furukawa Y., Kitamura K., "The role of nonstoichiometry in 180° domain switching of $LiNbO_3$ crystals," *Appl. Phys. Lett.* 72(16), 1981–1983 (1998).

Gu Y., Li M., Morozovska A. N., Wang Y., Eliseev E. A., Gopalan V., Chen L.-Q., "Flexoelectricity and ferroelectric domain wall structures: Phase-field modeling and DFT calculations," *Phys. Rev. B* 89, 174111 (2014).

Gu Z., Pandya S., Samanta A., Liu S., Xiao G., Meyers C. J. G., Damodaran A. R., Barak H., Dasgupta A., Saremi S., Polemi A., Wu L., Podpirka A. A., Will-Cole A., Hawley C. J., Davies P. K., York R. A., Grinberg I., Martin L. W., Spanier J. E., "Resonant domain-wall-enhanced tunable microwave ferroelectrics," *Nature* 560(7720), 622–627 (2018).

Guardia E., Skarmoutsos I., Masia M., "Hydrogen Bonding and related properties in liquid water: a car–parrinello molecular dynamics simulation study," *J. Phys. Chem. B* 119(29), 8926–8938 (2015).

Harvey M. A., Baggio S., Baggio R., "A new simplifying approach to molecular geometry description: the vectorial bond-valence model," *Acta Crystallogr. B* 62(6), 1038–1042 (2006).

Hehlen B., Al-Sabbagh M., Al-Zein A., Hlinka J., "Relaxor ferroelectrics: back to the single-soft-mode picture," *Phys. Rev. Lett.* 117, 155501 (2016).

Hirota K., Ye Z.-G., Wakimoto S., Gehring P. M., Shirane G., "Neutron diffuse scattering from polar nanoregions in the relaxor $Pb(Mg_{1/3}Nb_{2/3})O_3$," *Phys. Rev. B* 65, 104105 (2002).

Hlinka J., "Do we need the ether of polar nanoregions?," *J. Adv. Dielectrics* 02(02), 1241006 (2012).

Hoffmann M. M., Conradi M. S., "Are there hydrogen bonds in supercritical water?," *J. Am. Chem. Soc.* 119(16), 3811–3817 (1997).

Horsfield A. P., Bratkovsky A. M., Fearn M., Pettifor D. G., Aoki M., "Bond-order potentials: theory and implementation," *Phys. Rev. B* 53, 12694–12712 (1996).

Hu H.-L., Chen L.-Q., "Three-dimensional computer simulation of ferroelectric domain formation," *J. Am. Ceram. Soc.* 81(3), 492–500 (1998).

Ioffe L. B., Vinokur V. M., "Dynamics of interfaces and dislocations in disordered media," *J. Phys. C* 20(36), 6149 (1987).

Ishibashi Y., "The 90°-wall in the tetragonal phase of $BaTiO_3$-type ferroelectrics," *J. Phys. Soc. Jpn.* 62(3), 1044–1047 (1993).

Ishibashi Y., Salje E., "A theory of ferroelectric 90 degree domain wall," *J. Phys. Soc. Jpn.* 71(11), 2800–2803 (2002).

Jain A., Ong S. P., Hautier G., Chen W., Richards W. D., Dacek S., Cholia S., Gunter D., Skinner D., Ceder G., Persson K. A., "Commentary: The Materials Project: A materials genome approach to accelerating materials innovation," *APL Mater.* 1(1), 011002 (2013).

Jiang J., Bai Z. L., Chen Z. H., He L., Zhang D. W., Zhang Q. H., Shi J. A., Park M. H., Scott J. F., Hwang C. S., Jiang A. Q., "Temporary formation of highly conducting domain walls for non-destructive read-out of ferroelectric domain-wall resistance switching memories," *Nat. Mater.* 17(1), 49–56 (2017).

Jin L., Li F., Zhang S., "Decoding the fingerprint of ferroelectric loops: comprehension of the material properties and structures," *J. Am. Ceram. Soc.* 97(1), 1–27 (2014).

Jo J. Y., Yang S. M., Kim T. H., Lee H. N., Yoon J.-G., Park S., Jo Y., Jung M. H., Noh T. W., "Nonlinear dynamics of domain-wall propagation in epitaxial ferroelectric thin films," *Phys. Rev. Lett.* 102, 045701 (2009).

Kalinin S. V., Rodriguez B. J., Borisevich A. Y., Baddorf A. P., Balke N., Chang H. J., Chen L. Q., Choudhury S., Jesse S., Maksymovych P., Nikiforov M. P., Pennycook S. J., "Defect-mediated polarization switching in ferroelectrics and related materials: from mesoscopic mechanisms to atomistic control," *Adv. Mater.* 22(3), 314–322 (2010).

Kamba S., Kempa M., Bovtun V., Petzelt J., Brinkman K., Setter N., "Soft and central mode behavior in $PbMg_{1/3}Nb_{2/3}O_3$ relaxor ferroelectric," *J. Phys.: Condens. Mat.* 17(25), 3965–3974 (2005).

Kim S., Gopalan V., Gruverman A., "A case study in lithium niobate and lithium tantalite," *Appl. Phys. Lett.* 80(15), 2740–2742 (2002).

Kim S., Gopalan V., Kitamura K., Furukawa Y., "Domain reversal and nonstoichiometry in lithium tantalite," *J. Appl. Phys.* 90(6), 2949–2963 (2001).

King-Smith R. D., Vanderbilt D., "Theory of polarization of crystalline solids," *Phys. Rev. B* 47, 1651–1654 (1993).

King-Smith R. D., Vanderbilt D., "First-principles investigation of ferroelectricity in perovskite compounds," *Phys. Rev. B* 49, 5828–5844 (1994).

Kong L. B., Ma J., "PZT ceramics formed directly from oxides via reactive sintering," *Mater. Lett.* 51(2), 95–100 (2001).

Körbel S., Hlinka J., Sanvito S., "Electron trapping by neutral pristine ferroelectric domain walls in BiFeO$_3$," *Phys. Rev. B* 98, 100104 (2018).

Krogstad M. J., Gehring P. M., Rosenkranz S., Osborn R., Ye F., Liu Y., Ruff J. P. C., Chen W., Wozniak J. M., Luo H., Chmaissem O., Ye Z.-G., Phelan D., "The relation of local order to material properties in relaxor ferroelectrics," *Nat. Mater.* 17(8), 718–724 (2018).

Kumar A., Waghmare U. V., "First-principles free energies and Ginzburg-Landau theory of domains and ferroelectric phase transitions in BaTiO$_3$," *Phys. Rev. B* 82, 054117 (2010).

Kumar P., Buldyrev S. V., Stanley H. E., "A tetrahedral entropy for water," *Proc. Natl. Acad. Sci.* 106(52), 22130–22134 (2009).

Lente M. H., Eiras J. A., "90° domain reorientation and domain wall rearrangement in lead zirconate titanate ceramics characterized by transient current and hysteresis loop measurements," *J. Appl. Phys.* 89(9), 5093–5099 (2001).

Lente M. H., Picinin A., Rino J. P., Eiras J. A., "90° domain wall relaxation and frequency dependence of the coercive field in the ferroelectric switching process," *J. Appl. Phys.* 95(5), 2646–2653 (2004).

Li F., Cabral M. J., Xu B., Cheng Z., Dickey E. C., LeBeau J. M., Wang J., Luo J., Taylor S., Hackenberger W., Bellaiche L., Xu Z., Chen L. Q., Shrout T. R., Zhang S., "Giant piezoelectricity of Sm-doped Pb(Mg$_{1/3}$Nb$_{2/3}$)O$_3$-PbTiO$_3$ single crystals," *Science* 364(6437), 264–268 (2019).

Lines M. E., Glass A. M., *Principles and Applications of Ferroelectrics and Related Materials* (Clarendon Press, Oxford, 1977).

Liu G., Zhang S., Jiang W., Cao W., "Losses in ferroelectric materials," *Mater. Sci. Eng. R Rep.* 89, 1–48 (2015).

Liu J., Fernández-Serra M. V., Allen P. B., "First-principles study of pyroelectricity in GaN and ZnO," *Phys. Rev. B* 93, 081205 (2016a).

Liu S., Cohen R. E., "Origin of negative longitudinal piezoelectric effect," *Phys. Rev. Lett.* 119, 207601 (2017a).

Liu S., Cohen R. E., "Local atomic and magnetic structure of dilute magnetic semiconductor (Ba,K)(Zn,Mn)$_2$As$_2$," *Phys. Rev. B* 95, 094102 (2017b).

Liu S., Grinberg I., Rappe A. M., "Development of a bond-valence based interatomic potential for BiFeO$_3$ for accurate molecular dynamics simulations," *J. Phys.: Condens. Mat.* 25(10), 102202-1–6 (2013a).

Liu S., Grinberg I., Rappe A. M., "Intrinsic ferroelectric switching from first principles," *Nature* 534(7607), 360–363 (2016b).

Liu S., Grinberg I., Takenaka H., Rappe A. M., "Reinterpretation of the bond-valence model with bond-order formalism: an improved bond-valence-based interatomic potential for PbTiO$_3$," *Phys. Rev. B* 88, 104102 (2013b).

Matsuura M., Hirota K., Gehring P. M., Ye Z.-G., Chen W., Shirane G., "Composition dependence of the diffuse scattering in the relaxor ferroelectric compound (1−x)Pb(Mg$_{1/3}$Nb$_{2/3}$)O$_3$−xPbTiO$_3$ (0≤x≤0.40)," *Phys. Rev. B* 74, 144107 (2006).

Men T.-L., Thong H. C., Li J.-T., Li M., Zhang J., Zhong V., Luo J., Chu X., Wang K., Li J.-F., "Domain growth dynamics in (K, Na)NbO$_3$ ferroelectric thin films," *Ceram. Int.* 43(12), 9538–9542 (2017).

Metropolis N., Rosenbluth A. W., Rosenbluth M. N., Teller A. H., Teller E., "Equations of state calculations by fast computing machines," *J. Chem. Phys.* 21(6), 1087–1092 (1953).

Meyer B., Vanderbilt D., "Ab initio study of ferroelectric domain walls in PbTiO$_3$," *Phys. Rev. B* 65, 104111 (2002).

Miller R. C., Weinreich G., "Shell model simulations by adiabatic dynamics," *Phys. Rev.* 117, 1460–1466 (1960).

Mitchell P. J., Fincham D., "Shell model simulations by adiabatic dynamics," *J. Phys.: Condens. Mat.* 5(8), 1031 (1993).

Morozovska A. N., Ievlev A. V., Obukhovskii V. V., Fomichov Y., Varenyk O. V., Shur V. Y., Kalinin S. V., Eliseev E. A., "Self-consistent theory of nanodomain formation on nonpolar surfaces of ferroelectrics," *Phys. Rev. B* 93, 165439 (2016).

Nambu S., Sagala D. A., "Domain formation and elastic long-range interaction in ferroelectric perovskites," *Phys. Rev. B* 50, 5838–5847 (1994).

Nikolaeva E. V., Shur V. Y., Shishkin E. I., Sternberg A., "Nanoscale domain structure in relaxor PLZT x/65/35 ceramics," *Ferroelectrics* 340(1), 137–143 (2006).

Park S.-E., Shrout T. R., "Ultrahigh strain and piezoelectric behavior in relaxor based ferroelectric single crystals," *J. Appl. Phys.* 82(4), 1804–1811 (1997).

Paruch P., Giamarchi T., Tybell T., Triscone J.-M., "Nanoscale studies of domain wall motion in epitaxial ferroelectric thin films," *J. Appl. Phys.* 100(5), 0516081–10 (2006).

Paruch P., Guyonnet J., "Nanoscale studies of ferroelectric domain walls as pinned elastic interfaces," *C. R. Phys.* 14(8), 667–684 (2013).

Paściak M., Welberry T. R., Kulda J., Kempa M., Hlinka J., "Polar nanoregions and diffuse scattering in the relaxor ferroelectric $PbMg_{1/3}Nb_{2/3}O_3$," *Phys. Rev. B* 85, 224109 (2012).

Perdew J. P., Ruzsinszky A., Csonka G. I., Vydrov O. A., Scuseria G. E., Constantin L. A., Zhou X., Burke K., "Erratum: restoring the density-gradient expansion for exchange in solids and surfaces," *Phys. Rev. Lett.* 100, 136406 (2008).

Pertsev N. A., Petraru A., Kohlstedt H., Waser R., Bdikin I. K., Kiselev D., Kholkin A. L., "Dynamics of ferroelectric nanodomains in $BaTiO_3$ epitaxial thin films via piezoresponse force microscopy," *Nanotechnology* 19(37), 375703 (2008).

Pettifor D. G., Oleinik I. I., "Analytic bond-order potentials beyond Tersoff-Brenner. I. Theory," *Phys. Rev. B* 59(13), 8487–8499 (1999).

Pettifor D. G., Oleinik I. I., Nguyen-Manh D., Vitek V., "Bond-order potentials: bridging the electronic to atomistic modelling hierarchies," *Comp. Mat. Sci.* 23(1–4), 33–37 (2002).

Pramanick A., Prewitt A. D., Forrester J. S., Jones J. L., "Domains, domain walls and defects in perovskite ferroelectric oxides: a review of present understanding and recent contributions," *Crit. Rev. Solid State Mater. Sci.* 37(4), 243–275 (2012).

Rabe K. M., Ahn C. H., Triscone J.-M., *Physics of Ferro-Electrics: A Modern Perspective*, Volume 105 (Springer Science & Business Media, New York, 2007).

Rahman A., "Correlations in the motion of atoms in liquid argon," *Phys. Rev.* 136, A405–411 (1964).

Resta R., "Macroscopic electric polarization as a geometric quantum phase," *Europhys. Lett. (EPL)* 22(2), 133–138 (1993).

Resta R., "Macroscopic polarization in crystalline dielectrics: the geometric phase approach," *Rev. Mod. Phys.* 66, 899–915 (1994).

Sághi-Szabó G., Cohen R. E., Krakauer H., "First-principles study of piezoelectricity in $PbTiO_3$," *Phys. Rev. Lett.* 80, 4321–4324 (1998).

Scott J. F., "Applications of modern ferroelectrics," *Science* 315(5814), 954–959 (2007).

Semenovskaya S., Khachaturyan A. G., "Development of ferroelectric mixed states in a random field of static defects," *J. Appl. Phys.* 83(10), 5125–5136 (1998).

Sepliarsky M., Asthagiri A., Phillpot S. R., Stachiotti M. G., Migoni R. L., "Atomic-level simulation of ferroelectricity in oxide materials," *Curr. Opin. Solid State Mater. Sci.* 9(3), 107–113 (2005).

Sepliarsky M., Cohen R. E., "Development of a shell model potential for molecular dynamics for $PbTiO_3$ by fitting first-principles results," *AIP Conf. Proc.* 626, 36–44 (2002).

Sepliarsky M., Cohen R. E., *J. Phys.: Condens. Mat.* 2011(43), 435902 (2011).

Sepliarsky M., Stachiotti M. G., Migoni R. L., "Structural instabilities in $KTaO_3$ and $KNbO_3$ described by the nonlinear oxygen polarizability model," *Phys. Rev. B* 52, 4044–4049 (1995).

Shin Y.-H., Grinberg I., Chen I.-W., Rappe A. M., "Nucleation and growth mechanism of ferroelectric domain-wall motion," *Nature* 449, 881–886 (2007).

Shrout T. R., Zhang S. J., "Lead-free piezoelectric ceramics: alternatives for PZT?, *J. Electroceram.* 19(1), 113–126 (2007).

Slutsker J., Artemev A., Roytburd A., "Phase-field modeling of domain structure of confined nanoferroelectrics," *Phys. Rev. Lett.* 100, 087602 (2008).

Sutton A. P., *Electronic Structure of Materials* (Oxford University Press, Oxford, 1993).

Takenaka H., Grinberg I., Liu S., Rappe A. M., "Slush-like polar structures in single-crystal relaxors," *Nature* 546(7658), 391–395 (2017).

Takenaka H., Grinberg I., Rappe A. M., "Anisotropic local correlations and dynamics in a relaxor ferroelectric," *Phys. Rev. Lett.* 110, 147602 (2013).

Tan H., Takenaka H., Xu C., Duan W., Grinberg I., Rappe A. M., "First-principles studies of the local structure and relaxor behavior of $Pb(Mg_{1/3}Nb_{2/3})O_3-PbTiO_3$-derived ferroelectric perovskite solid solutions," *Phys. Rev. B* 97, 174101 (2018).

Tinte S., Stachiotti M. G., Phillpot S. R., Sepliarsky M., Wolf D., Migoni R. L., "Ferroelectric properties of $Ba_xSr_{1-x}TiO_3$ solid solutions obtained by molecular dynamics simulation," *J. Phys.: Condens. Mat.* 16(20), 3495–3506 (2004).

Tinte S., Stachiotti M. G., Sepliarsky M., Migoni R. L., Rodriguez C. O., "Atomistic modelling of $BaTiO_3$ based on first-principles calculations," *J. Phys.: Condens. Mat.* 11(48), 9679 (1999).

Tybell T., Paruch P., Giamarchi T., Triscone J.-M., "Domain wall creep in epitaxial ferroelectric $Pb(Zr_{0.2}Ti_{0.8})O_3$ thin films," *Phys. Rev. Lett.* 89, 097601 (2002).

Vielma J. M., Schneider G., "Shell model of $BaTiO_3$ derived from ab-initio total energy calculations," *J. Appl. Phys.* 114(17), 174108 (2013).

Völker B., Marton P., Elsässer C., Kamlah M., "Multiscale modeling for ferroelectric materials: A transition from the atomic level to phase-field modeling," *Contin. Mech. Thermodyn.* 23(5), 435–451 (2011).

Xiao Z., Poddar S., Ducharme S., Hong X., "Domain wall roughness and creep in nanoscale crystalline ferroelectric polymers," *Appl. Phys. Lett.* 103(11), 112903 (2013).

Xu G., Wen J., Stock C., Gehring P. M., "Phase instability induced by polar nanoregions in a relaxor ferroelectric system," *Nat. Mater.* 7(7), 562–566 (2008).

Xu R., Karthik J., Damodaran A. R., Martin L. W., "Néel-like domain walls in ferroelectric $Pb(Zr,Ti)O_3$ single crystals," *Nat. Commun.* 5, 3120 (2014).

Yadav A. K., Nelson C. T., Hsu S. L., Hong Z., Clarkson J. D., Schlepütz C. M., Damodaran A. R., Shafer P., Arenholz E., Dedon L. R., Chen D., Vishwanath A., Minor A. M., Chen L. Q., Scott J. F., Martin L. W., Ramesh R., "Observation of polar vortices in oxide superlattices," *Nature* 530(7589), 198–201 (2016).

Yadav A. K., Nguyen K. X., Hong Z., García-Fernández P., Aguado-Puente P., Nelson C. T., Das S., Prasad B., Kwon D., Cheema S., Khan A. I., Hu C., Íñiguez J., Junquera J., Chen L.-Q., Muller D. A., Ramesh R., Salahuddin S., "Spatially resolved steady-state negative capacitance," *Nature* 565(7740), 468–471 (2019).

Yang S. M., Jo J. Y., Kim T. H., Yoon J.-G., Song T. K., Lee H. N., Marton Z., Park S., Jo Y., Noh T. W., "AC dynamics of ferroelectric domains from an investigation of the frequency dependence of hysteresis loops," *Phys. Rev. B* 82(17), 174125–1–7 (2010).

Zhang C., Galli G., "Dipolar correlations in liquid water," *J. Chem. Phys.* 141(8), 084504 (2014).

Zhao Y., Truhlar D. G., "Construction of a generalized gradient approximation by restoring the density-gradient expansion and enforcing a tight Lieb–Oxford bound," *J. Chem. Phys.* 128(18), 184109 (2008).

15

Electronics Based on Domain Walls

J. Seidel[1,2] and R. Ramesh[3,4,5]

[1] School of Materials Science and Engineering, University of New South Wales, Sydney 2052, Australia
[2] ARC Centre of Excellence in Future Low-Energy Electronics Technologies, UNSW Sydney, Sydney NSW 2052, Australia
[3] Department of Materials Science and Engineering, University of California, Berkeley, CA, 94720, United States
[4] Materials Sciences Division, Lawrence Berkeley National Laboratory, Berkeley, CA, 94720, United States
[5] Department of Physics, University of California, Berkeley, CA, 94720, United States

15.1 Introduction

Domain walls in ferroelectrics and multiferroic systems have attracted a great deal of attention in recent years. The study of many new fundamental phenomena associated with these types of topological defects (Seidel 2019) have propelled them from textbook features mentioned in connection with ordered materials that break symmetry to functional nanostructures with great potential for novel electronics applications. These developments were inspired by earlier findings in magnetic materials systems that discussed domain walls within the context of racetrack geometries (Parkin et al. 2008) and now have evolved to more complex topological structures such as skyrmions (Milde et al. 2013), which have also been recently found in ferroelectric systems (Das et al. 2019). One of the pivotal developments in this regard has been the observation of electronic conductivity at domain walls in bismuth ferrite (Seidel at al. 2009), which has triggered a hunt for similar properties in other conventional ferroelectrics (Farokhipoor and Noheda 2011; Guyonnet et al. 2011; Maksymovych et al. 2011; Schröder et al. 2012; Oh et al. 2015), with many of them now confirmed to have comparable interesting properties. Based on these fundamental findings and the exploration of basic properties of domain walls, new concepts for nanoelectronic elements have been proposed recently. Changes in structure, symmetry, and solid-state chemistry uniquely confined within the wall (see Chapters 2 to 4) have driven home a fresh perspective that these are a distinct class of materials systems by themselves (Farokhipoor et al. 2014). This is the fundamental starting point for "domain wall nanoelectronics"—an emergent information technology where data is processed by and within the walls rather than the domains (Catalan et al. 2012). In this chapter, we will review some of these initial developments and recently introduced potential application concepts related to domain

J. Seidel and R. Ramesh, *Electronics Based on Domain Walls* In: *Domain Walls: From Fundamental Properties to Nanotechnology Concepts*. Edited by: Dennis Meier, Jan Seidel, Marty Gregg, and Ramamoorthy Ramesh, Oxford University Press (2020). © J. Seidel and R. Ramesh.
DOI: 10.1093/oso/9780198862499.003.0015

walls in ferroelectrics and multiferroics with a special (non-exclusive) focus on the heavily investigated bismuth ferrite $BiFeO_3$ system as one of the rare examples of a single-phase room-temperature multiferroic system that can be widely tailored in application-relevant epitaxial thin films.

15.2 Controlling Electronic Properties at Ferroelectric Domain Walls

Fundamental properties of domain walls that have been extensively investigated (see Figure 15.1) include domain wall conductivity, which is closely linked to the charge state of the wall. Initial work on control of conductivity also includes multi-level resistance states at walls and memristor concepts (Maksymovych et al. 2011). The interaction of

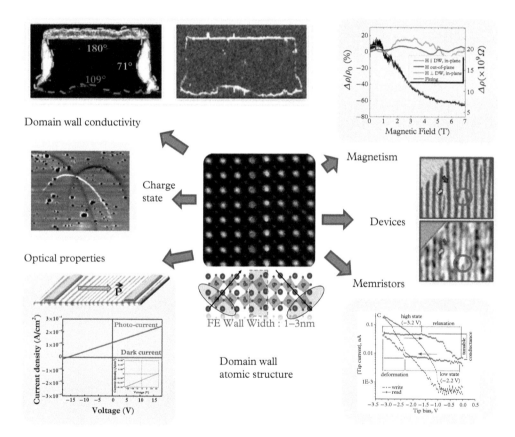

Figure 15.1 *Domain walls—atomic scale functional structures in multiferroics. Emerging functional properties arising at atomically sharp domain wall interfaces. Clockwise from the top: domain wall conductivity (Seidel et al. 2009), magnetoresistance at walls (He et al. 2012), exchange-bias-based devices (Chu et al. 2008), memristive behavior of walls (Maksymovych et al. 2011), atomically sharp nature of walls, photovoltaic properties (Yang et al. 2010), and charged domain wall properties (Meier et al. 2012).*

light with domain walls, i.e. their optical (Vats et al. 2019) and photovoltaic properties, is still being investigated and debated (see Chapter 9). All these findings are related to the atomic structure of domain walls, which has been visualized by electron microscopy, i.e. with HAADF-STEM measurements (Lubk et al. 2012, 2013) (see also Chapter 10). These measurements also reveal one of the most appealing aspects of domain walls in ferroelectrics and multiferroics: their atomic-scale thickness or size. The transition of the order parameter on length scales of the order of ~1 nm has been discussed as the ultimate functional feature in ordered solid-state systems (Seidel et al. 2012). A first demonstration of functionality includes domain-wall-controlled conductivity in prototype device structures (see Figure 15.2) (Seidel et al. 2009).

Domain wall geometry (Vasudevan et al. 2012; Stolichnov et al. 2015) and type (Li et al. 2016; Godau et al. 2017) have been shown to determine its charge state, which in turn influences its conduction properties. Typically, charged walls have a higher energy of formation than neutral domain walls and thus are unfavorable energetically. They can be stabilized, however, if sufficient mobile charges are available for screening the polarization discontinuity (Sluka et al. 2013). Stable charged domain walls can also be observed in systems with improper ferroelectric order (where ferroelectricity is not the primary order parameter, and indeed the magnitude of the spontaneous polarization is at least an order of magnitude smaller than that of the proper ferroelectrics such as $BiFeO_3$) and multiferroic systems (Balke et al. 2010). Initial findings, for example, in hexagonal manganites have shown that indeed this modulation of conductivity at walls can be observed experimentally (see also Chapter 6) (Meier et al. 2012).

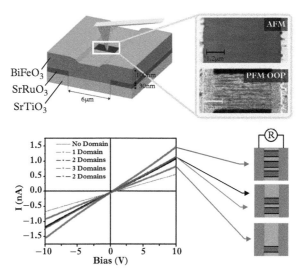

Figure 15.2 *Domain-wall-based functional device concept. Early demonstration of control of domain wall conductivity (Seidel et al. 2009). Artificially atomic force microscopy (AFM) tip-written domain wall pairs (0–3 pairs) lead to stepwise increase in current between two metal contacts. The associated ferroelectric domain structure is seen in piezoresponse force microscopy (PFM) out-of-plane (OOP) images.*

Of particular interest in multiferroic materials that exhibit some form of magnetic order is the magnetic state of domain walls. Little work has been done in this direction, mainly due to the bottleneck of revealing magnetic order at the atomic scale. Nevertheless, systematic approaches that involve arrays of engineered domain walls of the same type by heteroepitaxy, such as, e.g. the widely studied model system in bismuth ferrite (He et al. 2012; Domingo et al. 2017), have led to some insight. Work on exchange bias coupling in bismuth ferrite with a ferromagnetic overlayer (Heron et al. 2014) (e.g. FeCo) initially pointed to the influence of specific domain wall types on the magnetic state of such devices.

As shown in Figures 15.3(a,b), samples involving 71° or 109° walls have distinct magnetic signatures in SQUID-based magnetization measurements. These results have also been corroborated by magnetotransport (MR) measurements (He et al. 2012), which show a distinct and clear dependence of magnetoresistance on domain wall type, orientation, and applied magnetic field direction (Figures 15.3(c,d)). The magnetic field dependence of the transport behavior exhibited a markedly negative MR when both the magnetic field and transport were parallel to the walls. Negative MR values as high as 60% were obtained at a magnetic field of 7 T. When the magnetic field was applied perpendicular to the transport path or when the transport was perpendicular to the domain walls, little MR was observed, indicating that it is directly related to the domain walls in the material. Although the precise magnetic structure at these walls has remained elusive up to now, these measurements clearly show a distinct magnetic signature of the walls that should be further explored.

An overlooked or less investigated aspect with regard to domain walls is their nanoscale mechanical properties, which can be coupled to their electronic/magnetic structure. Some work has been reported on conventional and morphotropic ferroelectric systems (Heo et al. 2014; Sharma, et al. 2016a, 2016b, 2018; Alsubaie et al. 2017, 2018; Heo et al. 2017; Hu et al. 2017) (see also Chapter 1). The widely investigated mixed-phase structure in bismuth ferrite thin films presents a new concept, namely, that of hybrid domain walls, i.e. a transition of the order parameter coupled to a structural transition of the lattice of the material (Kim et al. 2014; Seidel et al. 2014; Lee et al. 2017). The study of the mechanical properties of nanoscale ferroelectrics using the above methods provides a unique tool to access and study information complementary to the otherwise already accessible electronic and structural properties of these materials with nanometer spatial resolution. The sensitivity and selectivity of the mechanical interaction opens new pathways for real space probing of multiple order parameters and phases of complex ferroic materials. Examples here include mechanical phase memory and mechanical control of magnetism by force in multiferroics, in analogy to electrical control of magnetism.

Some of the basic electronic properties discussed above together with ways to exploit coupling of the spin charge and the lattice properties of the material (Ramirez et al. 2009) have recently been utilized in prototype domain wall devices with the wall being the active element. These include memory and logic based on these nanoscale elements (Whyte et al. 2014; McGilly et al. 2015, 2016; Whyte and Gregg 2015; Li et al. 2016). The ability to precisely manipulate domain walls (Seidel et al. 2010; Vasudevan et al.

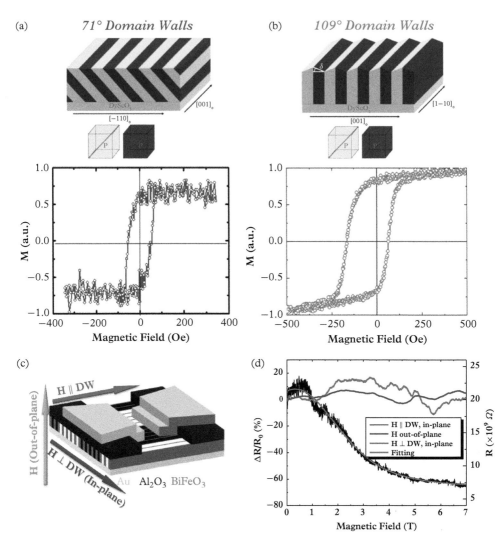

Figure 15.3 *Magnetism in multiferroic domain walls. Distinct magnetic signatures of ordered domain wall arrays in BFO on DyScO₃ substrates (blue and yellow indicate alternating ferroelectric domains with distinct orientation of the polarization P). (a,b) Magnetization measurements of samples containing predominantly 71° and 109° domain walls. (c,d) Magnetotransport measurements of domain walls found in samples involving arrays of engineered wall types in bismuth ferrite; magnetic field H applied in different directions leads to significant changes in domain-wall-mediated resistance (He et al. 2012).*

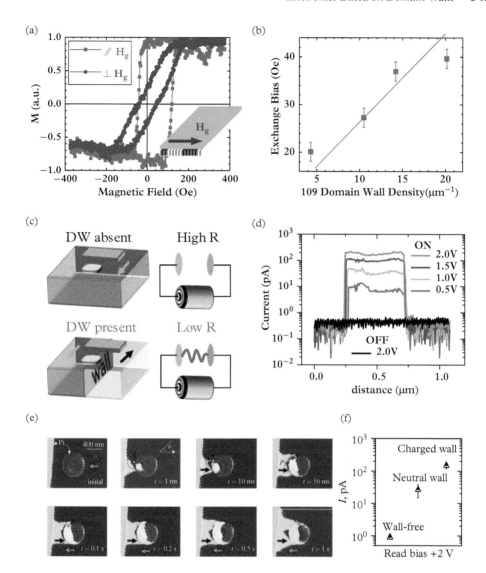

Figure 15.4 *Bismuth-ferrite-based domain wall device concepts for electronics and spintronics. (a,b) Domain-wall-controlled exchange bias. Magnetization M of the exchange layer depends on the domain wall orientation (magnetic field H parallel or perpendicular to walls), increasing density of 109° domain walls leads to a larger exchange bias (Chu et al. 2008). (c) Domain wall memory concept (Sharma et al. 2017), and demonstration with large device current on/off ratio. (d), (e) Domain wall conformational control, i.e. domain-wall-shape-dependent electronic conductivity. (f) memory cell device currents for different wall types (Sharma et al. 2019).*

2012; Sluka et al. 2013; Crassous et al. 2015; Stolichnov 2015), modulate and gate their electrical transport properties (Whyte et al. 2014; Whyte and Gregg 2015; Li et al. 2016), and create interconnects are important milestones toward domain wall nanoelectronics. In addition, one of the pivotal device developments in multiferroic systems that has been investigated for several years is electrical control of magnetism. One of the possible approaches involves interface-based magnetic coupling between a ferroelectric or multiferroic and a magnetic-exchange-bias-coupled layer in which the magnetism can be switched electrically. This is achieved through electrical control of the ferroelectric domain structure on the micro- or nanometer scale. Various approaches are based on carefully tailored domain geometries and orientations with respect to the applied electric fields. The magnetic state of domain walls in that regard is one possible means of controlling exchange bias coupling (Figures 15.4(a,b)) in magnetoelectric device architectures.

15.3 Ferroelectric Domain Wall Memory

The pursuit of domain wall memory has led to a recent demonstration of a nonvolatile prototype in bismuth ferrite thin films (Sharma et al. 2017). The controllable presence or absence of a domain wall (Figure 15.4(c)) was achieved in nanoscale devices based on electron-beam-patterned lateral electrodes in thin films combined with scanning probe microscopy base manipulation and readout of the ferroelectric and electrical transport properties of domain walls (Figure 15.4(d)). The nonvolatile nature of the memory combined with low-voltage operation and on-off ratios of 100–1000 hold promise for future developments in this area. The demonstrated memory cells can in principle be integrated in already well-established architectures such as cross-bar array structures that are widely used in commercial memory. For this to be possible, the switching energy and fatigue, scalability, low-energy operation, and memristive behavior and functionality need to be further studied and compared with existing technologies such as ferroelectric RAM (FeRAM). These results nevertheless provide a perspective to harness the special electronic properties of the functional ferroelectric domain walls in data storage devices.

The earlier discussed variation of the charge state of domain walls leads to an exciting additional possibility for memory applications: deterministically tuning the domain wall shape and hence charge state such that a solid-state device with a controlled variable output is obtained. The feasibility of this concept was recently demonstrated (Sharma et al. 2019). Morphological changes or conformational transformation of a domain wall (similar to conformational switches in organic molecules) has been shown to control its local charge state and hence electronic characteristics, which previously were not utilized in nanoelectronic devices. This is mainly because such behavior not only demands reliable site-specific injection and erasure of domain walls with controlled morphology, but also absolutely robust and precise control over their electronic states. The evolution of the electronic states of the wall concomitant with the conformational changes comes at a cost, that of the formation of charged DWs that are typically unstable (Bai et al.

2018; Jiang et al. 2018). Using piezoresponse force microscopy (PFM) images of the ferroelectric domain state acquired during stroboscopic imaging of domain wall pairs in bismuth-ferrite-based memory cells (Figure 15.4(e)), a distinct departure from the characteristic straight (parallel wall pair) neutral domain walls to a tilted domain wall geometry with charged walls (V-shaped wall pair) is visible. An analysis of the wall angle as a function of the electric pulse duration shows the controlled formation of a stable charged domain wall state alongside neutral domain walls. Therefore, by tuning the width (and/or height) of an applied electrical pulse, the domain wall configuration in such a memory cell can be transitioned between neutral and charged, demonstrating bias-controlled "folding" of the wall and tuning of the associated conduction state of the domain wall (Figure 15.4(f)).

15.4 Outlook

Where do we go from here? Domain walls as well as other topological structures show us new ways to novel tailored states of matter with a wide range of electronic properties. Domain wall electronics, particularly with ferroelectrics and multiferroics (Seidel 2012; Zhou et al. 2012; Seidel et al. 2013), provides new nanotechnological concepts for identifying, understanding, and designing new material properties. From a fundamental physics perspective, it would be exciting to be able to induce a metal–insulator transition right at the wall; some early work along this direction has been demonstrated with the WO_3 system (see work by Salje et al.). A true phase transition will require much more sophisticated control of the electronic structure at the wall in an otherwise insulating ferroelectric material. In this regard, there has been very little work done on controlling electronic correlations. $BiFeO_3$ itself is a wonderful model system; without correlations, the compound would likely have been a small bandgap insulator (Ederer and Spaldin). Electronic correlations open up a robust gap in this system. Thus, manipulating electronic correlations, especially at the domain walls, would be a valuable research direction. The quite diverse and wide range of ferroelectric materials also includes organic ferroelectrics and layered systems, which are amenable to inexpensive large-scale production by solution and spin-casting methods. This possibility offers a potential low-cost solution for mass production and their integration with flexible substrates. Also of special interest are the magnetic and optical properties of multiferroic systems which could enable the development of electrically tunable nano-spintronic devices based on domain walls. From an applications perspective, perhaps the biggest questions relate to signal margins and communication with the outside world. Regarding the signal margins, the current values of electronic current at domain walls is still smaller than what would be ideal, i.e. on the order of approximately microamperes. A related issue is the protocols required to write and read information within the walls. Will this involve an atomic force microscopy (AFM)-based (i.e. like a Millipede concept) approach or will it involve a more classical metallic wire contact? While the nanoscale is perhaps the biggest attraction of such domain walls, it is also likely to be the most difficult aspect when it comes to real applications. Another interesting aspect is the speed with which the transport can be manipulated and how robust the conducting state is to repeated reversals. Overall, we have seen considerable development of concepts and prototype electronics applications of domain walls recently and likely will see even more in the coming years.

....

REFERENCES

Alsubaie A., Sharma P., Liu G., Nagarajan V., Seidel J., "Mechanical stress-induced switching kinetics of ferroelectric thin films at the nanoscale," *Nanotechnology* 28, 075709 (2017).

Alsubaie A., Sharma P., Lee J. H., Kim J. Y., Yang C., Seidel J., "Uniaxial strain-controlled ferroelastic domain evolution in BiFeO$_3$," *ACS Appl. Mater. Interf.* 10, 11768 (2018).

Bai Z. L., Cheng X. X., Chen D. F., Zhang D. W., Chen L.-Q., Scott J. F., Hwang C. S., Jiang A. Q., "Hierarchical domain structure and extremely large wall current in epitaxial BiFeO$_3$ thin films," *Adv. Funct. Mater.* 28, 1801725 (2018).

Balke N., Gajek M., Tagantsev A. K., Martin L. W., Chu Y.-H., Ramesh R., Kalinin S. V., "Direct observation of capacitor switching using planar electrodes," *Adv. Funct. Mater.* 20, 3466 (2010).

Catalan G., Seidel J.Ramesh R., Scott J. F., "Domain wall nanoelectronics," *Rev. Mod. Phys.* 84, 119 (2012).

Chu Y.-H., Martin L. W., Holcomb M. B., Gajek M., Han S.-J., He Q., Balke N., Yang C.-H., Lee D., Hu W., Zhan Q., Yang P.-L., Fraile-rodríguez A., Scholl A., Wang S. X., Ramesh R., "Erratum: electric-field control of local ferromagnetism using a magnetoelectric multiferroic," *Nat. Mater.* 7, 678 (2008).

Crassous A., Sluka T., Tagantsev A. K., Setter N., "Polarization charge as a reconfigurable quasi-dopant in ferroelectric thin films," *Nat. Nanotechnol.* 10, 614–618 (2015).

Das S., Tang Y. L., Hong Z., Gonçalves M. A. P., McCarter M. R., Klewe C., Nguyen K. X., Gómez-Ortiz F., Shafer P., Arenholz E., Stoica V. A., Hsu S. L., Wang B., Ophus C., Liu J. F., Nelson C. T., Saremi S., Prasad B., Mei A. B., Schlom D. G., Íñiguez J., García-Fernández P., Muller D. A., Chen L. Q., Junquera J., Martin L. W., Ramesh R., "Observation of room-temperature polar skyrmions," *Nature* 568, 368 (2019).

Domingo N., Farokhipoor S., Santiso J., Noheda B., Catalan G., "Domain wall magnetoresistance in BiFeO$_3$ thin films measured by scanning probe microscopy," *J. Phys.: Condens. Mat.* 29, 334003 (2017).

Farokhipoor S., Noheda B., "Conduction through 71° domain walls in BiFeO$_3$ thin films," *Phys. Rev. Lett.* 107, 127601 (2011).

Farokhipoor S., Magén C., Venkatesan S., Íñiguez J., Daumont C. J. M., Rubi D., Snoeck E., Mostovoy M., de Graaf C., Müller A., Döblinger M., Scheu C., Noheda B., "Artificial chemical and magnetic structure at the domain walls of an epitaxial oxide," *Nature* 515, 379 (2014).

Godau C., Kämpfe T., Thiessen A., Eng L. M., Haußmann A., "Enhancing the domain wall conductivity in lithium niobate single crystals," *ACS Nano* 11, 4816 (2017).

Guyonnet J., Gaponenko I., Gariglio S., Paruch P., "Conduction at domain walls in insulating Pb(Zr$_{0.2}$Ti$_{0.8}$)O$_3$ thin films," *Adv. Mater.* 23, 5377–5382 (2011).

He Q., Yeh C.-H., Yang J.-C., Singh-Bhalla G., Liang C.-W., Chiu P.-W., Catalan G., Martin L. W., Chu Y.-H., Scott J. F., Ramesh R., "Magnetotransport at domain walls in BiFeO$_3$," *Phys. Rev. Lett.* 108, 067203 (2012).

Heo Y., Jang B. K., Kim S. J., Yang C. H., Seidel J., "Nanoscale Mechanical Softening of Morphotropic BiFeO$_3$," *Adv. Mater.* 26, 7568 (2014).

Heo Y., Hu S., Sharma P., Kim K.-E., Jang B.-K., Cazorla C., Yang C.-H., Seidel J., "Impact of isovalent and aliovalent doping on mechanical properties of mixed phase BiFeO$_3$," *ACS Nano* 11, 2805 (2017).

Heron J. T., Bosse J. L., He Q., Gao Y., Trassin M., Ye L., Clarkson J. D., Wang C., Liu J., Salahuddin S., Ralph D. C., Schlom D. G., Iñiguez J., Huey B. D., Ramesh R., "Deterministic switching of ferromagnetism at room temperature using an electric field," *Nature* 516, 370 (2014).

Hu S., Alsubaie A., Wang Y., Lee J. H., Kang K.-R., Yang C.-H., Seidel J., "Poisson's ratio of BiFeO$_3$ thin films," *Phys. Status Solidi A* 214, 1600356 (2017).

Jiang J., Bai Z. L., Chen Z. H., He L., Zhang D. W., Zhang Q. H., Shi J. A., Park M. H., Scott J. F., Hwang C. S., Jiang A. Q., "Temporary formation of highly conducting domain walls for non-destructive read-out of ferroelectric domain-wall resistance switching memories," *Nat. Mater.* 17, 49 (2018).

Kim K.-E., Jang B.-K., Heo Y., Lee J. H., Jeong M., Lee J. Y., Seidel J., Yang C.-H., "Electric control of straight stripe conductive mixed-phase nanostructures in La-doped BiFeO$_3$," *Nature Publishing Group Asia Mater.* 6, e81 (2014).

Lee J. H., Kim K.-E., Jang B.-K., Ünal A. A., Valencia S., Kronast F., Ko K.-T., Kowarik S., Seidel J., Yang C.-H., "Strain-gradient-induced magnetic anisotropy in straight-stripe mixed-phase bismuth ferrites: Insight into flexomagnetism," *Phys. Rev. B* 96, 064402 (2017).

Li L., Britson J., Jokisaari J. R., Zhang Y., Adamo C., Melville A., Schlom D. G., Chen L. Q., Pan X., "Giant resistive switching via control of ferroelectric charged domain walls," *Adv. Mater.* 28, 6574 (2016).

Lubk A., Rossell M. D., Seidel J., He Q., Yang S. Y., Chu Y. H., Ramesh R., Hÿtch M. J., Snoeck E., "Evidence of sharp and diffuse domain walls in BiFeO$_3$ by means of unit-cell-wise strain and polarization maps obtained with high resolution scanning transmission electron microscopy," *Phys. Rev. Lett.* 109, 047601 (2012).

Lubk A., Rossell M. D., Seidel J., Chu Y. H., Ramesh R., Hÿtch M. J., Snoeck E., "Electromechanical coupling among edge dislocations, domain walls, and nanodomains in BiFeO$_3$ revealed by unit-cell-wise strain and polarization maps," *Nano Lett.* 13, 1410 (2013).

Maksymovych P., Seidel J., Chu Y. H., Wu P., Baddorf A. P., Chen L. Q., Kalinin S. V., Ramesh R., "Dynamic conductivity of ferroelectric domain walls in BiFeO$_3$," *Nano Lett.* 11, 1906–1912 (2011).

Meier D., Seidel J., Cano A., Delaney K., Kumagai Y., Mostovoy M., Spaldin N. A., Ramesh R., Fiebig M., "Anisotropic conductance at improper ferroelectric domain walls," *Nat. Mater.* 11, 284 (2012).

McGilly L. J., Yudin P., Feigl L., Tagantsev A. K., Setter N., "Controlling domain wall motion in ferroelectric thin films," *Nat. Nanotechnol.* 10, 145–150 (2015).

Milde P., Köhler D., Seidel J., Eng L. M., Bauer A., Chacon A., Kindervater J., Mühlbauer S., Pfleiderer C., Buhrandt S., Schütte C., Rosch A., "Unwinding of a skyrmion lattice by magnetic monopoles," *Science* 340, 1076 (2013).

Oh Y. S., Luo X., Huang F.-T., Wang Y., Cheong S.-W., "Experimental demonstration of hybrid improper ferroelectricity and the presence of abundant charged walls in (Ca,Sr)$_3$Ti$_2$O$_7$ crystals," *Nat. Mater.* 14, 407–413 (2015).

Parkin S. S. P., Hayashi M., Thomas L., "Magnetic domain-wall racetrack memory," *Science* 320, 190 (2008).

Ramirez M., Ramirez M. O., Kumar A., Denev S. A., Chu Y. H., Seidel J., Martin L. W., Yang S.-Y., Rai R. C., Xue X. S., Ihlefeld J. F., Podraza N. J., Saiz E., Lee S., Klug J., Cheong S. W., Bedzyk M. J., Auciello O., Schlom D. G., Orenstein J., Ramesh R., Musfeldt J. L., Litvinchuk A. P., Gopalan V., "Spin-charge-lattice coupling through resonant multimagnon excitations in multiferroic $BiFeO_3$," *Appl. Phys. Lett.* 94, 161905 (2009).

Schröder M., Haußmann A., Thiessen A., Soergel E., Woike T., Eng L. M., "Conducting domain walls in lithium niobate single crystals," *Adv. Funct. Mater.* 22, 3936–3944 (2012).

Seidel J., "Domain walls as nanoscale functional elements," *J. Phys. Chem. Lett.* 3, 2905 (2012).

Seidel J., Martin L. W., He Q., Zhan Q., Chu Y.-H., Rother A., Hawkridge M. E., Maksymovych P., Yu P., Gajek M., Balke N., Kalinin S. V., Gemming S., Wang F., Catalan G., Scott J. F., Spaldin N. A., Orenstein J., Ramesh R., "Conduction at domain walls in oxide multiferroics," *Nat. Mater.* 8, 229–234.

Seidel J., Maksymovych P., Katan A. J., Batra Y., He Q., Baddorf A. P., Kalinin S. V., Yang C.-H., Yang J.-C., Chu Y.-H., Salje E. K. H., Wormeester H., Salmeron M., Ramesh R., "Domain wall conductivity in La-doped $BiFeO_3$," *Phys. Rev. Lett.* 105, 197603 (2010).

Seidel J., Singh-Bhalla G., He Q., Yang S.-Y., Chu Y.-H., Ramesh R., "Domain wall functionality in $BiFeO_3$," *Phase Transit.* 86, 53 (2013).

Seidel J., Trassin M., Zhang Y., Maksymovych P., Uhlig T., Milde P., Köhler D., Baddorf A. P., Kalinin S. V., Eng L. M., Pan X., Ramesh R., "Electronic properties of isosymmetric phase boundaries in highly strained Caâ doped $BiFeO_3$," *Adv. Mater.* 26, 4376 (2014).

Seidel J., "Nanoelectronics based on topological structures," *Nat. Mater.* 18, 188 (2019).

Sharma P., Heo Y., Jang B.-K., Liu Y. Y., Valanoor N., Li J., Yang C.-H., Seidel J., "Morphotropic phase elasticity of strained $BiFeO_3$," *Adv. Mater. Interf.* 3, 1600033 (2016a).

Sharma P., Kang K. R., Jang B.-K., Yang C.-H., Seidel J., "Unraveling elastic anomalies during morphotropic phase transitions," *Adv. Electron. Mater.* 2, 1600283 (2016b).

Sharma P., Zhang Q., Sando D., Lei C.-H., Liu Y., Li J.-Y., Nagarajan V., Seidel J., "Nonvolatile ferroelectric domain wall memory," *Sci. Adv.* 3, e1700512 (2017).

Sharma P., Kang K.-R., Liu Y.-Y., Jang B.-K., Li J.-Y., Yang C.-H., Seidel J., "Optimizing the electromechanical response in morphotropic $BiFeO_3$," *Nanotechnology* 29, 205703 (2018).

Sluka T., Tagantsev A. K., Bednyakov P., Setter N., "Free-electron gas at charged domain walls in insulating $BaTiO_3$," *Nat. Commun.* 4, 1808 (2013).

Stolichnov I., Feigl L., McGilly L. J., Sluka T., Wei X.-K., Colla E., Crassous A., Shapovalov K., Yudin P., Tagantsev A. K., Setter N., "Bent ferroelectric domain walls as reconfigurable metallic-like channels," *Nano Lett.* 15, 8049 (2015).

Vasudevan R. K., Morozovska A. N., Eliseev E. A., Britson J., Yang J.-C., Chu Y.-H., Maksymovych P., Chen L. Q., Nagarajan V., Kalinin S. V., "Domain wall geometry controls conduction in ferroelectrics," *Nano Lett.* 12, 5524 (2012).

Vats G., Bai Y., Zhang D., Juuti J., Seidel J., Optical control of ferroelectric domains: nanoscale insight into macroscopic observations," *Adv. Opt. Mater.* 7, 1800858 (2019).

Whyte J. R., Gregg J. M., "A diode for ferroelectric domain-wall motion," *Nat. Commun.* 6, 7361 (2015).

Whyte J. R., McQuaid R. G. P., Sharma P., Canalias C., Scott J. F., Gruverman A., Gregg J. M., "Ferroelectric domain wall injection," *Adv. Mater.* 26, 293–298 (2014).

Zhou J., Trassin M., He Q., Tamura N., Kunz M., Cheng C., Zhang J., Liang W.-I., Seidel J., Hsin C.-L., Wu J., "Directed assembly of nano-scale phase variants in highly strained $BiFeO_3$ thin films," *J. Appl. Phys.* 112, 064102 (2012).

Index